Teeth are one of the best sources of evidence for both identification and studies of demography, biological relationships and health in ancient human communities. This text introduces the complex biology of teeth and provides a practical guide to the:

excavation, cleaning, storage and recording of dental remains
identification of human teeth including those in a worn or fragmentary state
methods for studying variation in tooth morphology
study of microscopic internal and surface structure of dental tissues
estimation of age-at-death from dental development, wear and micro-
    structure
recording and interpretation of dental disease in archaeological and museum
    collections.

Dental Anthropology is *the* text for students and researchers in anthropology and archaeology, together with others interested in dental remains from archaeological sites, museum collections or forensic cases.

DENTAL ANTHROPOLOGY

DENTAL ANTHROPOLOGY

SIMON HILLSON

Institute of Archaeology
University College London

CAMBRIDGE
UNIVERSITY PRESS

PUBLISHED BY THE PRESS SYNDICATE OF THE UNIVERSITY OF CAMBRIDGE
The Pitt Building, Trumpington Street, Cambridge, United Kingdom

CAMBRIDGE UNIVERSITY PRESS
The Edinburgh Building, Cambridge CB2 2RU, UK
40 West 20th Street, New York, NY 10011–4211, USA
477 Williamstown Road, Port Melbourne, VIC 3207, Australia
Ruiz de Alarcón 13, 28014 Madrid, Spain
Dock House, The Waterfront, Cape Town 8001, South Africa

http://www.cambridge.org

First published 1996
Third printing 2002

Typeset in Times 11/14 pt.   *System* Miles33®   [wv]

A catalogue record for this book is avaialble from the British Library

Library of Congress Cataloguing in Publication data
Hillson, Simon.
Dental anthropology / Simon Hillson.
p. cm.
Includes bibliographical references and index.
ISBN 0 521 45194 9 (hardback). – ISBN 0 521 56439 5 (pbk.)
1. Dental anthropology.   I. Title.
[DNLM: 1. Tooth – anatomy & histology.   2. Anthropology, Physical.
3. Odontometry.   WU 101 H655d 1996]
GN209.H56 1996
611′.314–dc20
DNLM/DLC
for Library of Congress   95–52570 CIP

ISBN 0 521 45194 9 hardback
ISBN 0 521 56439 5 paperback

Transferred to digital printing 2003

To Kate, William, James and Harriet

Contents

Acknowledgements

To a large extent I have to thank Dr Jerry Rose, Professor of Anthropology at the University of Arkansas, for starting me on the path that led to this book. He put me up to it, encouraged me throughout and read drafts – although he cannot be held responsible for my views and interpretations, especially as we do not always agree. I have gained further impetus from discussions with many fellow members of the American Association of Physical Anthropologists and the Dental Anthropology Association during my visits to the USA. As always, I also acknowledge the support of colleagues at University College London, including Professors Alan Boyde, Sheila Jones, Don Brothwell (now at the University of York), Dr Chris Dean and Liz Pye. For many years they have helped with advice, discussion and facilities, and they taught me dental anatomy, histology and anthropology in the first place. Other colleagues at UCL have helped with my research in practical ways – Sandra Bond has put in many hours of laboratory preparation and scanning electron microscopy, Naomi Mott found wonderful specimens in our collections and Stuart Laidlaw assisted with photography. In addition, I acknowledge the inspiration of the unique collections of the Odontological Museum at the Royal College of Surgeons in London, and the unfailing help of Dr Caroline Grigson, who curates it. Further inspiration has come from the enormous collections from Roman and Medieval London held by the Museum of London Archaeology Service. In terms of the nuts and bolts of writing, I have once again greatly enjoyed working with Cambridge University Press, and I particularly thank my commissioning editor, Dr Tracey Sanderson. Above all, however, I could not have written this book without the support and active help of my family, who put up with my long hours of extra work over several years, during evenings, nights, weekends and holidays. My wife Kate and my eldest son William typed rows of numbers and corrections, whilst my younger children James and Harriet were stalwart companions during the crucial last days of the project.

Simon Hillson
February 1996

Abbreviations

ATP	Anderson, Thompson and Popovitch dental development data
B	buccal
BSE	SEM with back-scattered electron detector
CEJ	cement–enamel junction
D	distal
EDJ	enamel–dentine junction
ET	SEM with Everhart–Thornley dector
GCF	gingival crevice fluid
kDa	kilodalton (molecular weight)
La	labial
Li	lingual
M	mesial
MFH	Moorees, Fanning and Hunt dental development data
μm	micrometres (thousandths of a millimetre)
nm	nanometres (millionths of a millimetre)
O	occlusal
pkg	perikyma grooves
SEM	scanning electron microscope

1

Introduction to dental anthropology

'Show me your teeth and I will tell you who you are', Baron Georges Cuvier, the great eighteenth–nineteenth century zoologist and anatomist, is supposed to have said. This comment was really in the context of comparative anatomy, and refers to Cuvier's delight in reconstructing whole extinct animals from fossil fragments of their dentitions, but it will do just as well for human teeth. For anthropologists studying archaeological, fossil and forensic remains, the teeth are possibly the most valuable source of evidence in understanding the biology of ancient communities, following the course of evolution and identifying an individual from their fragmentary remains. Dental anthropology might therefore be defined as a study of people (and their close relatives) from the evidence provided by teeth. Teeth have a distinct anatomy and physiology, all their own and wholly different to the biology of the skeleton, and teeth are also unique amongst the resistant parts of archaeological and fossil remains in having been exposed on the surface of the body throughout life. Dental anthropology can therefore be studied in the mouths of living people, using much the same techniques as are employed for ancient remains. It is thus not surprising that practising dentists have always been prominent amongst dental anthropologists, with anatomists and other oral biologists from schools of dentistry, in addition to researchers whose training lies more in biological anthropology. The exposure of teeth in the living mouth is also very useful when training anthropologists, as everyone carries their own reference material with them – students can just open their mouths and look in a mirror.

One of the main themes of dental anthropology has been a study of variation in size and shape of the teeth, as recorded in casts of living mouths or seen in the skulls of archaeological and fossil collections. This work is founded on a series of classic 'odontographies' – dental studies of particular ethnic groups or fossil collections such as those of Robinson (1956) on the australopithecines

1

from South Africa, Weidenreich (1937) on the Chinese *Homo erectus*, Campbell (1925) on Aboriginal Australians, Moorrees (1957b) on the Aleuts and Pedersen (1949) on the Inuit. It was particularly brought to prominence during the second half of the twentieth century by the late Al Dahlberg, who amassed a large collection of casts from his work as a dentist with living Native Americans, and acted as a focus for the development of dental anthropology through his graduate training programme at the University of Chicago and his role in the establishment of a series of Dental Morphology Symposia (page 68). The widespread use of the phrase 'dental anthropology' probably dates to the forerunner of these symposia, a meeting held in London during 1958 (Brothwell, 1963a), which was celebrated 30 years later by a symposium of the American Association of Physical Anthropologists in Kansas City (Kelley & Larsen, 1991). In 1986, a Dental Anthropology Association was formed during a meeting of the AAPA at Albuquerque, New Mexico, and now numbers amongst its membership most of the active researchers in the field.

Dental anthropology is, however, a much wider subject than just morphology. It includes a study of the development of teeth in relation to age, their appearance in the mouth, and the processes of wear and other changes that occur once they are in place. It also includes the microscopic traces, preserved inside the tissues of the teeth, of the growth and ageing processes. Yet another area of interest is the study of dental diseases, in relation to diet and other factors, and the most recent development is a study of the biochemistry of dental tissues. These ideas and techniques have entered anthropology from oral biology, whose roots lie in odontology, a subject that is little mentioned nowadays but which has formed the scientific basis for modern dental surgery. Odontology has its origins in research at many centres during the eighteenth and nineteenth centuries (Hofman-Axthelm, 1981), but one major focus for its early development as a coherent discipline was the Royal College of Surgeons in London, whose museums were founded with the personal collections of the extraordinary surgeon polymath John Hunter (1771; 1778). Successive curators included Sir Richard Owen (1845), Sir John Tomes (1894) and Sir Frank Colyer (1936; Miles & Grigson, 1990) who, between them, published key texts on odontology. Many of their original specimens, illustrated in their great works, can still be seen at the Odontological Museum and Hunterian Museum of the Royal College of Surgeons and continue to act as an important resource.

The present book has been written mainly for biological anthropologists, amongst whom there are several different groups with an interest in teeth. One of the largest of these groups (known as bioarchaeologists in America) focuses on collections of human remains that have been excavated from archaeological sites, aiming to reconstruct the demography, biological affinities, diet, health and general way of life of past populations from a range of skeletal and dental evidence. The teeth are particularly resistant to the destructive effects of long burial in the ground, and thus occupy an important place in this work. Archaeological collections are often compared with similar studies of living people, and one further advantage of teeth is that direct comparisons can readily be made. Forensic anthropologists make up another group with an interest in teeth. Their aim, in most cases, is to identify very fragmentary remains and the teeth become important when the remains are so damaged as to make identification difficult by any other means. Forensic anthropology is usually considered to be distinct from forensic dentistry (often called forensic odontology), which concentrates particularly on such matters as bite marks, or matching dental records with evidence for dental surgery, so these areas have deliberately not been included here as there are several texts that deal with them in detail (Cottone & Standish, 1981; Whittaker & MacDonald, 1989; Clark, 1992). Anthropological methods come into their own where the remains have no evidence of dental treatment or it is not possible to find dental records for matching purposes – still a common enough occurrence in many parts of the world amongst those unable to afford treatment. Palaeoanthropologists make up another large group, with an interest in the fossil remains of (mostly extinct) primates, and they overlap with primatologists, who are concerned with the biology and behaviour of both living and fossil representatives of the primates. Full consideration of teeth in these fields would have expanded the book out of reasonable bounds, so the focus has been restricted to our own species *Homo sapiens* and our closest relatives. There are enough similarities within this group for descriptions of dental anatomy and physiology to cover them all, but their place within a broader range of primates is dealt with elsewhere (Swindler, 1976; Aiello & Dean, 1990).

We and our closest relatives are usually included in the family Hominidae (hominids), which is combined with the great apes or family Pongidae (pongids), into the super-family Hominoidea. There is a great deal of dispute about the correct way of dividing living and extinct hominids and pongids into families and species, and in any text some decision is needed about which terms and definitions to use (the terms used in this book are given in Table 1.1 and are an attempt to follow the most widespread practice). The australopithecines are a relatively well-defined group of African hominid fossils, and many researchers now separate the less heavily built of these into the genus *Australopithecus*, whilst placing the more robust into a separate genus *Paranthropus* (Grine, 1988). They have since been joined by the fossil remains

Table 1.1. *The family Hominidae*

Species	Sites	Stratigraphic division	Date ranges
Australopithecines			
Ardipithecus ramidus	East Africa	Pliocene	*c.* 4.4 Ma BP
Australopithecus anamensis	East Africa	Pliocene	4.2–3.9 Ma BP
Australopithecus afarensis	East Africa	Pliocene	3.75–2.8 Ma BP
Australopithecus africanus	South Africa	Pliocene	3–2.5 Ma BP
Paranthropus robustus	South Africa	Lower Pleistocene	1.8–1.5 Ma BP
Paranthropus boisei	East Africa	Pliocene, Lower Pleistocene	*c.* 2.6–1.2 Ma BP
Hominines			
Homo habilis	Africa (+ ?)	Pliocene, Lower Pleistocene	2.2–1.6 Ma BP
Homo erectus	Asia (+ Africa for earlier dates)	Middle Pleistocene	700–125 ka BP (1.9 Ma BP, 1.6 Ma BP)
Homo sapiens (archaic)	Africa + Europe	Middle Pleistocene	700–125 ka BP
Homo sapiens (Neanderthal)	Europe + West Asia	Upper Pleistocene	100–35 ka BP
Homo sapiens (anatomically modern)	Worldwide	Upper Pleistocene + Holocene	90 ka BP, 50 ka BP –present

BP, years before present; Ma, millions of years; ka, thousands of years; Holocene, 10 ka BP–present; Pleistocene, 2 Ma–10 ka BP; Pliocene, 5.1 Ma–2 Ma BP.

from Aramis in Ethiopia, which were originally labelled *Australopithecus* (White *et al.*, 1994), but have now been placed in a new genus *Ardipithecus* (White *et al.*, 1995). A further recent addition to the australopithecines is the new species *Australopithecus anamensis* (Leakey *et al.*, 1995), defined on fossil finds from Kenya. The other major genus within the Hominidae is *Homo* itself, and there is considerable argument about which of the earlier African fossils should be included in it. The species *Homo habilis*, as originally defined, includes a rather variable collection of specimens and it is now suggested that these may be better divided into two species (Wood, 1991). Similarly, whilst the core of *Homo erectus* is clearly defined as a group of Middle Pleistocene fossils from China and Java, earlier specimens referred to it are more controversial. The species *Homo sapiens* is frequently divided into three groups, and some researchers actually give these different species names too. Archaic *Homo sapiens* as used in this book includes a heterogeneous collection of Middle Pleistocene specimens from Africa and Europe. Neanderthals are a much more clearly defined group, with a core of specimens from Western Europe and material assigned from Eastern Europe and Western Asia. Anatomically modern *Homo sapiens* includes living people throughout the world and similar fossils from the Upper Pleistocene.

The organization of this book

Basic field and laboratory methods applying to the dental anthropology of archaeological, museum and forensic remains are outlined in Appendix A. Chapters 2, 3 and 4 all deal with morphology. The aim of Chapter 2 is to describe the basic anatomy of teeth and to summarize the criteria for identifying them that are most useful in anthropology, especially with fragmentary material. Chapter 3 is one of the largest sections, because it deals with morphological variation – the core of dental anthropology for many people – whilst Chapter 4 examines dental occlusion, or the way in which teeth fit together. Chapter 5 is concerned with the development of the teeth during childhood, concentrating on the important evidence that this provides for age-at-death in childrens' remains. Chapters 6, 7, 8 and 9 all cover different aspects of the microscopic structure of dental tissues, passing from enamel, to dentine and then cement, and the age estimation methods that are based upon them. Much of the discussion is concerned with images from dental microscopy, for which details are given in Appendix B for those unfamiliar with microscope work. Chapter 10 covers the relatively new field of the biochemistry of dental remains, which is likely to be the focus of much future research. Chapter 11 deals with the wear of teeth and the evidence it yields about age and diet, whilst Chapter 12 explores dental disease and its interpretation particularly in relation to the diet. The conclusion, Chapter 13, attempts to summarize the main achievements of dental anthropology, the problems and possible future directions. This book makes no attempt to quote the whole literature of dental anthropology – no book could – but it does attempt to provide an introduction to the main skills required, the major issues raised, with a pathway to the literature so that readers can follow up these arguments themselves. The bibliography is therefore one of the largest sections.

2

Dental anatomy

Terms and definitions

Further details for modern *Homo sapiens* are given in Jordan *et al.* (1992), Carlsen (1987), van Beek (1983) and Woelfl (1990), whereas details for other hominids are available in a variety of sources (Robinson, 1956; White *et al.*, 1981; Wood & Abbott, 1983; Wood *et al.*, 1983; Grine, 1985; Wood & Uytterschaut, 1987; Wood *et al.*, 1988; Wood & Engelman, 1988; Wood, 1991; White *et al.*, 1994), and the patterns of wear described in this chapter are based on the work of Murphy (1959a). Carlsen provided an alternative terminology of 'fundamental macroscopic units' for describing tooth morphology, which is not included here because it has not yet been employed in dental anthropology. For comparisons with non-human mammal teeth see publications by Hillson (1986a; 1992c) – the main potential confusions are with bear and pig molars, with incisors in deer, cattle and their relatives, and the great apes, which are described in detail, along with other primates, by Swindler (1976).

Labels for teeth

Each child has two dentitions. The deciduous (or milk) dentition is about half-formed by birth and erupts into the mouth during the next two years (page 124). It is replaced gradually by the permanent dentition, for which the first tooth starts to form just before birth, and the last tooth is finally completed in the early twenties. Each dentition is divided into four quadrants: upper left, upper right, lower left and lower right. Left and right quadrants are separated by the midline of the skull (the median sagittal plane) – so that the upper left quadrant mirrors the upper right and the lower left mirrors the lower right (Figure 2.1). Within each quadrant there are different classes of teeth – incisors (Latin *dentes incisivi*; cutting teeth), canines (Latin *dentes canini*; dog teeth),

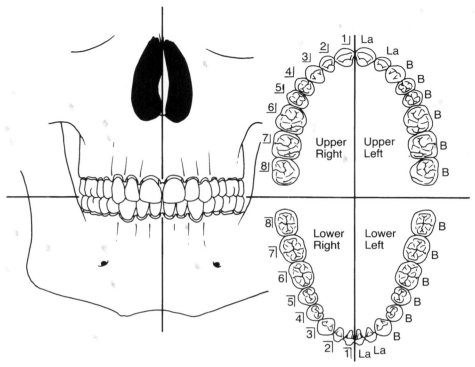

Figure 2.1 Quadrants of the permanent dentition (see Table 2.1). La, labial surfaces; B, buccal surfaces.

premolars and molars (Latin *dentes molares*; grinding teeth) – and incisors and canines are often described together as anterior teeth, whilst premolars and molars are called cheek teeth. In each quadrant of the permanent dentition there are normally two incisors, one canine, two premolars and three molars. Each quadrant of the deciduous dentition similarly comprises two incisors, one canine and two cheek teeth, which are normally called deciduous molars but which are, properly speaking, deciduous premolars. This problem arises because the traditional names in dentistry were adopted without reference to palaeontology. In the mammals as a whole, it is considered that deciduous dentitions consist only of incisors, canines and premolars, and there may be up to four premolars in each quadrant, depending on the species. In the case of human deciduous dentitions, the two 'molars' are equivalent to the third and fourth premolars of other mammals but, in spite of this, they continue to be called the first and second deciduous molars in human dentistry. There is a similar problem with the permanent premolars, which are described in (human) dental texts as the first and second premolars but are really third and fourth premolars. For consistency, however, the traditional dental terms are used in this book (Table 2.1).

Tooth names are cumbersome and several shorthand notations are used in dentistry (Table 2.1). The Zsigmondy system denotes the deciduous teeth of each quadrant by lower case letters (a – e), the permanent teeth by numbers (1–8), and the quadrants themselves by vertical and horizontal bars. Another common notation is the Fédération Dentaire Internationale (1971) two digit system, where the first digit indicates quadrant and dentition, and the second digit denotes the tooth. The FDI system is designed for rapid entry into computer databases and is ideal for anthropological recording of large collections.

Components and surfaces in tooth crowns and roots

Each tooth is divided into a crown and a root. The crown is the part that projects into the mouth and the root is embedded in the jaws. Dentine is the tissue that forms the core of the whole tooth, and the crown is coated with enamel whilst the root is coated with a thin layer of cement (Figure 2.3). The boundaries between these tissues are termed the enamel–dentine junction (EDJ), cement–dentine junction (CDJ), and cement–enamel junction (CEJ). The meeting point between the crown and the root is the *cervix* (Latin; neck – cervical is used as an adjective for this part of the tooth) and, for some reason, this formal anatomical name is retained today whilst the formal names *corona* (Latin; crown) and *radix* (Latin; root) are rarely used, even though their adjective derivatives coronal and radicular are often employed. The base of the crown is called the cervical margin and, girdling the cervical one-third of the crown, there is often a broad bulge called the *cingulum* (Latin; girdle). Inside the tooth is the pulp chamber, containing the soft tissue of the pulp, with small conical hollows (horns or diverticles) in its roof, and a floor which opens into a root canal, or canals. A tooth may have several roots, each with a root canal, and the point at which roots are divided is known as the root fork, or furcation.

The aspect of the crown (Figure 2.2) that faces teeth in the opposing jaw when the mouth closes is known as the occlusal aspect (Latin *facies occlusalis*; closed up face). In human molars and premolars there are broad crown surfaces that actually meet when the jaws shut, and these can truly be called occlusal surfaces, but incisors and canines are tall and spatulate with high crowns that do not normally meet edge-to-edge, and overlap instead (page 114). In anterior teeth, it is therefore clearer to call the occlusal extremity of the crown the incisal edge (Latin; *margo incisalis*). The complete opposite of occlusal is the aspect which contains the tips of the roots and, as the tip of each root is known as its apex, this is called the apical aspect.

The remaining four aspects of each tooth are labelled in relation to its

Table 2.1. *Tooth labelling systems*

Deciduous dentition

Tooth name	Zsigmondy system	FDI system
First incisor (often called 'central' incisor)	a	1
Second incisor (often called 'lateral' incisor)	b	2
Canine	c	3
First molar (more correctly the third premolar)	d	4
Second molar (more correctly the fourth premolar)	e	. 5

Permanent dentition

Tooth name	Zsigmondy system	FDI system
First incisor (often called 'central')	1	1
Second incisor (often called 'lateral')	2	2
Canine	3	3
First premolar (more correctly the third premolar)	4	4
Second premolar (more correctly the fourth premolar)	5	5
First molar	6	6
Second molar	7	7
Third molar	8	8

Zsigmondy system lines to denote jaw quadrants

	Right	Left
Upper	⌐	¬
Lower	⌐	¬

Examples of Zsigmondy system

|6̲ = permanent upper left first molar

c̄| = deciduous lower right canine

FDI system codes to denote quadrant

Permament dentition

	Right	Left
Upper	1	2
Lower	4	3

Deciduous dentition

	Right	Left
Upper	5	6
Lower	8	7

Examples of FDI system

26 = permanent upper left first molar

83 = deciduous lower right canine

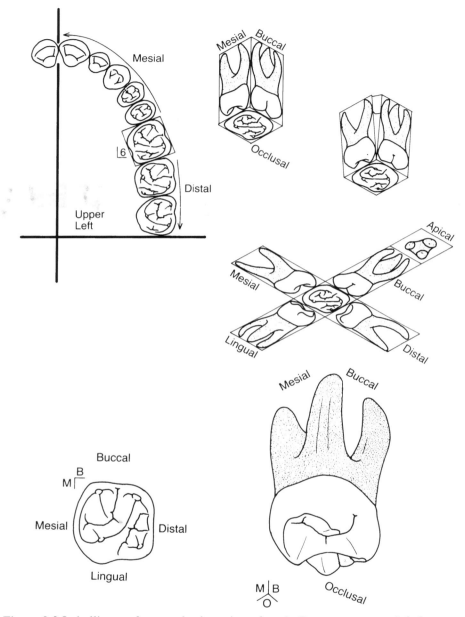

Figure 2.2 Labelling surfaces and orientation of teeth. Permanent upper left first
molar. Upper half of figure: the six aspects (See Table 2.2). Lower half of figure:
orthogonal projection of occlusal surface and isometric projection of mesial–
buccal–occlusal surface, showing the method of labelling orientation in Chapter
2 by marking one corner with abbreviations (M, mesial; D, distal; O, occlusal;
B, buccal; La, labial; Li, lingual).

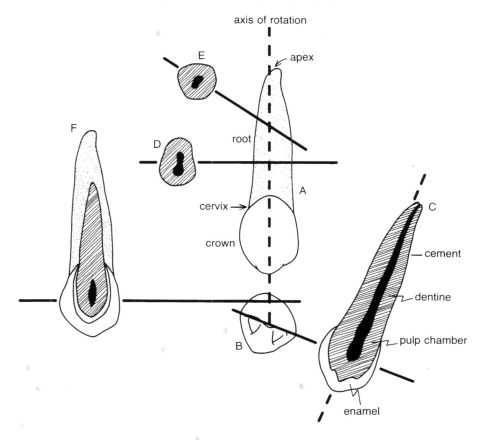

Figure 2.3 Planes of tooth section. Illustrated using the permanent upper right canine. A, buccal aspect; B, occlusal aspect; C, radial section (includes the axis of rotation of the tooth); D, transverse section of root (perpendicular to the axis of rotation); E and F, tangential sections of root and tooth (any other orientation).

position in the line of teeth, which curves around in an arch known as the dental arcade. One surface of each tooth faces along the arcade towards the median sagittal plane (page 6), so this surface is labelled mesial (Greek *mesos*; middle). The opposite surface faces along the arcade away from the median sagittal place, and is called distal (a word made up from the 'dist' part of *distant* to be the opposite of mesial). When pairs of neighbouring teeth in the same jaw are being considered, it is usual to talk about their adjoining sides (mesial facing distal) as approximal surfaces. The surface that faces the tongue is known as lingual (Latin *facies lingualis*; tongue face) but in the upper jaw, where it faces the palate, the same surface is often termed palatal (Osborn,

Table 2.2. *Surfaces of teeth*

Mesial – the surface of all teeth that faces along the dental arcade towards the median sagittal plane.
Distal – the surface of all teeth that faces along the dental arcade away from the median sagittal plane.
Lingual – the surface of all teeth that faces the tongue.
Labial – the surface of incisors facing the lips.
Buccal – the surface of canines, premolars and molars facing the cheeks.
Occlusal – the surface of all teeth that parallels the plane in which upper and lower dentitions meet. In particular, it is used to describe the surface of premolars and molars that meets in normal occlusion.
Incisal – the cutting edge of incisors and canines (equivalent to occlusal in the teeth that overlap during normal occlusion).
Apical – the surfaces of a tooth that face towards the apex of the roots.

1981). Finally, the surface that faces outside the dental arcade, towards the cheeks and lips, is called variously buccal (Latin *facies buccalis*; cheek face), labial (Latin *facies labialis*; lip face), or vestibular (after the vestibule, which is the space formed between the teeth, lips and cheeks). The word 'labial' is usually reserved for incisors, although some authors use it for the canines as well (Woelfl, 1990; Jordan *et al.*, 1992). Similarly 'buccal' is usually used to describe canines, premolars and molars, whereas 'vestibular' may be used for all teeth and is really a much more useful term. This part of the nomenclature is thus in rather a mess, but the terms in this book (Table 2.2) probably represent the most common usage worldwide.

Measurements of crown size

Only measurements of the crown are of much help for identification purposes, and definitions for the mesiodistal and buccolingual (or labiolingual) diameters are given on pages 70 and 71. In general usage, mesiodistal diameter is described as the crown length, and buccolingual diameter as the crown breadth, whereas crown height is the distance from CEJ to cusp tips. Root length similarly describes the distance from CEJ to apex.

General note on drawings

The drawings in this chapter are designed purely to illustrate points in the text and tables. They are best seen as caricatures, which intentionally emphasize features and try to distill the key elements for identification into simple lines, without complicating the picture by shading. Most use an orthogonal projection, with each tooth viewed from one of the six aspects, seen as if its outlines were traced on a piece of paper held directly above. To give an impression of modelling to complex three-dimensional structures, some figures use an isometric projection. Here, the tooth is tilted so that three aspects can be viewed at once from the corner between them, and the two main orientations are views from the mesial–buccal–occlusal corner and from the distal–lingual–occlusal corner. In all cases, the orientation of the tooth in the view is shown by labelling one corner with M (mesial), D (distal), O (occlusal), La (labial), Li (lingual) or B (buccal). All teeth are reproduced at twice life size, using average dimensions as the base. Root (cement) surfaces are stippled, and exposed dentine surfaces are indicated by hatching where cut sections or worn surfaces are shown (Figure 2.3). Some figures are designed to show the structure of the pulp chamber and roots in section (Figures 2.5, 2.10, 2.15, 2.23 and 2.30) and three planes of section are used – radial, transverse and tangential – defined in Figure 2.3. Other figures are designed particularly to aid in orientating specimens (Figures 2.6, 2.7, 2.11, 2.12, 2.16, 2.24 and 2.31). To use them, the tooth should initially be held right-way-up with the lingual surface facing the observer and held over the figure to match against the quadrants of dentition shown.

General appearance of teeth in x-rays

In an x-ray (page 306) heavily mineralized enamel shows as a more radio-opaque layer than the dentine or bone (which have roughly equivalent densities), and it is possible to identify the EDJ clearly under the sides of the crown (Figure 2.4). Cement cannot usually be distinguished, because it has insufficient density contrast with dentine, and forms a very thin layer in any case. The pulp chamber and root canal are seen as radiolucencies that are clearly visible in the cervical part of the crown and the root, but are masked by the convoluted enamel of the occlusal crown. Sometimes, diffuse radiolucencies can be seen at the mesial and distal borders of the cervical part of the root. These are called cervical burnout and are a feature of normal radiographic anatomy.

The tooth sockets, with their lining of alveolar bone (page 260), show as a thin radio-opaque layer known as the lamina dura in radiographs, separated from the dentine and cement of the tooth roots by a narrow radiolucent space representing the periodontal ligament (page 260). The radiographic appearance of the lamina dura is highly variable, and it may seem patchy and diffuse even in a normal jaw. The crest of the alveolar process (the part of each jaw that holds the dental arcade) is represented in radiographs as a radio-opaque line which joins the lamina dura, but which is often itself poorly defined.

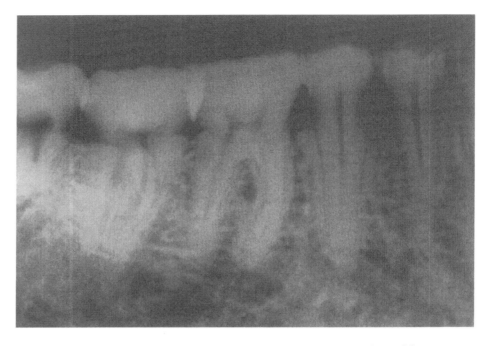

Figure 2.4 Radiograph of lower permanent molars and premolars with supporting bone, in a post-Medieval archaeological specimen from London. See text
for explanation of features. The lamina dura is most strongly marked in the first
molar and second premolar.

Incisors

See Figures 2.5 to 2.9 and Tables 2.3 to 2.6.

Incisor crowns

Incisor crowns of all hominids are similar; spatulate, with a sharp incisal edge,
convex labial surface and concave lingual surface. The lingual surface (Figure
2.6) has marginal ridges spreading out occlusally from a cingulum bulge (the
tuberculum) at the cervix. Two shallow depressions, running down from the
incisal edge on both labial and lingual sides, divide the tooth into three lobes

Figure 2.5 Permanent and deciduous left first and second incisors. Top row,
permanent upper incisors; second row, permanent lower incisors; third row,
deciduous upper incisors; bottom row, deciduous lower incisors. A, labial
aspects of first and second incisors; B, mesial aspect of first incisor; C, transverse
root section looking down to apical; D, radial tooth section in labiolingual plane;
E, radial tooth section of first incisor in mesiodistal plane. Approximately twice
life-size.

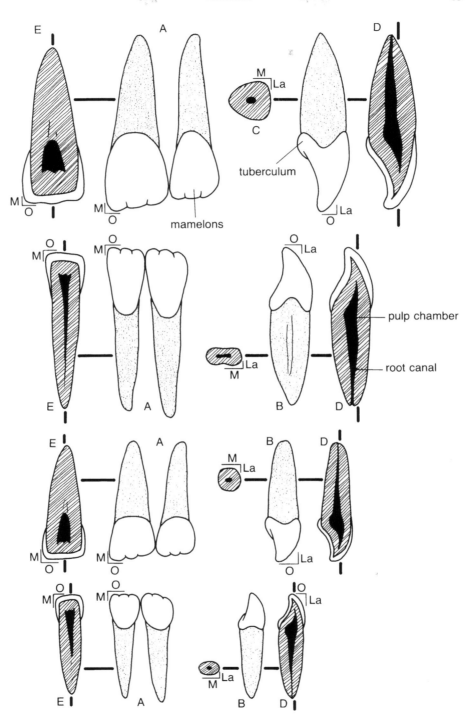

tuberculum

mamelons

pulp chamber

root canal

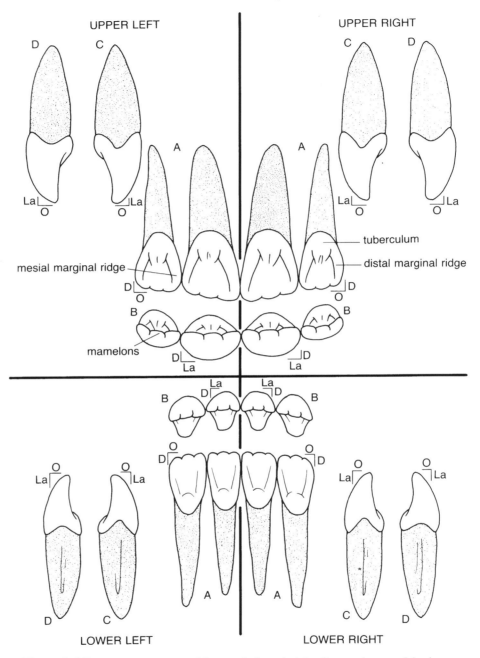

Figure 2.6 Permanent upper and lower, left and right first and second incisors. Upper left, upper right, lower right and lower left quadrants. A, lingual aspects of first and second incisors (first placed to mesial of second); B, incisal aspects; C, mesial aspect of first incisors; D, distal aspect of first incisors. Approximately twice life-size.

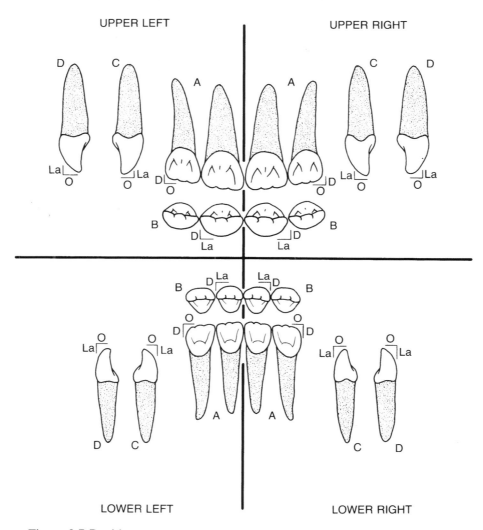

Figure 2.7 Deciduous upper and lower, left and right first and second incisors. Upper left, upper right, lower right and lower left quadrants. A, lingual aspects of first and second incisors (first placed to mesial of second); B, incisal aspects; C, mesial aspect of first incisors; D, distal aspect of first incisors. Approximately twice life-size.

whose tips are marked by three low cusplets (mamelons) along the incisal edge of unworn incisors. Where the lingual marginal ridges are strongly developed, and reach the incisal edge, the mesial and/or distal mamelons may be further sub-divided to produce four or five in total (Figure 2.9). The incisal edge slopes from mesial down to distal, and the mesial surface is flatter and

Dental anatomy

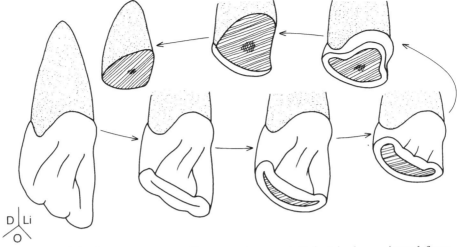

Figure 2.8 Incisor wear stages. Permanent upper left first incisor, viewed from the distal–lingual–incisal corner. Approximately twice life-size.

more vertically arranged, whilst the distal surface bulges out above the cervix (Figures 2.5 and 2.6). The cervical margin curves strongly down to apical on the labial and lingual sides but on the mesial and distal sides there is a correspondingly strong occlusal curve, which is more pronounced on the mesial side (Figure 2.6).

For both permanent and deciduous dentitions, the upper incisor crowns are always larger than those of lower incisors, and the upper first incisor crown is larger than the second, whereas the two lower incisors are about the same size as one another. These contrasts hold good within all hominid taxa but, for each incisor type, there are considerable overlaps between taxa in the range of crown dimensions.

Figure 2.9 Permanent upper incisor variants. A, three mamelons (and lingual rootlet); B, four mamelons; C, five mamelons; D, double shovelling (A to D are all upper left first incisors viewed from mesial–labial–incisal corner). E to G, labial outline of upper left first incisors (tapering, square and ovoid); H to K, labial outline of upper left second incisors (normal, distal element reduced, peg-shaped and distal indentation). L, two tubercle extensions (and cement extension); M, three tubercle extensions; N, shovelling; O, shovelling with tubercle furrow (L to O are all upper left first incisors viewed the distal–lingual–incisal corner). P, normal incisor; Q barrel-shaped incisor (P and Q are upper left second incisors viewed from the distal–lingual–incisal corner). R, crown root groove (lingual view of upper left second incisor). Approximately twice life-size.

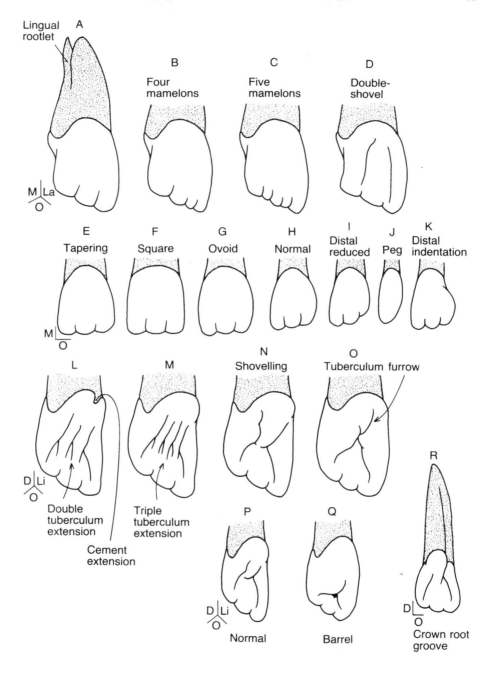

A — Lingual rootlet
M | La
O

B — Four mamelons
C — Five mamelons
D — Double-shovel

E — Tapering
F — Square
G — Ovoid
H — Normal
I — Distal reduced
J — Peg
K — Distal indentation
M
O

L
Double tuberculum extension
Cement extension
D | Li
O

M
Triple tuberculum extension

N — Shovelling

O — Tuberculum furrow

P — Normal
D | Li
O

Q — Barrel

R
D
O
Crown root groove

Table 2.3. *General orientating features for incisors*

1. The labial surface is convex, whereas the lingual surface is concave.
2. The lingual surface has marginal ridges and a tuberculum.
3. The mesial surface is more upright and the distal surface more bulging to distal, giving a distally skewed outline in labial view.
4. The incisal edge slopes from mesial to distal; similarly the worn occlusal surface.
5. Higher occlusal curve of cervical margin on mesial side than on distal (more in permanent than deciduous).
6. The labial element of the root is often bulkier than the lingual.
7. Slight distal curve of root apex.

See Figure 2.6.

Table 2.4. *Key features for distinguishing upper and lower incisors*

1. Upper incisors are larger than lower.
2. Upper crowns are more distinctly shovel-shaped, with a broad and strongly convex labial surface, large tuberculum and strong marginal ridges.
3. Lower incisors are more chisel-shaped, with flatter labial surface, narrow but prominent tuberculum and weak marginal ridges.
4. Upper crowns are longer than broad; lower crowns are broader than long.
5. Upper crowns are more markedly asymmetrical than lower, with a sharp mesioincisal corner and rounded, bulging distoincisal corner.
6. Upper roots are stout and rounded triangular in section; lower roots are compressed mesial–distal. More in permanent than deciduous.

See Figures 2.5–2.7.

Incisor roots and pulp chamber

Incisors normally have one root, with flattened sides, or broad, shallow grooves running down its mesial and distal surfaces, dividing it into labial and lingual elements, with the labial element being bulkier than the lingual (Figure 2.5). The apical one-third of the root is usually, but not always, tilted slightly to distal. Incisor pulp chambers are located inside the cervix of the tooth and have three diverticles (one for each mamelon) but are not clearly separated from the root canal, so that a funnel- or slot-like entrance forms the floor. There is usually one root canal, although there may sometimes be two, but in such cases the division occurs within the root and there is still only one entrance from the pulp chamber.

Table 2.5. *Key features for distinguishing first and second incisors*

Upper

1. First incisors are larger than second.
2. First incisor crowns are longer (in relation to breadth) than second.
3. Second incisor crowns are more asymmetrical than first, with a more rounded distoincisal corner.
4. Second incisors are more variable than first.

Lower

1. First incisor crowns are almost symmetrical; second incisor crowns are clearly asymmetrical.
2. The incisal edge is rotated (see in occlusal view) in second incisors.
3. Second incisor roots have a slightly greater distal curve than first.

See Figures 2.5–2.7.

Incisor wear and fragmentation

Wear starts at the tips of the mamelons, which are rapidly worn away to leave a smooth incisal edge. A thin line of dentine is exposed (Figure 2.8) along its length, with heavier wear sometimes exposing a wider line at the distal end. As wear progresses, differences even out as a broader band of dentine is exposed and, if the marginal ridges of the crown are prominent, the mesial and distal ends of this band are swollen. When wear reaches the tuberculum, a lingual bulge is added to create a broad dentine area with an enamel rim and, eventually, the rim is breached as wear proceeds down the root.

Incisors often fall out from their sockets in dry skulls (unless they have abnormally curved roots). The crown may also break away at the cervix, and then fracture along a labiolingual radial plane. Fracturing along a mesiodistal plane is less common but, if wear has exposed a line of dentine at the incisal edge, this may act as a line of weakness. The resulting crown halves or quadrants are still clearly identifiable as incisors but, where crown enamel is destroyed in cremations, the remaining roots are fractured into drum-like elements which are less easily identified (although the compressed roots of lower incisors may be distinguishable).

Identifying incisors

All incisors are most readily orientated (Table 2.3) from their asymmetry in labial or lingual view. This is particularly marked in second incisors, especially the upper, and is notable in *Paranthropus* and (for lower second incisors) in *Australopithecus africanus*. Lower incisors are distinguished (Table 2.4) by

Table 2.6. *Permanent upper incisor variants*

General outline

1. In labial outline first incisors may be square, ovoid or tapering[a].
2. The distal lobe of second incisors may be reduced, sometimes enough to produce a peg-shaped form[a].
3. The crown may be hook-shaped in mesial or distal view.

Mamelons

1. There are usually three mamelons, but sometimes there are four or five[a].
2. The mesial mamelon is usually highest but, in second incisors, it may be the central mamelon.

Marginal ridges

1. Lingual marginal ridges vary in bulk; when pronounced, they produce *shovel-shaped* incisors[a].
2. The labial surface may also have marginal ridges (*double-shovel* form)[a].
3. The cervical part of marginal ridges may be swollen in second incisors, creating a deep fissure at the tuberculum[a].
4. The grooves lining the marginal ridges vary in prominence and may be fissure-like.

Tuberculum

1. The tuberculum varies in prominence, especially in second incisors; in *barrel-shaped* forms, it is almost full crown height and encloses a deep pit[a].
2. There may be one, two (the normal number) or three tuberculum projections – small ridges extending into the dished lingual surface[a].

Concavities and grooves

1. The mesial side of second incisor crowns may be indented part way down[a], so that the mesial marginal ridge is curved or sharply cut by a groove.
2. There may be a groove running down the mesiolingual corner of the crown, crossing the cervical margin and extending down the root (*crown-root groove*)[a].

Cervical margin

1. A *cement extension* may cut the first incisor labial cervical margin[a].
2. The mesial or distal cervical margins may curve in a gentle 'U' or a sharp 'V'.

Root

1. There may be an additional groove down the labial side of the root.
2. The root may divide at its apical end to produce labial and lingual rootlets[a].
3. There may be two root canals.

[a] See Figure 2.9.

Table 2.7. *Key features for deciduous incisors*

1. Deciduous incisors are much smaller than permanent.
2. Deciduous roots are resorbed prior to exfoliation (page 138).
3. The root is short (even in its unresorbed state) relative to crown height.
4. Deciduous upper incisor crowns are long relative to their height.
5. Wear proceeds rapidly through the thin enamel, accentuating the long and low appearance of the upper incisor crown in labial view.
6. There are three mamelons along the unworn incisal edge and the central is markedly smaller than the mesial or distal.
7. The pulp chamber has less pronounced diverticles than are present in permanent teeth, the central diverticle being the smallest.
8. The root canal is large in diameter relative to the outside root diameter.

See Figures 2.5 and 2.7.

their short, chisel-like crowns, in which the contrast between first and second incisor (Table 2.5) is much less marked than in the upper teeth. Upper incisors show considerable variation in form (Table 2.6 and Figure 2.9), particularly the second incisor, which is the most variable tooth in the dentition after the third molar (page 75). Lower incisors show relatively little variation of this type, although the central mamelon may sometimes be smaller than the others, especially in *Paranthropus*. The deciduous incisors are, in many ways, like scaled-down versions of the permanent teeth, but their squat, low, bulging crowns are distinctive (Table 2.7).

Canines

See Figures 2.10 to 2.14 and Tables 2.8 to 2.12.

Canine crowns

Hominid canine crowns are broadly spatulate, with three coalesced elements similar to incisors, but with the main difference that the central element is always dominant. There is a single main, central cusp, with ridges running down the incisal edge to mesial and distal – the mesial ridge being shorter than the distal (except in the deciduous upper canine) – and with a prominent buttress that runs down the lingual surface, to meet the cingulum bulge of the tuberculum (Figures 2.11 and 2.13). Marginal ridges, outlined by grooves, are prominently developed on the lingual surface, with the mesial marginal ridge bulkier and more vertically arranged than the distal, and reaching slightly higher to occlusal. For most canines, the crown is therefore asymmetrical in buccal or lingual outline, with a more vertical mesial side and more bulging distal side, but the deciduous upper canine is arranged in exactly the opposite way (Figures 2.10 and 2.12). The cervical margin (Figure 2.11) in all canines curves down smoothly to apical on the buccal and lingual sides and up to occlusal on the mesial and distal sides (the mesial curve being the more prominent in permanent canines, as in incisors).

Canine roots and pulp chamber

Canines normally have one long root with, in permanent canines, broad, shallow grooves running down its mesial and distal sides, and its apical third curving to distal (although this is quite variable). The pulp chamber lies inside the cervix, with one large diverticle corresponding to the main cusp (Figure 2.10), and it grades into the root canal, without a clear floor. Inside the root there is usually one large canal, of compressed section, but sometimes this divides within the root into two canals – a buccal and a lingual.

Canine wear and fragmentation

In most hominids, the tip of the canine main cusp comes into wear first and a dot of dentine is exposed (Figure 2.13). Soon afterwards, the mesial and/or distal ridges of the incisal edge are worn enough to expose a line of dentine. With further wear, the main labial bulge of the central cusp element comes into the worn facet, together with the lingual buttress and marginal ridges, to produce an area of dentine with an outline a little like a heavy moustache. As wear approaches the cervix, the dentine area opens out into a diamond shape, bounded by a higher rim of enamel and, in the final stages, this rim is breached as wear proceeds down the root. The wear facet often slopes to distal on lower canines, but *Australopithecus afarensis* shows characteristically asymmetrical wear facets, sloping strongly to distal on upper canines and to mesial on lower. Approximal wear is initiated at the contact points of the canines on the most prominent mesial and distal bulges of the crown sides (the distal contact point is usually more cervically placed than the mesial). Wear proceeds slowly at these contact points, but eventually produces facets.

Canines usually fall out of their sockets in dry skulls (unless they have an abnormally curved root). The crowns also frequently break away from the root

Figure 2.10 Permanent and deciduous left canines. Top row, permanent upper canine; second row, permanent lower canine; third row, deciduous upper canine; bottom row, deciduous lower canine. A, buccal aspect; B, mesial aspect; C, transverse root section looking down to apical; D, radial tooth section in labiolingual plane; E, radial tooth section in mesiodistal plane. Approximately twice life-size.

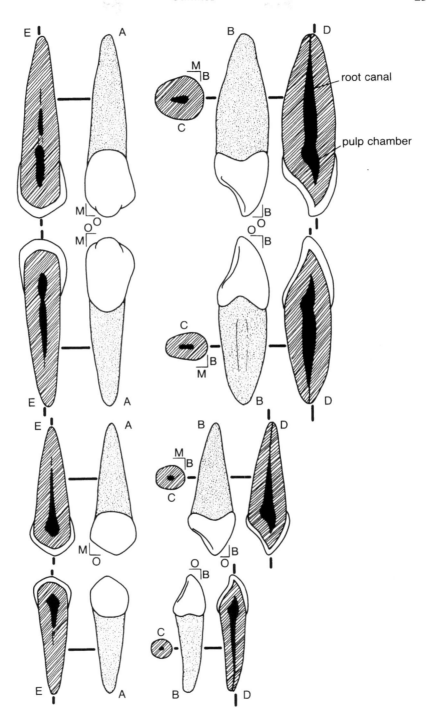

root canal

pulp chamber

Dental anatomy

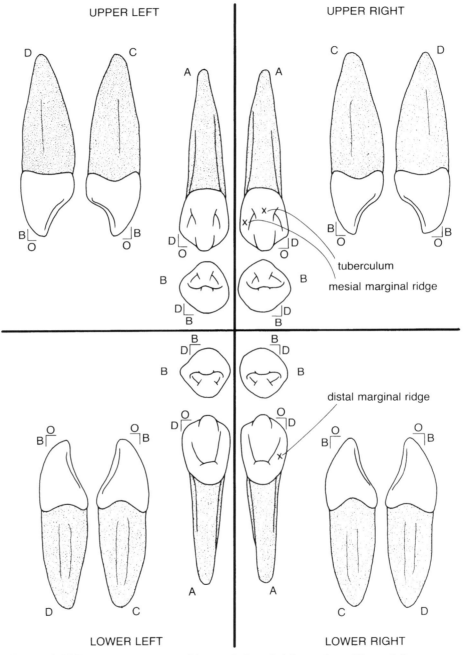

Figure 2.11 Permanent upper and lower, left and right canines. Upper left, upper right, lower right and lower left quadrants. A, lingual aspect; B, incisal aspect; C, mesial aspect; D, distal aspect. Approximately twice life-size.

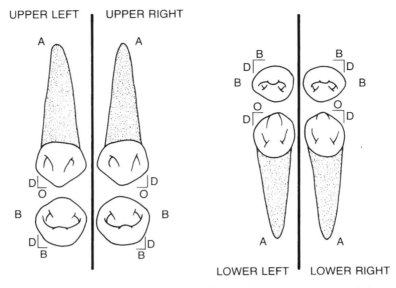

Figure 2.12 Deciduous upper and lower, left and right canines. Upper left, upper right, lower right and lower left quadrants. A, lingual aspect; B, incisal aspect. Approximately twice life-size.

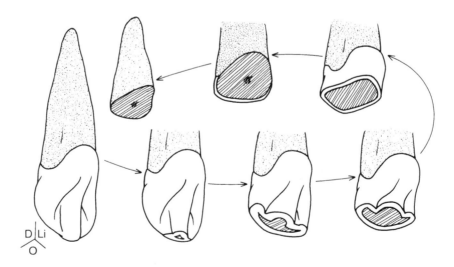

Figure 2.13 Canine wear stages. Permanent upper left canine, viewed from the distal–lingual–incisal corner. Approximately twice life-size.

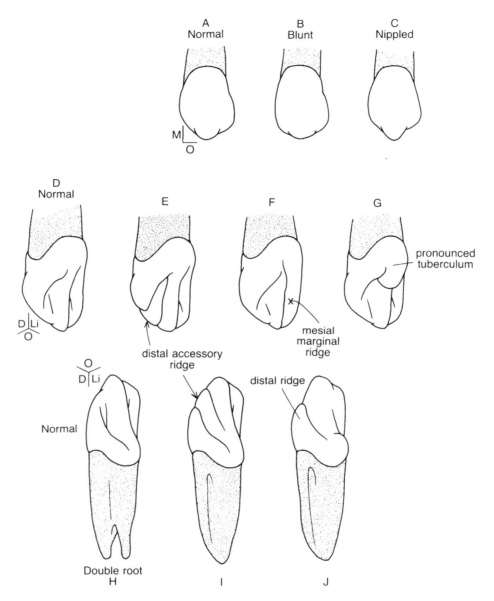

Figure 2.14 Permanent canine variants. A to C, buccal outline of upper left canine (normal, blunt, nippled). D, normal ridges and tuberculum; E, distal accessory ridge; F, canine mesial ridge; G, lingual tubercle (D to G are all upper left canines viewed from the distal–lingual–incisal corner. H, double-rooted lower canine; I, distal accessory ridge; J, distal ridge (H to J are all lower left canines viewed from the distal–lingual–incisal corner). Approximately twice life-size.

Table 2.8. *Key features for distinguishing incisors and canines*

1. Unworn canines have a clear central cusp on their incisal edge, whereas incisors have three (usually) similar-sized mamelons.
2. The central cusp is supported by a prominent buttress, which runs down the lingual surface and joins the tuberculum, although this buttress is less prominent in some lower canines. The buttress is never present in incisors.
3. Canine crowns are taller than incisor crowns, and their roots are longer.
4. Worn canines tend to have a diamond-shaped area of exposed dentine, whereas worn incisors show a line of dentine.
5. One common error is to confuse worn lower canines (with a less prominent buttress than normal) with upper incisors. In such cases, the most useful distinctions are the higher, shorter (mesiodistally) canine crown, and more prominent mesial and distal root grooves in lower canines.

Table 2.9. *Key features for distinguishing upper and lower canines*

1. Upper canines are stouter, broader teeth than lower.
2. In buccal outline, upper canines bulge to mesial and to distal much more markedly than lower.
3. In buccal outline, the mesial crown margin of a lower canine is often so upright that it lies almost in a straight line with the mesial root side.
4. The lingual buttress, tuberculum and marginal ridges are less strongly developed in lower canines than upper, and uppers are more concave lingually.
5. The grooves outlining the marginal ridges and tuberculum are often deep and even fissure-like in upper canines, but never so in lower.
6. The upper canine sometimes has tuberculum extensions, never the lower.
7. The lower canine root is often more compressed in section than the upper.
8. Lower canines sometimes have an accessory root, never seen in the upper.

See Figures 2.10–2.12.

at the cervical margin and then fracture again along a buccal–lingual radial plane through the main cusp tip, but are still easily identifiable. Cremated remains produce fractured drum-like segments that may be difficult to distinguish from those of upper incisors.

Identifying canines

Canines are most easily orientated by their asymmetry (Table 2.10, Figures 2.11 and 2.12), as seen in buccal or lingual view (NB deciduous upper canines

Table 2.10. *Key features for orientating canines*

1. The buccal surface is broadly convex; the lingual surface is concave.
2. The crown, seen in buccal view, leans over to distal in all canines except the deciduous upper canines, which have exactly the opposite asymmetry.
3. The mesial marginal ridge on the lingual surface is more prominent than the distal, again in all canines except deciduous upper.
4. The incisal curve of the cervical margin is deeper on the mesial side than the distal (the difference is not pronounced in deciduous canines).
5. If facets have been worn at the contact points, they are usually more cervically placed on the distal side (on all except deciduous upper canines).

See Figures 2.10–2.12.

Table 2.11. *Variation in permanent canine form*

1. The slope of the incisal ridges varies, as does the form of the unworn main cusp tip; 'normal', 'blunt', 'nippled'[a].
2. A groove may separate the central and distal crown elements on the lingual side, and it sometimes outlines a prominent *distal accessory ridge* (DAR)[a].
3. The tip of the distal marginal ridge sometimes reaches onto the incisal edge[a]. This is particularly marked in strongly asymmetrical *Australopithecus* lower canines and is a separate feature to the DAR.
4. In upper canines, there may be a similar mesial ridge[a].
5. The relative prominence of the mesial and distal marginal ridges varies. Both may be strongly developed in the upper canine, especially in *Paranthropus*, where the central buttress of the lingual surface is also weakly developed.
6. The prominence of the tuberculum varies in the upper canine. Sometimes it is swollen into a separate cusplet[a].
7. The number and size of tuberculum extensions in the upper canines varies (as for upper incisors).
8. The depth of grooving down the root varies: lower canines may be split into two rootlets[a].
9. The lower canine may have a buccal or lingually placed accessory root.
10. The direction and degree of root curvature varies.

[a] See Figure 2.14.

have an opposite direction of asymmetry to all other anterior teeth). Upper canines are more robust, bulging teeth than lower canines (Table 2.9), in both permanent and deciduous dentitions, although there is little difference in their average crown dimensions and, whereas permanent canines vary in overall form and detailed morphology (Table 2.11 and Figure 2.14), they show few

Table 2.12. *Key features for deciduous canines*

1. Deciduous canines are much smaller than permanent.
2. Deciduous roots are resorbed prior to exfoliation (page 138).
3. Deciduous crowns are low, long and chubby – particularly the upper canine.
4. Deciduous canines are strongly constricted at the cervix and the crown flares out above this to mesial and to distal, with a marked labial convexity and tuberculum.
5. The root is rounded and markedly conical in form.
6. The deciduous upper canine incisal edge contrasts with all other canines in having its mesial ridge longer than the distal.

See Figure 2.12.

consistent size differences between hominid taxa. Worn lower canines in all hominid taxa may take on a superficial resemblance to upper incisors as their main cusp is lost, and care is needed in identification (Table 2.8). Deciduous canines have many features in common with their permanent successors, but their crowns are distinctively long and low and their roots are markedly rounded and conical (Table 2.12).

Permanent upper premolars

See Figures 2.15 to 2.19 and Tables 2.13 to 2.15.

Upper premolar crown features

Upper premolar crowns have two main cusps; buccal and lingual (Figure 2.15). The buccal cusp is larger and taller, with its tip on the crown midline, whereas the smaller lingual cusp tip is set slightly to mesial (Figure 2.16). The cusps are joined at their sides by prominent mesial and distal marginal ridges – the mesial being slightly higher than the distal. Between the two main cusps is a mesial–distal groove, usually with deeper pits at its ends (mesial fossa and distal fossa), which are separated by buttresses running down from the tips of the cusps towards the centre of the groove (Figure 2.16). Smaller grooves extend from the mesial fossa inside the mesial marginal ridge, and another (especially in the first premolar) often extends over the ridge crest (developmental groove; Figures 2.16 and 2.18). Similar small grooves extend out from the distal fossa, but less often cross the marginal ridge.

The contrast in size between the buccal and lingual cusps is less marked amongst the australopithecines than it is in *Homo*, and the crown has a more

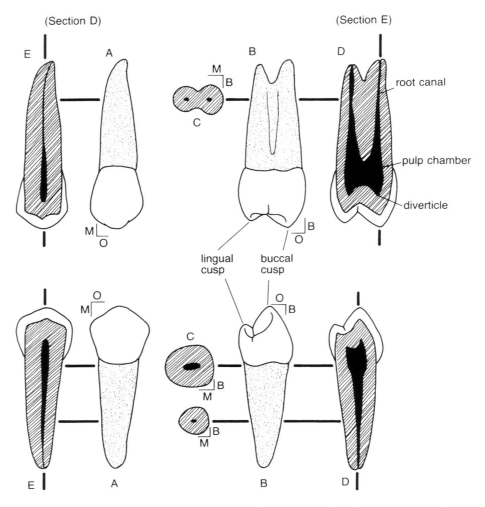

Figure 2.15 Permanent upper and lower left first premolars. Top row, upper first premolar; lower row, lower first premolar. A, buccal aspect; B, mesial aspect; C, transverse root sections looking down to apical; D radial tooth section in buccolingual plane; E, radial/tangential tooth section in mesiodistal plane. Approximately twice life-size.

massive, bulging appearance, especially in *Paranthropus* (Figure 2.19). The occlusal outline of the crown is thus a regular trapezoid in *Homo*, but more oval amongst the australopithecines and, furthermore, australopithecine upper premolars bear a prominent cingulum bulge, which is only modestly developed in *Homo*. In all hominids, the mesial and distal sides flare out above the cervical margin, up to the marginal ridges. The buccal and lingual cervical

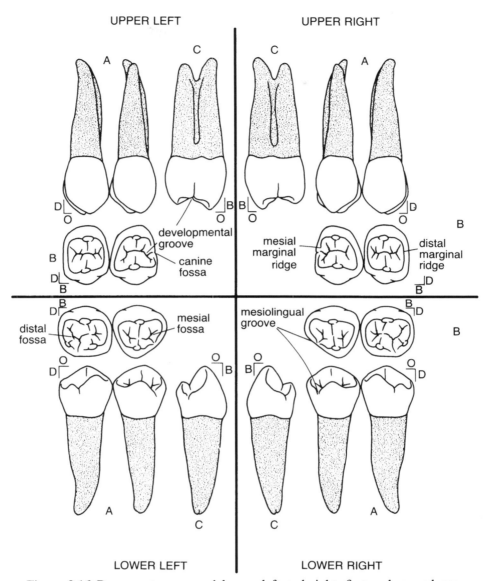

UPPER LEFT   UPPER RIGHT

developmental groove

canine fossa

mesial marginal ridge

distal marginal ridge

distal fossa

mesial fossa

mesiolingual groove

LOWER LEFT   LOWER RIGHT

Figure 2.16 Permanent upper and lower, left and right, first and second premolars. Upper left, upper right, lower right and lower left quadrants. A, lingual aspects of first and second premolars (first placed to mesial of second); B, occlusal aspects; C, mesial aspect of first premolar. Approximately twice life-size.

margins are at roughly similar levels, whereas the distal (and especially the mesial) margins curve slightly to occlusal (Figures 2.15 and 2.16).

In living *Homo*, average crown diameters are about the same for both first and second premolars, or in some populations smaller for the second premolar (Figure 3.1). Within one individual, the second premolar is usually the smaller

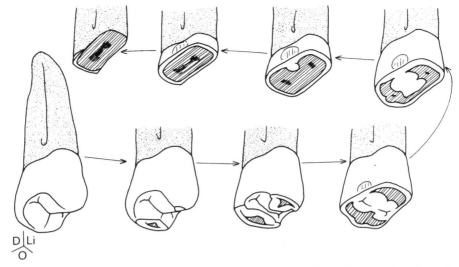

Figure 2.17 Permanent upper premolar wear stages. Second premolar viewed from distal–lingual–occlusal corner. Approximately twice life-size.

tooth but, in australopithecines, especially *Paranthropus*, the second tends to be the larger tooth. *Paranthropus* premolars are on average larger than those of *Australopithecus*, which are in turn larger than those of *Homo*, although there is considerable overlap in range.

Upper premolar roots and pulp chamber

Upper premolars may have one, two or three roots. A fully two-rooted tooth has a larger buccal and a smaller lingual root (Figure 2.18). The one-rooted form has grooves down the mesial and/or distal sides, dividing the root into two elements with various degrees of separation – often just the apices are separated. The three-rooted form is particularly common in *Paranthropus* (Figure 2.19), but is also occasionally found in *Homo*, and has mesiobuccal, distobuccal and lingual roots. The two buccal roots are smaller in diameter than the lingual root but are usually longer and, again, various degrees of separation exist, with the furcation for the buccal roots usually to apical of their furcation with the lingual root.

The pulp chamber occupies the root trunk and cervical part of the crown (Figure 2.15). There are two diverticles under the main cusps, the buccal diverticle being much more prominent than the distal. Both one-rooted and two-rooted premolars always have two root canals, and two funnel shaped entrances open from the buccal and lingual corners of the pulp chamber floor. In three-rooted forms, two smaller funnel shaped entrances open from the buccal corners, in addition to the lingual opening.

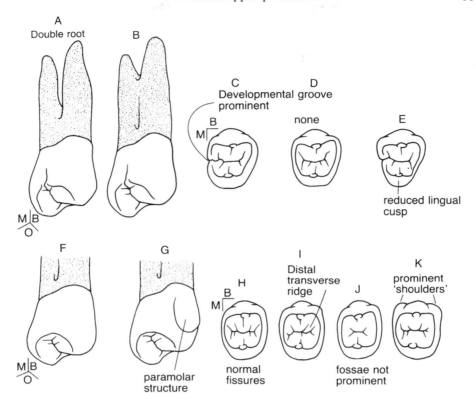

Figure 2.18 Permanent upper premolar variants. A and B, upper left first pre-molars viewed from the mesial–buccal–occlusal corner (double-rooted and normal). C, prominent developmental groove; D, no developmental groove; E, reduced lingual cusp (C to E are all upper left first premolars seen in occlusal view). F and G, upper left second premolars viewed from the mesial–buccal–occlusal corner (normal and paramolar structure). H, normal fissures; I, distal transverse ridge; J, fossae not prominent; K, prominent shoulders (H to K are all upper left second premolars seen in occlusal view). Approximately twice life-size.

Permanent upper premolar wear and fragmentation

Occlusal wear is initiated at the tips of the cusps, normally exposing a dot of dentine in the buccal cusp first (Figure 2.17). After both cusp tips have been worn away to expose areas of dentine, the crest of one of the marginal ridges is breached to link these areas, and leaves a peninsula of enamel in the central groove that is eventually reduced, with further wear, to a small dot or 'island'. When this island is lost, the softer dentine forms a dished area in which the darker secondary dentine is often exposed, surrounded by the enamel rim.

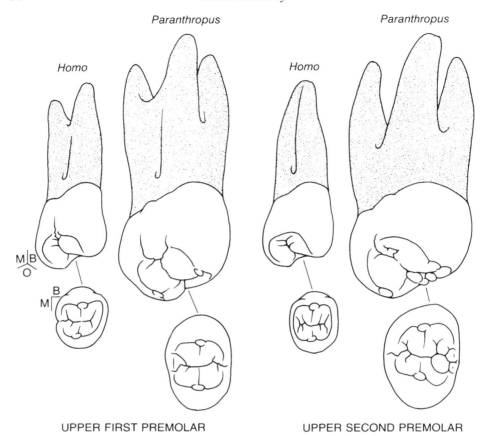

Figure 2.19 Permanent upper left first and second premolars in *Homo* and *Par-anthropus* (reconstructions based on material from Swartkrans), viewed from mesial–buccal–occlusal corner and occlusal aspect. Approximately twice life-size.

Finally the rim is breached, usually on the lingual side first, and the root starts to wear down. The worn occlusal surface slopes down from buccal to lingual at all stages.

The distal contact point is usually positioned slightly more to buccal than the mesial contact point in both upper premolars, and is higher up the crown than the mesial point in the first premolar, but lower in the second. At each contact point, an approximal wear facet develops and grows in size, until it eventually breaches the worn enamel rim of the occlusal surface.

Upper premolars in dry skulls quite often fall out of their sockets, although the two rooted forms remain in place more frequently. The crown commonly fractures away at the cervical margin and, when this occurs, tends to break

Table 2.13 *Key features for orientating permanent upper premolars*

1. The unworn buccal cusp is higher than the lingual.
2. Worn occlusal surfaces tilt from buccal to lingual.
3. The lingual cusp tip is skewed to mesial.
4. The mesial marginal ridge is higher than the distal.
5. A canine fossa is sometimes present on the first premolar mesial side.
6. Buccal root elements are bulkier than lingual.
7. The buccal diverticle of the pulp chamber is more prominent than the lingual.
8. The apical third of roots usually bends to distal.

See Figures 2.15 and 2.16.

Table 2.14. *Key features for distinguishing first and second permanent upper premolars*

1. The occlusal outline in first premolars is often almost triangular – it is square or oval in second premolars.
2. Size contrast between the buccal and lingual cusps is more marked in first premolars.
3. The mesial skew of the lingual cusp is more marked in first premolars.
4. The central groove is longer and fossae are more marked in first premolars (often reduced to a short slot in second).
5. The marginal ridges are bulkier in second premolars.
6. A 'developmental groove' cuts the crest of the mesial marginal ridge in first premolars (occasionally present in second).
7. A 'canine fossa' is present in first premolars – this is a broad concavity in the mesial crown side.
8. First premolars are normally two-rooted (or at least has two apices) in *Homo*; second premolars are usually one-rooted.

See Figure 2.16.

into two halves along the central groove, often with further fracturing along a buccolingual plane to create quadrants that are usually still identifiable. The roots are similarly distinctive and may even be identifiable from cremations, where they tend to fracture into flattened sections with hourglass outlines.

Identifying upper premolars

Upper premolars are readily distinguished from other tooth classes by their double cusps and tendency to double roots. They are most easily orientated (Table 2.13) from the mesial skew of the lingual cusp, but it can be difficult

Table 2.15. *Variation in permanent upper premolar form*

Cusps

1. The buccal cusp varies in pointedness, and in the bulge of its shoulders[a] where they join the marginal ridges.
2. The lingual cusp may be reduced in first premolars[a].
3. An *accessory cusplet* may appear in the centre of the occlusal area, arising from the buccal cusp element.
4. A *paramolar structure* is occasionally present, arising from the cingulum of the buccal crown side[a], but is never prominent.

Ridges and grooves

1. The *marginal ridges* are often interrupted by a groove extending from the fossa[a]. Such grooves are sometimes fissure-like.
2. One or two small cusplets may be present on the marginal ridge crests (especially on the distal marginal ridge of *Paranthropus*[b], creating a 'molarized' appearance).
3. The prominence of the main primary fissures varies[a], and secondary fissures and grooves may be superimposed over the primary pattern, particularly in australopithecine premolars[b].
4. An additional distal transverse ridge[a] may extend into such irregular fissure patterns.

Roots

1. In *Homo*, first premolars have two roots[a] (sometimes one and rarely three), whereas second premolars have one (sometimes two and rarely three).
2. In *Australopithecus*, both premolars generally have two roots (sometimes three and rarely one).
3. In *Paranthropus*, both premolars have three roots[b] (sometimes two and never one).

[a] See Figure 2.18.
[b] See Figure 2.19.

to separate first from second premolars (Table 2.14), especially in isolated and worn teeth. The best clue is the more triangular occlusal outline of most first premolars, which contrasts with the squarer second premolar outline (Figure 2.16). Some variation in form is commonly shown (Table 2.15 and Figure 2.18), and there are particular differences between hominid taxa, associated with the increased 'molarization' of premolars in australopithecines (Figure 2.19).

Permanent lower premolars

See Figures 2.15, and 2.20 to 2.22, with Tables 2.16 to 2.18.

Lower premolar crown features

Lower premolars are usually two-cusped or three-cusped, with one dominant, buccal, cusp and one or two smaller lingual cusps (Figures 2.15, 2.16 and 2.21). Marginal ridges run down from the main buccal cusp to mesial and to distal, connecting it with the lingual cusps (the mesial marginal ridge slopes down more sharply than the distal). A mesial–distal groove separates the cusps in the centre of the occlusal surface and a buttress runs down the buccal cusp into this groove, dividing it into two depressions – the mesial and distal fossae (Figure 2.16). A groove may extend lingually from the mesial fossa over the mesial marginal ridge and, in three-cusped forms, there is a groove between the two lingual cusps that creates an additional central fossa. The deepest parts of these grooves are often fissure-like.

The occlusal outline of lower premolars is round in two-cusped forms, to squarish in three-cusped (Figure 2.16). Buccal and lingual crown sides bulge somewhat in most hominids, whereas the mesial and distal sides are more flaring, out from the cervical margin up to the marginal ridges. This flare is more pronounced in first premolars than second. The cervical margin runs approximately level around the crown, but slightly lower on the buccal side, and with a very weak occlusal curve on the mesial and distal sides, particularly in the first premolar.

In living *Homo* the first premolar crown is roughly the same size as the

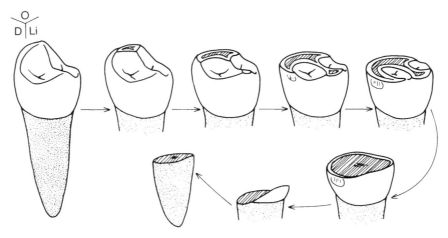

Figure 2.20 Permanent lower premolar wear stages. First premolar viewed from distal–lingual–occlusal corner. Approximately twice life-size.

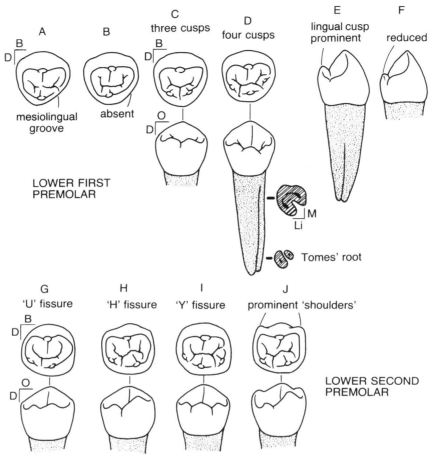

Figure 2.21 Permanent lower premolar variants. A and B, lower left first pre-molars seen in occlusal view (prominent mesiolingual groove, no mesiolingual groove). C and D, lower left first premolars seen in occlusal and lingual view (three-cusped form, four-cusped form and Tomes' root). E and F, lower left first premolars seen in mesial view (prominent lingual cusp, reduced lingual cusp). G, 'U' form fissures; H, 'H' form fissures; I, 'Y' form fissures (with four cusps); J, prominent 'shoulders' (G to J are all lower left second premolars seen in occlusal and lingual views). Approximately twice life-size.

Figure 2.22 Permanent lower left first and second premolars in *Homo*, *Austral-opithecus* and *Paranthropus* (reconstructions based on material from Sterkfont-ein and Swartkrans). Top row, lower first premolar viewed from mesial–buccal–occlusal corner and occlusal aspect. Middle row, lower second premolar viewed from mesial–buccal–occlusal corner and occlusal aspect. Bottom row, transverse root sections of hominid lower premolars, showing three-groove and four-groove forms (labelled as in text). Approximately twice life-size.

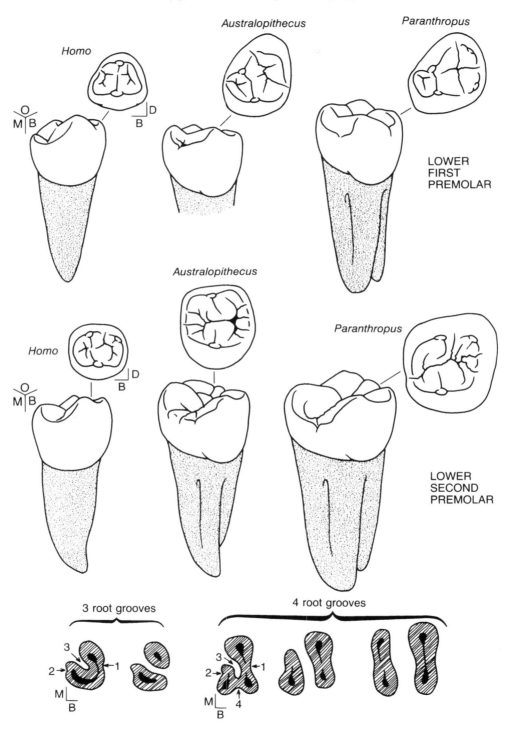

LOWER
FIRST
PREMOLAR

LOWER
SECOND
PREMOLAR

3 root grooves   4 root grooves

Table 2.16. *Key features for orientating permanent lower premolars*

1. The buccal cusp is largest.
2. The mesial marginal ridge slopes down more markedly to lingual than the distal.
3. The distal approximal wear facet is positioned more to lingual than the mesial facet.
4. The occlusal wear facet slopes to buccal.
5. In first premolars a prominent groove runs from the mesial fossa over the mesial marginal ridge.
6. In three-cusped premolars, the mesiolingual cusp is more prominent than the distolingual.
7. The apical third of the root is slightly curved to distal.

See Figure 2.16.

Table 2.17. *Key features for distinguishing first and second permanent lower premolars in* Homo

1. First premolars are usually two-cusped, and second premolars three-cusped.
2. The occlusal outline in first premolars is circular, but squarer in second premolars.
3. The mesial/distal flare of the crown is more pronounced in first premolars than second.
4. There is a marked contrast in size between buccal and lingual cusps in first premolars, with marginal ridges running steeply down.
5. The smaller difference between buccal and lingual cusps in second premolars makes the crown bulkier, with a central pit.
6. A prominent mesiolingual groove is present in many first premolar crowns.
7. Double (or partly divided) roots are much more common in first premolars than second premolars.

See Figure 2.16.

second whereas, amongst the australopithecines, the second premolar is usually largest (Figure 3.1). This difference is most marked in *Paranthropus*, where the premolars form part of the expanded cheek tooth row. The average size of both premolars increases in series from *Homo* to *Australopithecus* to *Paranthropus*, which has an expanded distal marginal ridge, decorated with cusps to form a much more molar-like tooth (Figure 2.22).

Table 2.18. *Variation in permanent lower premolar form*

First premolar main cusps

1. In living *Homo* first premolars are normally two-cusped, with the lingual cusp much smaller than the buccal[a], although there may be a total of three or four cusps.
2. In australopithecines (especially *Paranthropus*), the lingual cusp is more nearly equal to the buccal, giving a bulkier crown[b]. Prominent occlusal buttresses connect these cusps, separating the mesial and distal fossae.

Second premolar main cusps

1. Three-cusped forms dominate in living *Homo* and *Australopithecus*, but two-cusped forms can be found, in which the distolingual cusp is lost, and there may even be four cusps[a].
2. *Paranthropus* usually has two main cusps, the lingual being similar in size to the buccal[b].

Marginal ridges and other cusps

1. The distal marginal ridge is strongly developed in *Paranthropus*, to form an expanded occlusal area[b].
2. Both marginal ridges (especially the distal) may bear cusplets along their crests[b]. These are prominent in australopithecines (notably *Paranthropus*).
3. A paramolar structure – groove, pit or cusp – may be present on the buccal side, but is usually poorly developed.

Grooves and fissures

1. The strength of the first premolar mesiolingual groove varies[a].
2. Fissures in second premolars of living *Homo* may take on a 'U', 'H' or 'Y' form[a].

Roots

1. Living *Homo* premolars usually have one root. Occasionally Tomes' type double roots (mesiobuccal and distolingual) are present and, more rarely, treble roots (mesiobuccal, distobuccal and lingual). Multiple roots of any kind are very rare in second premolars[ab].
2. Second premolars in australopithecines usually bear two flattened roots (mesial and distal), although the roots are not always visible[b].
3. First premolar roots in *A. africanus* are mostly of Tomes' type but, because they are not always fully visible, it is not clear whether this condition often includes two separate roots. *A. africanus* has double roots.
4. Most *P. robustus* first premolars have Tomes' type double roots, but a few have large, flattened, mesial and distal roots. All *P. boisei* first premolars are of the latter type.

[a] See Fgure 2.21.
[b] See Figure 2.22.

Lower premolar roots and pulp chamber

Lower premolars in living *Homo* normally have a single conical root, of round section (Figure 2.15). Many, however, show a slight flattening of the mesiolingual and/or the distal surfaces, to give a more triangular section, and sometimes these flattened areas are deepened into grooves (Figure 2.22):

1. Distal groove – usually shallow.
2. Midline mesial groove – usually confined to the apical third of the root, but occasionally deep enough to split it into two rootlets (buccal and lingual).
3. Mesiolingual groove – most commonly found in the first premolar, but rarely pronounced, and often confined to the apical third of the root. Sometimes, it forms a deep cleft to produce a 'C'-shaped root section (Tomes' root; Figures 2.21 and 2.22) and, in some cases, the cleft is deep enough to split the root into two (mesiobuccal and distolingual rootlets).
4. Buccal groove – very rare in modern *Homo* but sometimes deep enough to divide the root. Double-rooted teeth of this type are, however, common in australopithecines and the 'molarised' premolars of *Paranthropus* usually have two large, flattened roots.

In most one-rooted forms, there is just one root canal of oval section but, in the multi-rooted forms, there may be two or more (Figure 2.22). In living *Homo*, the division usually occurs within the root, with only one entrance from the pulp chamber, whereas in australopithecines there are often several entrances, as in molars. The pulp chamber (Figure 2.15) is thus not clearly demarcated from the root canal in living *Homo*, as the root canal entrance takes the place of the pulp chamber floor, although the roof has diverticles associated with the main cusps (with the buccal being much the most prominent).

Lower premolar wear and fragmentation

Lower premolars start to wear at their most prominent point, the tip of the buccal cusp (Figure 2.20). This is where the first dot of dentine is exposed and, later, a smaller dot also appears at the tip of the most prominent lingual cusp. The pattern of wear varies after this, with trails of the yellower dentine appearing where the crests of ridges are breached and, finally, only a rim of enamel remains, with a dished area of dentine inside. Throughout the process, the worn facet tilts to buccal and the enamel rim is usually breached first on this side. The initial contact points of the lower premolars with their neighbouring teeth are between the middle and occlusal third of the crown sides, but lower and more to lingual on the distal side than the mesial. Approximal wear produces facets around these points, which may aid in identifying slightly worn teeth.

In dry jaws, single-rooted lower premolars frequently fall out. The crown may break away at the cervical margin, and then fracture again along a buccol-ingual radial plane through the main cusp tip. In cremations, the roots tend to break up into drum-shaped fragments, which are often difficult to identify.

Identifying permanent lower premolars

The general form of lower premolars in living *Homo* is distinctive with their round occlusal outline, dominant buccal cusp and small lingual cusps, but the division between first and second premolars (Table 2.17), left and right (Table 2.16), relies heavily on details of the occlusal surface, and worn teeth may be difficult to identify. Considerable variation (Table 2.18 and Figure 2.21) is seen particularly in the first premolar and, together, the lower premolars show some of the clearest morphological differences between hominid taxa (Figure 2.22).

Permanent upper molars

See Figures 2.23 to 2.27 and 2.30, and Tables 2.18 to 2.20.

Permanent upper molar crowns

Upper molar crowns have four main cusps (Figure 2.24). The three largest (mesiolingual, mesiobuccal and distobuccal) form a raised triangle (Figures 2.24 and 2.26), with the two mesial cusps linked by a marginal ridge, and the distobuccal and mesiolingual cusps linked by a strong oblique ridge (or *crista obliqua*). At the centre of this cusp triangle is a prominent central fossa (or *fovea centralis*), from which three fissures extend (Figure 2.24). One runs between the mesiobuccal and distobuccal cusps onto the buccal surface (the buccal fissure), and another indents the oblique ridge. The last extends to mesial, ending in a small mesial fossa (or *fovea anterior*) just inside the mar-ginal ridge. The fourth cusp, the distolingual, is less prominent than the others and is connected to the distobuccal cusp by the distal marginal ridge, but is otherwise separated from the main cusp triangle by the distal fossa (or *fovea posterior*). From this (Figures 2.24 and 2.26), a marked fissure runs to lingual (the lingual fissure) and another indents the side of the oblique ridge, whilst two more indent the distal marginal ridge, passing over onto the distal side and usually isolating a small cusplet (the metaconule). Both buccal cusp tips rise higher than either lingual cusp, and the lingual crown side bulges out more strongly than the buccal side (Figure 2.23). The cervical margin runs

UPPER FIRST MOLAR

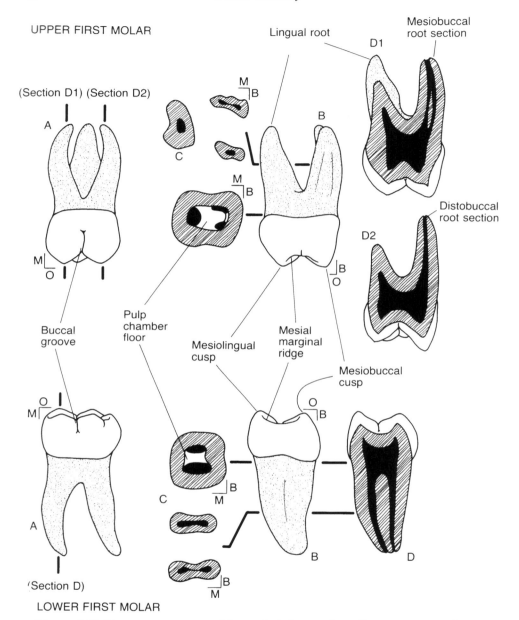

(Section D1) (Section D2)

Lingual root

Mesiobuccal root section

D1

B

Distobuccal root section

D2

Buccal groove

Pulp chamber floor

Mesiolingual cusp

Mesial marginal ridge

Mesiobuccal cusp

(Section D)

LOWER FIRST MOLAR

Figure 2.23 Permanent upper and lower left first molars. Top row, upper first molar; lower row, lower first molar. A, buccal aspect; B, mesial aspect; C, transverse sections of roots and root trunk looking down to apical; D, tangential tooth sections in a buccolingual plane. Approximately twice life-size.

UPPER FIRST MOLAR

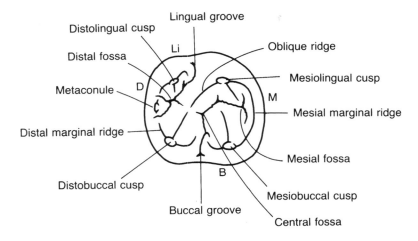

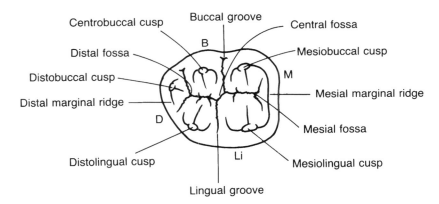

LOWER FIRST MOLAR

Figure 2.24 Permanent upper and lower left first molars. Occlusal views of upper left first molar (top) and lower first molar (bottom). Four times life-size.

straight along the buccal and lingual sides, but curves slightly to occlusal on the mesial and distal sides (especially mesial).

In living *Homo*, the first molar crowns of any one individual are usually largest, followed by the second and then the third (Figure 3.1). An individual australopithecine or early *Homo* jaw by contrast normally has larger second molars than first molars but, whilst the third molar crown is generally larger than the first molar, it may be either larger or smaller than the second (it is always larger in *Paranthropus*). The mean diameters of upper molar crowns

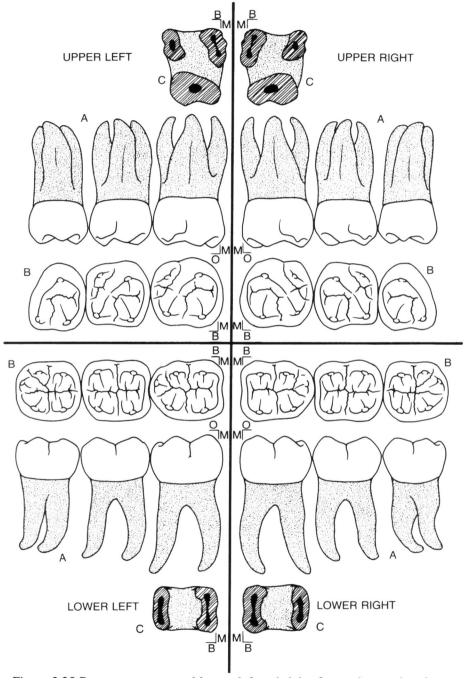

Figure 2.25 Permanent upper and lower, left and right, first and second molars. Upper left, upper right, lower right and lower left quadrants. A, lingual aspects (with first molars closest to mesial and third to distal); B, occlusal aspect (similarly organized); C, transverse root section of first molar looking up to cervical. Approximately twice life-size.

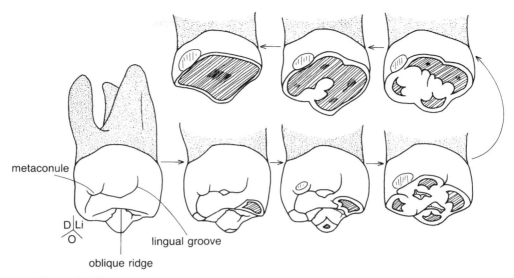

metaconule

D|Li
O

lingual groove

oblique ridge

Figure 2.26 Permanent upper molar wear stages. First molar viewed from distal–lingual–occlusal corner. Approximately twice life-size.

increase in a sequence from *Homo sapiens*, to *H. erectus*, and early *Homo*, *Australopithecus*, *Paranthropus robustus* and *P. boisei*.

Permanent upper molar roots and pulp chamber

Upper molars normally have three roots, one (the largest) lingual and two buccal (Figure 2.23). The tips of the buccal roots curve towards one another, whereas the lingual root is directed to lingual. The fork, or furcation, between the buccal roots is closer to the cervical margin than the mesial furcation, which is in turn closer than the distal furcation. One shallow groove runs down the lingual surface of the lingual root. Two grooves (mesial and distal) run down the distobuccal root, and three grooves (two mesial and one distal) similarly run down the flattened mesiobuccal root.

The pulp chamber lies within the base of the crown and root trunk (Figure 2.23). It has four diverticles, corresponding to the four main cusps, and the diverticles underlying the buccal cusps are the most prominent. Openings to the root canals lie in the corners of the pulp chamber floor. The opening for the mesiobuccal root is slot-shaped, leading to two fine root canals (or one canal of flattened section), whereas the openings for the distobuccal and lingual roots are rounder and more funnel-shaped, leading to one canal each.

Permanent upper molar wear and fragmentation

Usually, a dot of dentine is exposed under the mesiolingual cusp first, followed by the buccal and distolingual cusps (Figure 2.26). The order is, however,

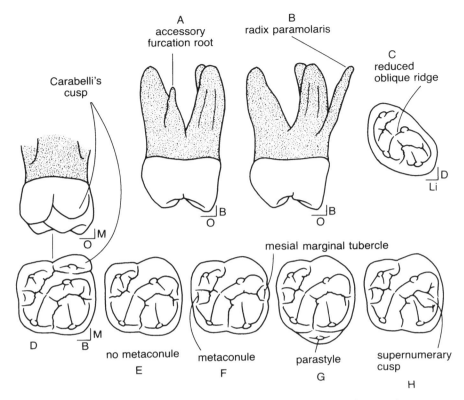

Figure 2.27 Permanent upper molar variants. A and B, upper first molars seen in distal view (accessory furcation root, radix paramolaris). C, upper left molar with reduced oblique ridge seen in occlusal view. D, cusp of Carabelli in upper left first molar, seen in occlusal and lingual views. E to H, upper left molars seen in occlusal view (metaconule missing, mesial tubercle and metaconule both present, parastyle and supernumerary cusp). Approximately twice life-size.

variable. Wear proceeds most rapidly on the lingual cusps and the enamel between them is breached to expose a large 'island' of dentine. Following this, the enamel between both lingual and buccal cusps is breached, leaving a ridge of enamel midway along the buccal side. Eventually, this too is lost, to create a single dished area of dentine with an enamel rim, which ultimately breaks down as the cervical margin is reached, surviving longest on the buccal side. Wear may continue into the root trunk, eventually reaching the furcations and extending down individual roots.

The initial contact points are one-third the way down the crown sides from the occlusal surface. Contact between first molar and second premolar crowns normally occurs just to buccal of the tooth midline. Contact between first and second molars is around the midline, but contact between second and third

Table 2.19. *Key features for orientating upper molars*

1. The main cusp triangle has two higher buccal cusps and one lower, broader cusp on its lingual corner.
2. The distolingual cusp is clearly separate from the main triangle, and lower.
3. The buccal crown side is relatively vertical, whereas the lingual side bulges.
4. If a Carabelli feature is present, it is sited on mesiolingual corner (Figure 2.27).
5. Occlusal wear facets slope to lingual.
6. The lingual root is stouter than the buccal roots, and well separated from them.
7. The mesiobuccal root has a flattened section and often two root canals.
8. The root trunk is tilted to distal, but the relative levels of mesial and distal furcations is variable.
9. The buccal pulp chamber diverticles are more strongly developed than the lingual; and mesial more than distal.
10. A Carabelli cusp is accompanied by an extra mesiolingual diverticle.

See Figures 2.23–2.25.

molars is highly variable. With approximal wear, a facet appears around the point of initial contact confined, to start with, to the enamel but cutting through the rim in the later stages of wear.

The three roots of upper molars are usually locked firmly into their sockets, and it is common for the roots to remain *in situ* whilst the crown fractures away. The crowns often break up into quadrants, along the fissures that separate the main cusps; the quadrants of upper molar crowns being more irregular than those of lower molars. In cremations, the enamel flakes away from the dentine to expose peaks of the EDJ under the cusps, and the roots break into characteristic short, drum-like lengths. With practice it is often possible to identify such fragments by using the variation in root canal form.

Identifying permanent upper molars

Upper molars are distinguished by their trapezoidal to triangular occlusal outline, their main triangle of cusps (the so-called trigon), and their triangle of roots, together providing clear orientating features (Table 2.19), even in worn teeth, which are found in all hominid taxa. There is a large range of variation (Table 2.21 and Figure 2.27), and the third molar is the most variable tooth in

Table 2.20. *Key features for distinguishing first, second and third permanent upper molars in* Homo

1. First molar crown diameters are larger than second, and second larger than third.
2. First molar occlusal outline is trapezoidal, second squarer and third triangular (but variable).
3. The distolingual cusp is usually large in first molars, reduced in second and small or absent in third.
4. First and second molars have mesial and distal approximal wear facets; third molars have a mesial facet only.
5. The roots are most divergent in first molars, less in second and least in third (often fused).
6. The root canal entrances in the pulp chamber floor are closest in first molars, and further apart in third.

See Figures 2.23–2.25.

the mouth. The differences between first, second and third molars are differences of degree (Table 2.20 and Figure 2.25), and it may be difficult to distinguish them from one another even in unworn specimens, particularly first and second molars.

Permanent lower molars

See Figures 2.23 to 2.25, 2.28 to 2.30, and Tables 2.22 to 2.24.

Permanent lower molar crowns

The occlusal outline of the lower molar crown is roughly rectangular in *Homo* (Figure 2.25) and *Australopithecus*, but more oval in *Paranthropus* where the sides of the crown bulge strongly (Figure 2.30). Four main cusps are arranged in the corners of the rectangle and divided by a cross of fissures, which meet in a central depression or fossa (Figure 2.24). The two fissures separating mesial from distal cusps curve over onto the buccal and lingual crown sides – the buccal fissure being the deeper of the two (Figure 2.24). The other two fissures end in the mesial and distal fossae (anterior and posterior foveae) and, in most lower molars, the mesial fossa is bounded by a continuous mesial marginal ridge, whereas a fissure from the distal fossa cuts through the distal marginal ridge (Figure 2.29). All four main cusps are similar in height (Figure 2.23), but the more bulging buccal crown side, with its lower cervical margin, makes the buccal cusps appear broader and lower than the lingual cusps.

Table 2.21. *Variation in permanent upper molar form*

Cusps and ridges

1. Unworn cusps may be high and pointed, or lower and more rounded. *Paranthropus* shows this low rounded form strongly, with bulging crown sides[b].
2. The distolingual cusp varies widely – becoming smaller and more variable from first, to second to third molar. In third molars it is usually absent.
3. The distobuccal cusp is sometimes reduced, or absent, in third molars.
4. The oblique ridge is occasionally breached by a deep fissure connecting the central and the distal fossae. In a few second and third molars, it is so shortened that mesiolingual and distobuccal cusps fuse[a].
5. An additional cusp (Carabelli's cusp) may arise from the lingual crown side below the mesiolingual cusp[a]. The cusp varies greatly in size, and may be represented by a small furrow in the same position. All these features are most common and pronounced on the first molar.
6. The distal marginal ridge usually has two grooves crossing it, separating off a small cusplet (the metaconule) but, occasionally, there is only one groove[a]. Metaconules are most common in first molars, followed by second. The mesial marginal ridge may bear one or more small mesial marginal tubercles[a].
7. A small supernumerary cusplet may be present on the mesiolingual corner of the mesiobuccal cusp[a].
8. A groove, sometimes with a cusp (the parastyle), may be present on the buccal surface at the base of the mesiobuccal cusp[a]. These features vary in prominence and, although potentially found in any upper molar, are most common in the third and least common in the first.

Fissures and fossae

1. The buccal and lingual fissures vary in prominence – some fading out into a shallow furrow, and some terminating sharply, or in a pit. Some buccal pits have a raised rim, or are underlined on their cervical side by a ridge (common in *Paranthropus*[b]).
2. The mesial fossa may be absent, with the mesial fissure running without interruption through the mesial margin (common in *Paranthropus*).
3. Accessory occlusal wrinkles are common, especially in third molars and in *Paranthropus*. Teeth with many such fissures are described as 'crenellated'.

Roots and cervical margin

1. Roots sometimes coalesce; joined by a radicular plate, or as a single fused root. Mesiobuccal and lingual roots fuse most often in third molars, followed by second, then first. Any third molar roots may be fused, often coalescing into one large, conical root.
2. An accessory furcation root may arise between mesiobuccal and lingual main roots[a].
3. In second and third molars, a large accessory root (the radix paramolaris) may arise from the main root trunk on its buccal side[a].
4. The mesiobuccal root usually has two narrow canals, but may have one or three. Distobuccal and lingual roots usually have one canal, but may contain two.

[a] See Figure 2.27.
[b] See Figure 2.30.

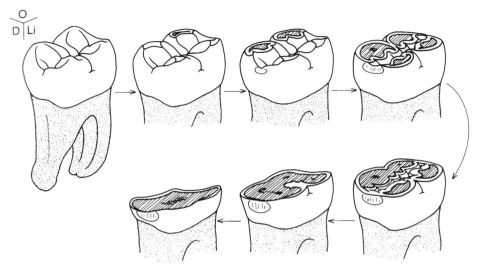

Figure 2.28 Permanent lower molar wear stages. Second molar viewed from distal–lingual–occlusal corner. Approximately twice life-size.

Lower molars in living *Homo* commonly have three, four or five cusps (Figure 2.29). In the three-cusped form, the distolingual cusp is absent. In the five-cusped form, a further distobuccal cusp is added to the basic rectangle – smaller than the others, and separated from its buccal neighbour by an additional fissure, running onto the buccal crown side. There may also be an additional distal cusp (designated cusp 6) or lingual cusp (cusp 7). In living *Homo*, the first lower molar is usually five-cusped, the second four-cusped and the third either five- or four-cusped (Figure 2.25) but, in early *Homo*, *Australopithecus* and *Paranthropus*, all lower molars are usually five- or six-cusped.

Average crown diameters (Figure 3.1) in living *Homo* are largest in first molars, followed by second and then third and, in any one individual dentition, this is also the normal pattern (just occasionally the second molar is smallest). In *Australopithecus africanus*, the second molar is the largest whereas, in

Figure 2.29 Permanent lower molar variants. A to I, lower left molars seen in occlusal view (three-cusped form, four-cusped form, five-cusped form, cusp 6, cusp 7, 'Y' fissures, '+' fissures, 'X' fissures, deflecting wrinkle). J to L, mesial marginal ridge in lower left molar crowns seen in mesial view (cut by a groove, continuous crest, tubercle). M to O, lower left molars viewed from the mesial–buccal–occlusal corner (radix ectomolaris, furcation root, protostylid). P, radix endomolaris in a lower left molar viewed from the distal–lingual–occlusal corner. Approximately twice life-size.

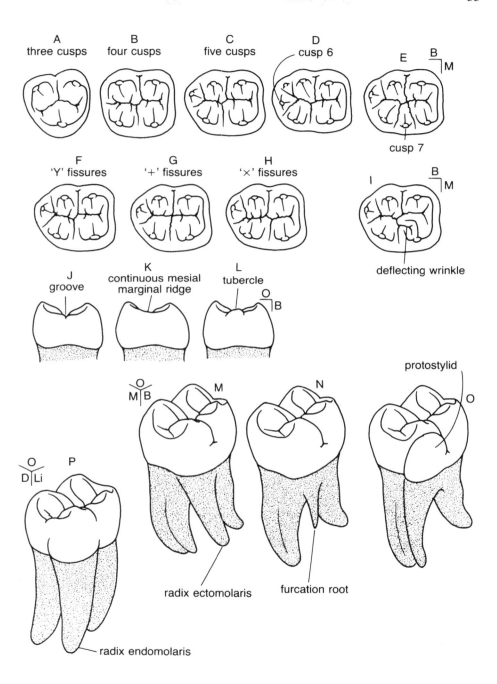

A three cusps

B four cusps

C five cusps

D cusp 6

E

B | M

cusp 7

F 'Y' fissures

G '+' fissures

H 'x' fissures

I

B | M

deflecting wrinkle

J groove

K continuous mesial marginal ridge

L tubercle

O | B

O M | B

M

N

protostylid

O

O D | Li

P

radix ectomolaris

furcation root

radix endomolaris

individual *Paranthropus* and *A. afarensis* dentitions, the teeth increase regularly in size from first molar, to second and then third. As with other cheek teeth, lower molar diameters increase in series from *Homo sapiens*, to *H. erectus*, and early *Homo, Australopithecus, Paranthropus robustus* and *P. boisei*.

Permanent lower molar roots and pulp chamber

Lower molars normally have two roots, mesial and distal, of which the mesial is the larger and more strongly grooved (Figures 2.23 and 2.25). Both roots are flattened in section and slope to distal, with the fork, or furcation, dividing them much closer to the cervical margin on the buccal side than on the lingual side. The pulp chamber lies inside the base of the crown and main root trunk, and it is longer than broad, with three, four or five diverticles in its roof corresponding to the main cusps. Crevice-like openings at the mesial and distal corners of its floor run into the root canals – usually two canals for the mesial and one for the distal root.

Permanent lower molar wear and fragmentation

In many teeth, wear exposes a dot of dentine on the mesiobuccal cusp first, followed by the distobuccal and then lingual cusps (Figure 2.28). Wear continues more rapidly over the buccal cusps and, eventually, the enamel between them is breached to expose a large area of dentine. Next, the enamel between the lingual and buccal cusps is breached, leaving a ridge of enamel midway along the lingual side. Eventually, this too is lost, to expose a dished area of dentine with an enamel rim, which ultimately breaks down as the buccal cervical margin is reached. Wear may continue into the root trunk, eventually reaching the furcation and extending down the individual roots.

The initial contact points between teeth are over one-third the way down the crown sides, and just to buccal of the tooth midline. The contact point between first and second molars is often positioned lower than that between the first molar and second premolar, but contact between second and third molars is highly variable. With approximal wear a facet appears at the contact point and grows slowly, sometimes breaking through the enamel rim before the lingual side is worn away.

The double roots of lower molars are usually locked firmly into their sockets, and in archaeological material it is common for them to remain *in situ* whilst the crown fractures away. The crowns break into quadrants along the fissures – lower molar quadrants being more regular than those of upper molars. In cremations, the enamel flakes away from the dentine to expose the peaks of the EDJ under the cusps, and the roots break into characteristic short,

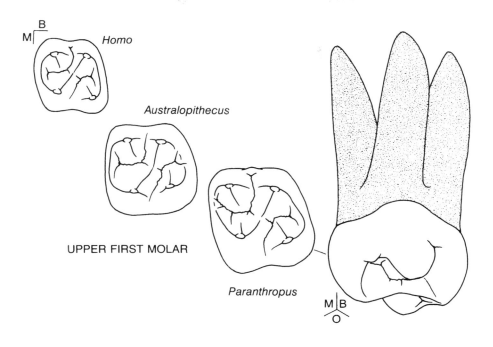

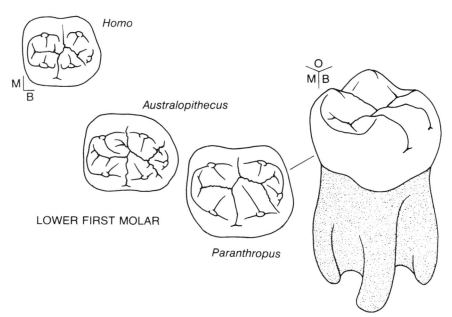

Figure 2.30 Permanent upper and lower left first molars in *Homo*, *Australopithecus* and *Paranthropus*, viewed from the occlusal aspect and mesial–buccal–occlusal corner (reconstructions based on material from Sterkfontein and Swartkrans). Top row, upper molars; lower row, lower molars. Approximately twice life-size.

Table 2.22. *Key features for orientating lower molars*

1. The mesial end of the occlusal outline is flat or concave; the distal end is bulging (especially in five-cusped crowns).
2. The fifth cusp (when present) is in a distobuccal position.
3. The buccal crown side is more bulging, with a lower cervical margin, than the lingual.
4. Occlusal wear facets slope to buccal.
5. Approximal wear facets lie just to buccal of the crown midline.
6. The root trunk is angled to distal, and the buccal root furcation is closer to the crown cervical margin than the lingual furcation.
7. The mesial root is longer and with a larger cross-section than the distal.
8. The pulp chamber is broader, and with higher diverticles, at its mesial end.
9. Mesial roots usually have two canals; distal only one.

See Figures 2.23–2.25 and 2.28.

Table 2.23. *Key features for distinguishing first, second and third permanent lower molars*

1. First molars are usually five-cusped, with a long occlusal outline, bearing a triangular distal extension.
2. In *Homo*, second molars are usually four-cusped, with a regular rectangular occlusal outline.
3. In *Homo*, third molars are variable and often irregular, with five, four or three cusps, and triangular, rectangular or oval outline.
4. First and second molars have both mesial and distal approximal wear facets; the third has only the mesial facet.
5. The occlusal curve of mesial/distal cervical margin is slightly more marked in first molars, followed by second, then third.
6. Root divergence decreases from first molars, to second, to third (frequently fused in third molars).

See Figure 2.25.

flattened drum-like lengths which, with practice, can be reconstructed and identified from their root canal form.

Identifying permanent lower molars

Most lower molars are distinguished by their four- or five-cusped crown outline, with two flattened roots. In modern *Homo*, the lower first molar is a distinctive tooth (Table 2.23) with its regular five-cusped form and, although third molars often have five cusps too, they are much less regular. Second

Table 2.24. *Variation in permanent lower molar form*

Cusps and ridges

1. In some teeth, the unworn cusps are high and pointed; in others, they are lower and more rounded. *Paranthropus* molars have an especially low and bulging crown[b].
2. Any lower molar may have five, four or three main cusps[a]. Five-cusped forms are normal in first molars, and common in all molars of *Australopithecus* and *Paranthropus*.
3. The tuberculum sextum or cusp 6 arises in the distal margin of the crown, to lingual of the distobuccal cusp in a five-cusped form[a]. It is rare in *Homo*, but almost universal in the first molars of *Paranthropus*.
4. The tuberculum intermedium or cusp 7 occurs rarely between mesiolingual and distolingual cusps, most often in first molars[a].
5. The mesiolingual cusp is sometimes indented on its distal side by a furrow – the deflecting wrinkle[a].
6. Both mesial and distal marginal ridges may take three forms: a simple continuous ridge, a small tubercle or a gap in the ridge where a fissure passes through[a].

Furrows and fissures

1. In some molars, the fissures forming the main cross in the central fossa meet at precisely the same point (the '+' pattern)[a]. In the 'Y' pattern, there are two meeting points connected by a short length of fissure, so that lingual and distal arms meet at one point, and buccal and mesial arms meet at the other. In the 'X' pattern, the lingual and mesial arms meet at one point, whilst the buccal and distal arms meet at the other.
2. The buccal furrow may be weakly or strongly indented and often ends in a buccal pit (foramen caecum). Some are strongly developed, with a raised rim, or a tubercle on their cervical side. In five-cusped forms, the more distal buccal furrow may also occasionally end in a pit.
3. In all lower molars (especially the third), a prominent groove may arise from the buccal furrow (sometimes at an additional pit), with or without an associated cusp, on the buccal crown surface at the base of the mesiobuccal cusp[a]. The cusp is called the protostylid. It is particularly common in the first molars of *Australopithecus africanus*.

Roots

1. A small furcation root may arise from the buccal furcation[a].
2. The radix ectomolaris, a large additional root, may arise at the buccal side of the main mesial root[a].
3. The radix endomolaris, another additional root, may arise to lingual of the main distal root[a].

[a] See Figure 2.29.
[b] See Figure 2.30.

molars are normally four-cusped and have a regular rectangular occlusal outline. In all three molars, orientation (Table 2.22 and Figure 2.25) is best carried
out from the flattened mesial end of the occlusal outline, the bulging buccal
crown side and the distal tilt of the roots, all features that can be used even
when the crown is well worn. Lower molars show considerable variation in
form (Table 2.24 and Figure 2.29).

Deciduous molars

See Figures 2.31 to 2.34, and Tables 2.25 to 2.26.

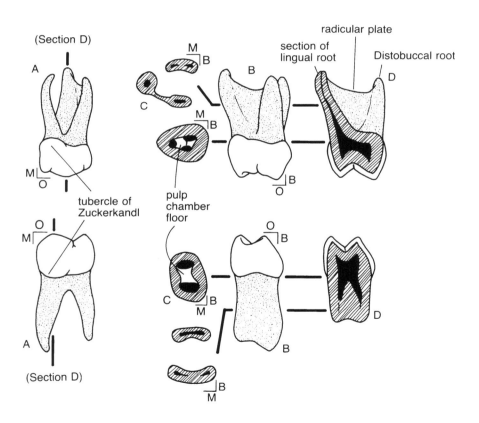

Figure 2.31 Deciduous upper and lower left first molars. Top row, upper molar;
lower row, lower molar. A, buccal aspect; B, mesial aspect; C, transverse sections of roots and root trunk looking down to apical; D, radial/tangential tooth
sections in buccolingual plane. Approximately twice life-size.

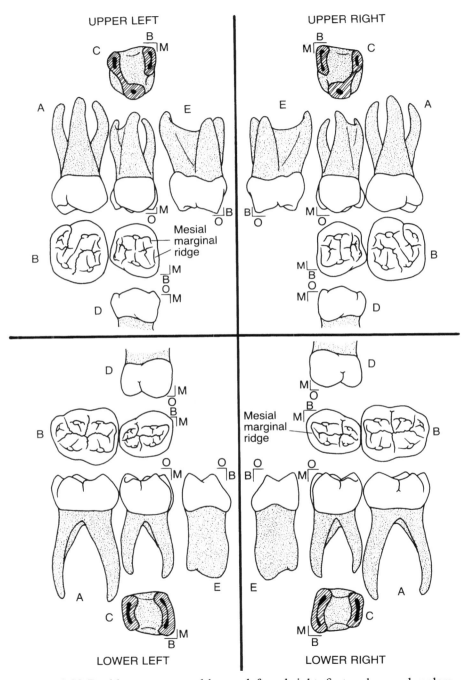

UPPER LEFT

UPPER RIGHT

Mesial marginal ridge

Mesial marginal ridge

LOWER LEFT

LOWER RIGHT

Figure 2.32 Deciduous upper and lower, left and right, first and second molars. Upper left, upper right, lower right and lower left quadrants. A, lingual aspects (with first to mesial of second); B, occlusal aspect (similarly arranged); C, transverse root section of first molar looking up to cervical; D, buccal aspect of first molar; E, mesial aspect of first molar. Approximately twice life-size.

DECIDUOUS UPPER FIRST MOLAR

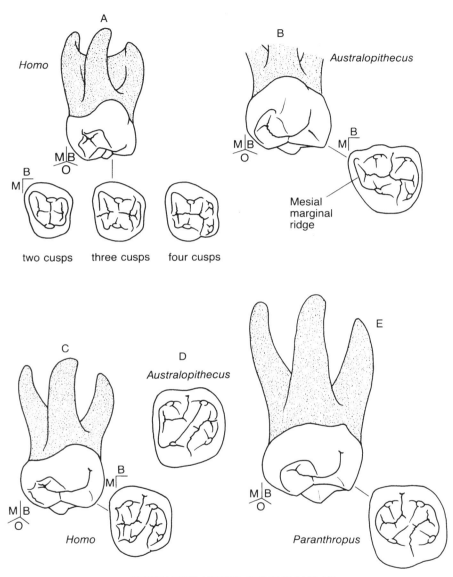

DECIDUOUS UPPER SECOND MOLAR

Figure 2.33 Deciduous upper first and second molars in *Homo, Australopithecus* and *Paranthropus,* viewed from the occlusal aspect and mesial–buccal–occlusal corner (reconstructions based on material from Sterkfontein and Swartkrans). Upper row, first molars in *Homo* (with two-, three- and four-cusped forms in occlusal view) and *Australopithecus.* Bottom row, second molars in *Homo, Australopithecus* (occlusal view only) and *Paranthropus.* Approximately twice life-size.

DECIDUOUS LOWER FIRST MOLAR

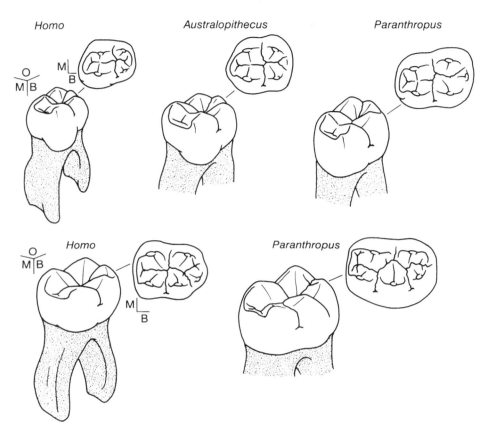

DECIDUOUS LOWER SECOND MOLAR

Figure 2.34 Deciduous lower first and second molars in *Homo*, *Australopithecus* and *Paranthropus*, viewed from the occlusal aspect and mesial–buccal–occlusal corner (reconstructions based on material from Sterkfontein and Swartkrans). Upper row, first molars in *Homo*, *Australopithecus* and *Paranthropus*. Bottom row, second molars in *Homo* and *Paranthropus*. Approximately twice life-size.

Deciduous upper molar crowns

Second deciduous molar crowns (Figure 2.32) are very similar to those of both permanent molars but, although first deciduous molars retain the main cusp triangle, their distobuccal cusp is usually much reduced in living *Homo*, and the oblique ridge is often breached by a fissure extending from the central fossa. A few first molars retain a distolingual cusp, however, and this cusp is normally present in australopithecines (Figure 2.33). In both first and second deciduous molars, the mesial marginal ridge is prominent and normally has

Table 2.25. *Key features for identifying deciduous molars*

1. The buccal and lingual crown sides both bear a markedly bulging cingulum.
2. The mesial and distal crown sides flare out strongly from the cervix.
3. First deciduous molars bear a tuberculum molare, or molar tubercle of Zuckerkandl – an additional cingulum swelling on the mesiobuccal crown side.
4. Lower molar crowns have a rather trough-like appearance, due to the close spacing of the buccal and lingual cusp rows, and the prominent distal groove.
5. First molars (but not second) have a distinctively angled mesial marginal ridge.
6. Upper first molars in living *Homo* usually have three cusps only.
7. Second deciduous molars have a marked resemblance to permanent first molars.
8. The enamel on deciduous molars is thin relative to the permanent teeth and wears rapidly.
9. The root trunk is narrow.
10. Roots are spread widely, with inwardly curving apices (grasping claw-like).
11. Roots and root canals have a flattened section.
12. Upper molars often bear radicular plates.
13. Roots in well-worn crowns are usually resorbed, for exfoliation (page 138).
14. The pulp chamber is large relative to the crown and root trunk, with prominent diverticles.

See Figures 2.31 and 2.32.

two tubercles at its crest, with the more buccal of these tubercles strongly developed in the first deciduous molar, to give an oblique set to the mesial marginal ridge, especially in *Australopithecus* (Figures 2.32 and 2.33). Four-cusped first and second molars usually have a distal marginal ridge, but this is often breached by a groove extending from the distal fossa. Second deciduous molars are always larger than first and, on average in living *Homo,* their crowns are about the same size as third permanent molars. Australopithecine deciduous molars are, on average, only slightly larger than those of modern *Homo* (Figure 2.33).

Deciduous upper molar roots and pulp chamber

Like upper permanent molars, upper deciduous molars normally have three roots – lingual, mesiobuccal and distobuccal (Figures 2.31 and 2.32). Each

Table 2.26. *Variation in deciduous upper molar form*

Cusps

1. Most upper first deciduous molars in living *Homo* are three-cusped, but the distolingual cusp is sometimes prominent, to give four cusps[a]. Four cusps are normal in *Australopithecus*. Two-cusped (both distolingual and distobuccal cusps reduced) first molars are also found.
2. Second deciduous molars may also have reduced distolingual cusps.
3. The mesiobuccal cusp is larger than the distobuccal in *Homo* or *Australopithecus* first deciduous molars, whereas these two cusps are more equal in size amongst *Paranthropus*.
4. The size and form of the mesiolingual cusp varies characteristically throughout the hominids in both first and second deciduous molars. Its lingual surface slopes markedly (lingual bevel) in *Australopithecus afarensis*, but in *Paranthropus* the lingual surface is swollen[a].
5. Carabelli structures occur on both first and second deciduous molars, but are most common in the second. Proper cusps of Carabelli occur only in the second molar.
6. The mesiobuccal cusp may bear a supernumerary cusplet.

Ridges and crown sides

1. The distal marginal ridge varies in prominence in *Homo* first deciduous molars, but is strongly developed in *Australopithecus africanus* and *Paranthropus*. It occasionally bears a small cusplet.
2. The tubercle of Zuckerkandl is prominent in *Homo*, but is weak or absent in *Paranthropus*[a].
3. *Paranthropus boisei* second deciduous molars protrude on the mesiobuccal side, giving an asymmetrical occlusal outline. These teeth may also have a unique accessory cusp in the lingual groove.
4. The parastyle, or its associated groove/pit, may occur in first or second deciduous molars, but is more common and pronounced in the first.

Roots

It is normal for distobuccal and lingual roots to be joined by radicular plates, but their form and extent vary. Occasionally, radicular plates are absent, or join other root pairs.

[a] See Figure 2.33.
See also Table 2.21.

root has a flattened section, with sides indented by grooves, and they diverge markedly, to curve in again at the apex. All three roots are larger, and more divergent, in the second deciduous molar than the first, with the distobuccal roots smallest in both teeth. Radicular plates are common, thin sheets of root tissue joining the distobuccal and lingual roots in most first and second molars. The pulp chamber occupies a large proportion of crown base and root trunk relative to those of permanent molars, with more prominent diverticles (Figure 2.31). As in the upper permanent molars, the entrance to the lingual root canal is roughly circular and funnel shaped, whereas the entrances to the two buccal roots are more crevice-like. The mesiobuccal root commonly has two canals running down its length, and the distobuccal and lingual roots commonly have one each. The canals are often flattened in section and emerge at the apices as slots rather than round foramina.

Deciduous lower molar crowns

There are five cusps on most lower second deciduous molar crowns, arranged in a similar way to those of the first permanent molar (Figures 2.31, 2.32, 2.34). Lower first deciduous molars commonly have either four or five cusps, arranged in buccal and lingual rows that are pressed rather closely together, and a mesial marginal ridge that slopes in a characteristic manner down from the mesiobuccal cusp to the mesiolingual cusp, and is angled in occlusal view. In the second molar, the mesial marginal ridge is also present but has no such distinctive features. The distal marginal ridge in both teeth is usually reduced, and cut by a groove, to give lower deciduous molar crowns a somewhat trough-like appearance. The crowns of second deciduous molars are always larger than first and, in living *Homo*, are on average just a little smaller than those of the third permanent molar.

Deciduous lower molar roots and pulp chamber

There are two roots, mesial and distal, of which the mesial is always largest, and the second deciduous molar always has larger roots than the first (Figure 2.32). Both roots are compressed in section and broadly grooved, more deeply so on their facing sides, and they diverge strongly (especially in second molars), with incurved apices. The pulp chamber is large relative to the overall crown size, with four or five prominent diverticles, depending upon the number of cusps (Figure 2.31). Two crevice-shaped entrances open from the pulp chamber floor into rather flattened root canals, and there may be one, two or three canals running down each root, with the larger numbers occurring in the mesial root.

Deciduous molar root resorption and fragmentation

The pattern of wear and fragmentation is similar to that of permanent molars but, with their thinner enamel, wear proceeds more rapidly. One unique feature of deciduous teeth is extensive root resorption, leaving a rough, scalloped surface. By the time deciduous molars are lost (exfoliated), there is little left but remnants of the root trunk.

Identifying deciduous molars

Deciduous molars can readily be distinguished from permanent molars (Table 2.25), even in worn teeth, where the thinner enamel and resorbed roots (page 138) are still distinctive. Similar orientation criteria to those outlined for permanent molars can be used (Tables 2.19 and 2.22, and Figure 2.32). First deciduous molars differ from second (Table 2.25) particularly in the form of the mesial marginal ridge and they also bear, on their mesiobuccal crown sides, a prominent cingulum bulge called the molar tubercle of Zuckerkandl (Figures 2.31 and 2.32), which makes an excellent feature for orientation. This is much less prominent in australopithecines than it is in *Homo*. Both upper and lower deciduous molars show considerable variation in form (Tables 2.26 and 2.27).

Table 2.27. *Variation in deciduous lower molar form*

1. Four or five cusps may be present, due to variation in the distobuccal cusp. In *Homo* first molars, distobuccal and distolingual cusps may be reduced, giving a three-cusped crown.
2. In first deciduous molars of *Australopithecus*, and especially *Paranthropus*, the distal cusps and marginal ridge are expanded relative to the mesial cusps and marginal ridge[a].
3. The mesiobuccal cusp is larger than, and placed to mesial of, the mesiolingual cusp in both first and second molars of *Homo* and *Australopithecus*. In *Paranthropus* the two cusps are more often level[a].
4. The mesial marginal ridge of the first deciduous molar slopes strongly from buccal to lingual in living *Homo* and *Australopithecus*, but runs level in *Paranthropus*[a]. The mesial marginal ridge in the second molar, and the distal marginal ridge in either tooth, are often much reduced.
5. The buccal groove is often fissure-like. This is more often the case in the second than in the first deciduous molar for living *Homo*, where the second molar may also bear a buccal pit. In *Paranthropus*, however, first molars have a deep buccal groove which, in *P. boisei*, runs down to indent the cervical margin.
6. The tubercle of Zuckerkandl varies in size, but is generally well developed in *Homo* and *Australopithecus*, and weak or absent in *Paranthropus*[a].
7. The protostylid and/or associated grooves may occur in either first or second molars, but is more prominent in the first.
8. Second molars may bear a deflecting wrinkle.
9. A radix endomolaris may occur in either first or second molars. A radix ectomolaris is only ever found in first molars.

[a] See Figure 2.34.
See also Table 2.24.

3

Variation in size and shape of teeth

Introduction

In Chapter 2, the different classes of teeth were described by defining an average or idealized form and then by outlining the range of variation, to facilitate identification of teeth even when their form is slightly aberrant. In this chapter, the aim is to define variation in form more closely and examine its distribution amongst the Hominidae. This subject has generated a larger literature than any other aspect of dental anthropology, and has inspired a series of International Symposia on Dental Morphology, including Fredensborg 1965 (Pedersen *et al.*, 1967), London 1968 (Dahlberg, 1971), Cambridge 1974 (Butler & Joysey, 1978), Turku 1979 (Kurtén, 1982), Paris 1986 (Russell *et al.*, 1988), Jerusalem 1989 (Smith & Tchernov, 1992), Florence 1992 (Moggi-Cecchi, 1995) and Berlin 1995 (Radlanski & Renz, 1995). One reason for this interest is that tooth crowns are formed full-size during childhood, so that their morphology can be studied in mixed collections of individuals of varying ages, whereas studies of skeletal morphology are based on fully grown adults to eliminate the factor of growth. Furthermore, living people (with known biological affinities) can be compared directly with ancient dental remains simply by taking dental impressions from their mouths.

Defining size and shape

Human eyes and brain are unsurpassed in discerning tiny differences between objects compared side by side, but it is much more difficult to define a scheme for recording size and shape in such a way that comparisons can be made between hundreds of such objects. This problem is particularly acute in teeth, which have few clearly defined points between which simple measurements can be made. In addition, any measurements must record homologous structures – present in all individuals and all species studied, with the same function and the same relationships with neighbouring structures. This is particularly difficult to achieve for many aspects of tooth morphology and, for this reason, many studies distinguish between metrical and non-metrical variation. 'Metrical' implies features measured directly, whereas 'non-metrical' implies features that are scored visually in terms of presence, absence, degree of development, or form. In fact, non-metrical features are more complex than their label implies and increasing numbers of studies have attempted to measure them as well (page 103).

Species, sexes and populations

A species is a group of animals whose members resemble one another, and breed together. They do not under normal conditions breed with members of other species and this reproductive boundary allows the species to be defined by a particular collection of genes and a particular collection of structural features. Members of the species vary amongst themselves, but can all be recognized by a particular combination of distinctive features, as living primates are distinguished by their overall size and shape, the colour and form of their fur coats, and by the form of their teeth (James, 1960; Swindler, 1976). When fossils represent extinct animals, it is difficult to apply the species concept because the remains are fragmentary, and can be dated only to the nearest few thousands or millions of years. A species becomes no more than a group of specimens that resemble one another and come from deposits of similar date. The dividing lines between such species are necessarily arbitrary, and different researchers combine the available fossil specimens in different ways to produce different groupings. One particular difficulty is the question of sexual dimorphism. Most species of living mammals show overall differences between the sexes, but the extent of difference varies between species. For example, male and female gibbons are a similar size, whereas male gorillas are much larger than the females. Living humans lie between these extremes and the status of extinct hominids is difficult to estimate, because it is not clear how much sexual dimorphism should be allowed for when defining species.

Living species are further divided into populations, which are groups of individuals that tend to breed together and are therefore also represented by a common collection of genes. The reproductive boundaries surrounding them are not so pronounced as those surrounding the species as a whole, and may be maintained by geography, behaviour or social factors. In the modern world, with convenient and rapid travel, human populations are becoming more diffuse but, in many parts of the world, boundaries are vigorously maintained by particular social and economic status, or by physical conflict. It is often assumed that, in the past, most human populations were clearly defined units and dental morphology has been used to show differences between population groupings (pages 83 and 101). With archaeological material such studies assume that each collection is a sample of a once living population and, for this to be true, the process of sampling must have been random. A cemetery collection is more properly regarded as a death assemblage with a very different composition, in terms of ages and conditions of people, to a living population. The death assemblage contains predominantly the very young, the old and the infirm, whilst the living population is mostly composed of older children, the middle-aged and the healthy. Furthermore, not all components of the death assemblage necessarily find their way equally into a cemetery, and it is quite common for juvenile remains to be excluded, for example. Add to this the selective nature of preservation and it becomes clear that the inclusion of individuals in a cemetery collection is a far from random process. This does not necessarily preclude the definition of broad population groupings, or the assessment of general biological relationships, but it inevitably reduces the expectations for such approaches.

Overall tooth dimensions

Remane (1930) defined many dental measurements, particularly for the molars, but the maximum dimensions of the teeth are much the most commonly used and in effect describe the minimum size of box into which the tooth would fit.

Definitions of dental measurements

Mesiodistal diameter (length) of crown

Many definitions have been given for this measurement, but the most often quoted is that of Moorrees (1957b; Moorrees *et al.*, 1957) – the greatest mesiodistal dimension, taken parallel to the occlusal and labial/buccal surfaces of the tooth crown. If this definition is followed strictly, there is a difficulty when malocclusion causes teeth to be rotated or displaced out of the dental arcade (page 112) and the most mesial and most distal points are then in different positions on the crown (i.e. they are no longer homologous points). Other authors have therefore used a definition that allows for this possibility by measuring the diameter between the mesial and distal contact points of each crown, as they would be if the crown were in normal occlusion (Goose, 1963; Wolpoff, 1971). This is currently the recommended procedure, but care must be taken to determine the definition used when interpreting published reports. The measurement is made using callipers with arms machined to fine points (page 308) in order to fit between teeth still held in the jaw. There is a difficulty with approximal wear (page 242), which produces a facet on which no clear measurement point is defined and, in any case, reduces the original mesiodistal diameter. For these reasons, most researchers exclude teeth with marked approximal attrition (Kieser, 1990), whilst others have attempted to make a correction (Hrdlicka, 1952; Doran & Freeman, 1974; Grine, 1981). Occlusal attrition (page 233) must also affect the mesiodistal diameter, because it eventually removes the original contact points and again, most researchers would exclude heavily worn teeth.

Buccolingual crown diameter (breadth)

This measurement is easily defined once a particular definition of the mesiodistal diameter has been accepted. It is the greatest distance between the buccal/labial and lingual surfaces of the crown, taken at right angles to the plane in which the mesiodistal diameter was taken. Callipers are used with their beam parallel to the occlusal surface of the tooth and their arms applied to the most prominent buccal and lingual bulges of the crown sides. It is easier if the calliper arms have broad flat surfaces. Buccolingual diameters are unaffected by approximal wear, but are modified by marked occlusal attrition.

Other dental measurements

Crown height, defined by Moorrees (1957b) for molars, is the distance in a plane parallel to the vertical axis of the tooth, between the most apical point of the cervical margin on the buccal side and the tip of the mesiobuccal cusp (the crown must be unworn). Root length (Garn *et al.*, 1979a) is the maximum length from the cervical margin of the crown to the apex of the roots. One suggestion for measurements that are unaffected by wear is to take mesiodistal and buccolingual diameters at the cervical margin (Goose, 1963). These give similar results to the normal crown diameters (Falk & Corruccini 1982; Colby, 1996), and it is surprising that more use has not been made of them.

Errors of measurement

One observer, repeating a measurement several times on the same teeth, is liable to obtain a range of different results, and different observers repeating the measurements produce an even greater range (Kieser, 1990). These are known as intra-observer and inter-observer errors respectively, and the usual procedure of measuring to the nearest 0.1 mm reflects the expected level of error. Assessment of observer error should however be built into any study, and it is very important to quote precise measurement definitions. Measurements can be taken both on original teeth and models prepared from dental

Table 3.1. *Dental indices*

Crown module = (mesiodistal diameter + buccolingual diameter)/2
Crown index = (100 × buccolingual diameter)/mesiodistal diameter
Robustness index (crown area) = mesiodistal diameter × buccolingual diameter
Summary tooth size (TS) = (summed robustness indices for all tooth classes)/ (number of tooth classes included)

impressions (page 299). Casting inevitably introduces errors, but Hunter and Priest (1960) found less than 0.1 mm difference between casts and originals, whilst Hollinger and colleagues (1984) stated that differences amounted to < 0.82% of measurements. Tooth wear is inevitably a source of error, although it can be identified even in casts. Dental calculus (page 255) may also affect measurements, but is readily recognized on direct examination, although it is not necessarily apparent in casts and therefore should be noted when impressions are taken.

Dental indices

Mesiodistal and buccolingual diameters are often combined as indices intended to summarize occlusal form (Table 3.1). The crown module is the average diameter of the crown in a particular tooth class, whereas the crown index is the buccolingual diameter expressed as a percentage of the mesiodistal diameter. The robustness index is effectively the area that would be enclosed by the occlusal surface if it were a perfect rectangle, and summary tooth size is the summed mean area for all tooth classes in the group under study.

Distributions and relationships of tooth crown diameters in living people

For one population and a single sex, the mesiodistal and buccolingual diameters of each tooth type have normal (Gaussian) distributions. This is usual for the dimensions of anatomical structures in adults, and is true of skeletal measurements as well. Archaeological and museum collections may well diverge from the ideal of normality, but this is to be expected with small numbers of individuals and their uncertain derivation.

In living *Homo*, mesiodistal diameters show moderate correlations with buccolingual diameters of the same crowns (Garn *et al.*, 1968b), with correlations slightly greater in females than males, in upper than lower teeth, and in cheek teeth than anterior teeth. On average, the mesiodistal diameter is larger than the buccolingual in the upper incisors, but the reverse is true in the lower incisors (Garn *et al.*, 1967c) and, in canines, the diameters are approximately equal. Cheek teeth all have larger buccolingual diameters than mesiodistal, with the exception of lower molars.

Correlations for diameters between different teeth in the same jaw are also moderate, both for the permanent and deciduous dentitions (Moorrees & Reed, 1964; Garn *et al.*, 1965a; Harris & Bailit, 1988). If a tooth from one part of a jaw is large then teeth from other parts tend to be large also but, when the anterior teeth are treated as a group, their crown diameters are inversely related to those of the cheek teeth, also treated as a group (Kieser, 1990). Thus, individuals that have larger than normal anterior tooth crowns have correspondingly smaller than normal cheek teeth and *vice versa*. Most studies show low correlations between deciduous teeth diameters and their permanent successors (Moorrees & Chadha, 1962; Clinch, 1963; Arya *et al.*, 1974), although high correlations for the permanent first molar with the deciduous second molar were noted by Moorrees and Reed (1964).

Patterns of mean crown diameter within a population or group are summarized (see top row in Figures 3.1 and 3.2) by crown size profiles (Garn *et al.*, 1968d, 1968c, 1969). In the upper teeth, the mean buccolingual diameter increases steadily to distal along the tooth row, from incisors, to canines, premolars and then molars. For the mesiodistal diameter, the upper molars characteristically have the highest values and the premolars the lowest, with upper first incisors and canines in between (separated by a low value for second incisors). In the lower dentition, both mesiodistal and buccolingual profiles are similar, with the molars distinguished by their much larger diameters and a gradual increase from first to second incisors, to canines and then premolars. These general patterns can be seen in most living human populations, from any part of the world.

Variability profiles (Kieser & Groenveld, 1988) chart the values of the coefficient of variation (Figure 3.3). In the upper dentition, the second incisors and third molars are the most variable teeth, followed (quite closely in some populations) by the second premolar. By contrast, the least variable teeth are the first molars, first incisors, and canines. The lower dentition has a less consistent pattern, but the premolars are often amongst the more variable teeth.

Explanations for variability in crown diameters

Most discussion of these patterns (Kieser, 1990) has centred around two theories:

1. Field theory. Butler (1939) suggested that tooth form was controlled by factors called morphogens, and defined three fields – incisor, canine and molar – each with a morphogen diffusing in a gradient around a polar tooth

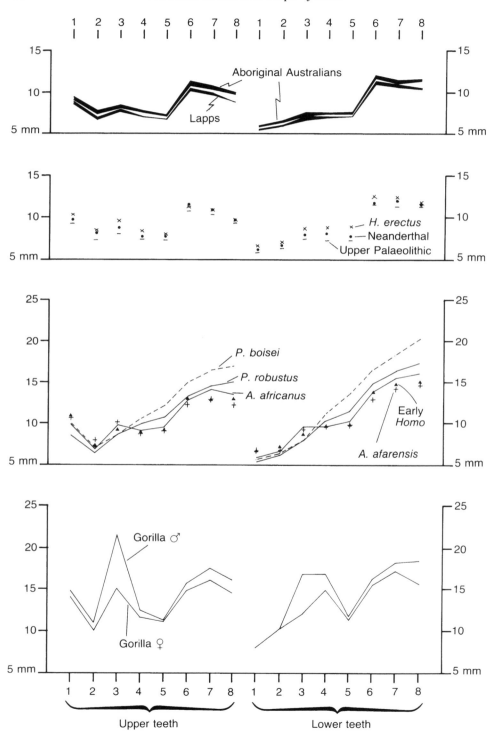

within the growing jaws. Tooth germs (page 118) closest to this polar tooth would thus be more strictly controlled than those further away, and it is true that upper second incisors are more variable than first incisors, whereas molars increase in variability from first to third. Lombardi (1978) found supporting evidence in a factor analysis of crown diameters.

2. The clone theory. Osborn (1978) suggested that the dental papilla (page 118) might develop by division from different populations (clones) of mesenchymal cells. The molar clone would start with first deciduous molars, second deciduous molars, first, second and then third permanent molars. Successive teeth in the series would involve more cell divisions, leading to an accumulation of variability along the series.

In terms of the relationships between dental measurements, there is little to choose between the two theories, because both generate the same pattern, but the clone theory has additional support from laboratory experiments.

Asymmetry in crown diameters

Within one dentition, the diameters of a left side crown differ slightly (by 0.1–0.4 mm) from those of its equivalent or antimere on the right side (Kieser, 1990; Bocklage, 1992; Harris, 1992). Asymmetry of this kind has two components:

1. Directional asymmetry, or the tendency for one side to be consistently larger than the other within one population. This is expressed by subtracting the population mean of the right-side crown diameter from that of the left. If there is no consistent difference, then the result will be zero, but if the right is generally larger than the left then the difference is negative (or positive for left larger than right). A component of directional asymmetry (averaging ± 0.06 mm) is common in human dentitions, the level varying between teeth not only in the extent of asymmetry, but also its direction. Thus, canines may show left larger than right, whilst first premolars show right larger than left. Different populations show different patterns within one jaw, although often the direction of asymmetry in an upper tooth is

---

Figure 3.1 Plots of mean mesiodistal diameter for upper and lower permanent dentition. Top row, aboriginal Australians and Lapps, men and women (lower edge of thick black line is female means and upper edge, male). Second row, Middle Upper Palaeolithic dentitions, Neanderthalers and *Homo erectus*. Third row, early *Homo*, *Australopithecus africanus*, *Australopithecus afarensis*, *Paranthropus robustus* and *Paranthropus boisei*. Bottom row, gorilla, male and female.

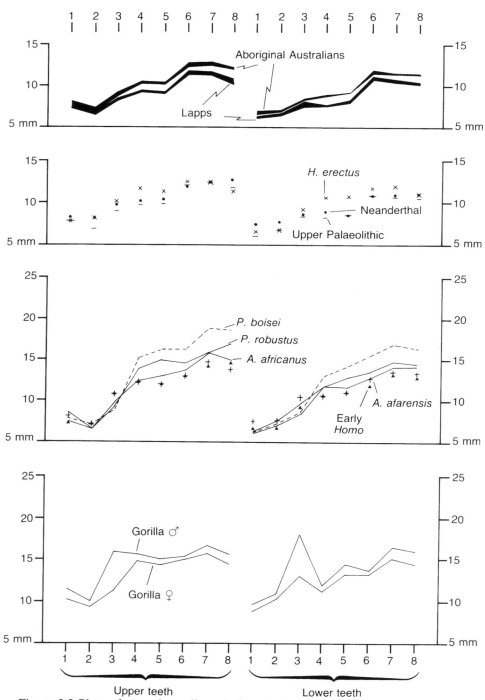

Figure 3.2 Plots of mean buccolingual diameter for upper and lower permanent dentition. Rows as for Figure 3.1.

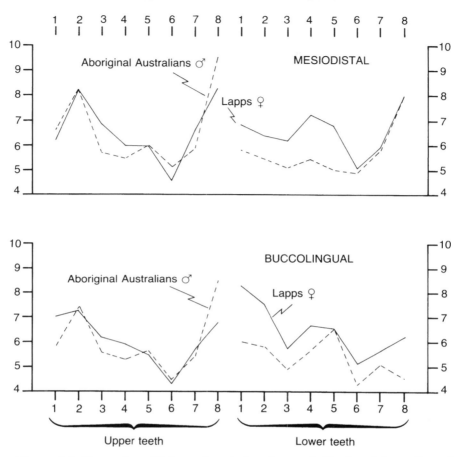

Figure 3.3 Plots of coefficient of variation for mesiodistal and buccolingual diameter in aboriginal Australian men and Lapp women (chosen as examples respectively amongst the largest and smallest of teeth in *Homo sapiens*). Top, mesiodistal diameter. Bottom, buccolingual diameter.

opposite in the equivalent lower tooth, so that, when upper left canines are on average larger than right, lower right canines tend to be larger than left. The reason for such patterning is not clear, and it potentially causes diffi- culties for population studies of dental measurements because most of these assume asymmetry to be random, and therefore measure only one side of the dentition. Fortunately, directional asymmetry is usually small in relation to the precision with which measurements can be taken.

2. Random, or fluctuating asymmetry, where the largest side varies between individuals. This is the random variation of left minus right differences about the mean value for the difference, usually expressed as a standard

deviation (known as root mean square, or RMS, asymmetry). Garn and colleagues (1966d; 1967a) found RMS asymmetries of 0.15–0.42 for both mesiodistal and buccolingual diameters in the Fels Growth Study (page 131); slightly higher for the upper dentition than the lower, and for boys than girls. Within one class of teeth, the more distal members were more asymmetrical than the mesial members (see field theory, page 75). Similar patterns were found in Aboriginal Australians (Townsend & Brown, 1980), and Ticuna Indians from Colombia (Harris & Nweeia, 1980). The fluctuating asymmetry component is usually larger than the directional, but the actual differences in diameter are still not large when compared with intra- and inter-observer error and the precision of measurement.

Greene (1984) indeed suggested that measurement error could be an important effect. Mean and variance of asymmetry would themselves only be affected if the error had sidedness, but this may be a serious consideration, as practically nothing is known about the effects of the observer's handedness on measurement error. Another way to measure asymmetry is to calculate correlations between diameters on the left and right sides (Bailit *et al.*, 1970). A correlation coefficient of $r = 1$ would indicate perfect symmetry, but common values range from $r = 0.63$ to 0.96 (Moorrees & Reed, 1964; Garn *et al.*, 1967a; Arya *et al.*, 1974; Perzigian, 1977). The difficulty with this approach is that it would be directly affected by measurement error (Greene, 1984). Furthermore, it registers any departure from symmetry, and thus does not distinguish between directional and fluctuating components (Bocklage, 1992).

There has been considerable discussion of the role of Selyeian stress (page 177) on asymmetry. Experimental studies with laboratory rats and mice have shown that cold, heat, noise or protein-deficient diet increase fluctuating asymmetry (Sciulli *et al.*, 1979), and twin studies (Potter & Nance, 1976; Potter *et al.*, 1976) suggest that inherited factors have much less effect on dental asymmetry than environmental factors. Bailit and colleagues (1970) devised a project to test the controlling role of environmental stresses at a population level. They compared crown diameters for the inhabitants of Tristan da Cunha, the Nasioi of Bougainville, the Kwaio of Malaita and children from Boston. The Tristanites had most asymmetry and the Boston children the least, with the Nasioi and the Kwaio variably in between. These results matched the anticipated pattern, in relation to the climatic, nutritional and health differences between the groups. High levels of asymmetry in Neanderthalers (Suarez, 1974) are also matched by high levels amongst recent Inuit (Doyle & Johnston, 1977; Mayhall & Saunders, 1986), which might suggest similarly high levels of stress. In the same way, differences amongst ancient Native American groups and living Americans have been interpreted as stress related (Perzigian, 1977), but Black (1980) found levels of asymmetry as large as the most asymmetrical ancient Native Americans in a group of living children from Michigan, presumably with lower levels of stress. DiBennardo and Bailit (1978) encountered similar problems in a large group of Japanese children, and the relationship between asymmetry and stress is thus still unclear.

Inheritance of crown diameters

A phenotype is the combination of physical and physiological characteristics displayed by an individual, resulting from an interaction between that individual's genetic material with the environment in which it grew and developed. In this context, environment means any factor, such as disease, diet or behaviour, which comes from outside the individual and affects the process of development. Phenotypic characteristics such as the crown diameters, which can take any value within a given range for one population, are known as continuous variants. Three groups of factors may be involved in their generation:

1. Variation in the genetic material, or genotype.
2. Environmental factors, shared in common between members of the same family.
3. Environmental effects not shared within the family – impacting differently on each member.

It is difficult to disentangle the relative contributions of these factors. In any case, these contributions vary between individuals, sexes and populations, and probably vary also between different parts of the dentition or even different parts of a single tooth crown.

Most studies estimate the relative importance of genotypic and environmental factors by calculating a statistic known as heritability (h^2). This is the slope of a regression line relating the measurements of fathers with those of sons or daughters, which indicates the extent to which children follow their parent. Several studies, in a range of living human populations, have found high heritabilities for crown diameters (Goose, 1971; Townsend & Brown, 1978; Townsend, 1980), and for the pattern of crown diameters shown in size profiles (Garn *et al.*, 1968d), although different teeth in the dentition may have different heritabilities (Alvesalo & Tigerstedt, 1974). These results are supported by the evidence of twin studies. Monozygous (MZ or identical) twins share the same genotype and environment, and are compared with dizygous (DZ or fraternal) twins, in which the genotype is different but the prenatal and early postnatal environment is the same. DZ twins show a much greater variance in crown diameters than MZ (Osborne *et al.*, 1958; Lundström, 1963), which suggests it has a strong genetic component.

The pattern of correlations between parents and siblings also led Goose (1971), and Potter *et al.* (1968; 1976) to suggest that the mode of inheritance was multifactorial, additive and without dominance – many genes acting together, each with a small but mutually enhancing effect. In the Fels Growth Study (page 131), correlations in sister/sister pairs were highest, followed by brother/brother, and then sister/brother, suggesting that at least some of the genes controlling crown diameter resided on the X chromosome (Garn *et al.*, 1965b). Some studies have duplicated these results (Lewis & Grainger, 1967), whilst others have yielded different patterns of correlation to those expected for sex-linked genes (Potter *et al.*, 1968; Goose, 1971). In several large studies, Alvesalo and colleagues (1971; Townsend & Alvesalo, 1985) concluded that genes affecting crown diameters were resident on both X and Y chromosomes, in addition to loci on other chromosomes. Overall, it is clear that the size of the dentition is part of a complex of features, related to a variety of genes (Potter & Nance, 1976; Potter *et al.*, 1976) on several chromosomes.

The shared environment of children from within one family includes factors of the prenatal environment relating to the mother's physiology, or common immunological and dietary factors in the mother's breast milk supply. Furthermore, a shared diet, infectious diseases and behavioural patterns may also affect all children in the family. Garn and colleagues (1979b) found a clear relationship between childrens' crown diameters and the mother's health during pregnancy, implying that their heritability includes shared environmental as well as genetic factors. Potter *et al.* (1983) applied a path analysis approach to family studies of mesiodistal diameters, and confirmed that complex transmissible and non-transmissible environmental factors were involved, varying between different teeth.

Sexual dimorphism in crown diameters

If the average body weight difference between males and females is expressed as a percentage of the average female weight, then the overall sexual dimorphism of gorillas is 95% (i.e. males are nearly twice the size of females). Similarly, orang-utans have 75% dimorphism and chimpanzees 19%, and amongst the Old World monkeys there are even higher levels of dimorphism, up to 117% in mandrills. In living people, body size dimorphism averages 10% but, as we are the only surviving members of the family Hominidae, it is unclear how representative this level is for the extinct members. This is an important consideration for interpretation of fossil remains, and is an area where dental studies have a useful role.

The highest levels of dental dimorphism in primate permanent teeth (70% or higher) are seen in the diameters of baboon canines and lower first premolars. In other Old World monkeys, these teeth show around 50% dimorphism. Gibbon canines are a similar size in males and females, whereas gorillas display around 40% dimorphism, followed by orang-utan and chimpanzee at 20–30%. Human dental dimorphism is much less pronounced but, as in the apes, it centres on the canines. In the large Fels Growth Study (page 131), mesiodistal diameter dimorphism (Garn *et al.*, 1964) varied between 6% (canines) and 3% (incisors), with slightly higher figures for buccolingual dimorphism (Garn *et al.*, 1966e), giving men somewhat squarer crowns in occlusal outline than women (Garn *et al.*, 1967b). In most living human populations (Garn *et al.*, 1967d), lower canines show the greatest dimorphism (up to 7.3%), followed by the upper canine, whereas premolars are the least dimorphic. The Fels Study group are therefore amongst the most dentally dimorphic populations, and the Ticuna Indians of Columbia are the least dentally dimorphic group yet published (Harris & Nweeia, 1980). Dimorphism in the permanent dentition is variable, and its distribution within Fels Study families suggests strongly that it has a substantial inherited component. The dimensions of the crown at the cervical margin (page 71) are at least as dimorphic as the more usual diameters (Colby, 1996). The deciduous dentition is also dimorphic (Moss & Moss-Salentijn, 1977; Black, 1978; De Vito & Saunders, 1990), with largest differences shown in molars and canines (up to 7% dimorphism).

The complex pattern of male/female differences in crown diameters through the dentition suggests that a multivariate approach might form the basis of a method for sex determination. Garn and colleagues (1967b; 1977; 1979a), were able to sex correctly up to 87% of individuals studied from their permanent tooth diameters. In other studies, 60–90% of children have been correctly assigned to male or female from their deciduous crown diameters (Black, 1978; De Vito & Saunders, 1990). Ditch and Rose (1972) and Rösing (1983) tested dental discrimination in archaeological material, sexed independently from a study of the pelvis, and obtained correct classifications in 90% of cases. Similar methods have potential for studies of fossil hominids. Bermudez de Castro and colleagues (1993) used a reference group of archaeological material from Gran Canaria Island to classify teeth from the Middle Pleistocene site of Sima de los Huesos (Sierra de Atapuerca, Spain) and the Eastern European Neanderthal group from Krapina in Croatia. Using a single collection of archaeological material, sexed by skeletal features, to classify much more ancient and completely unrelated material is fraught with difficulty even though the Krapina material does show two quite clearly separated groups in tooth size. In a large collection of archaeological material, it would be better to establish a baseline group, in which the sex was established by reliable indicators in the pelvis, and to base classification of unknown specimens on the pattern of crown diameters seen in this group. It must also be remembered that the actual size differences between sexes in individual teeth are very small (about 0.4 to 0.5 mm), so that intra- and inter-observer error may well be an important factor.

Tooth size and body size

Amongst the primates as a whole, there is a high positive correlation between body size and crown size (Gingerich, 1977), at least in males (Lucas, 1982), and sexual dimorphism in body size is also correlated with crown diameter dimorphism (Leutenegger & Kelly, 1977). Within living humans, the correlation between crown diameters and body size is low (Garn *et al.*, 1968a; Henderson & Corruccini, 1976; Lavelle, 1977; Perzigian, 1981).

Differences in crown diameters amongst living and recent human populations

The individual published record for crown size in living people comes from two members of the US Marine Corps (Keene, 1967), but the largest teeth, on average, today are those of the Aboriginal Australians and the smallest crowns are generally found amongst Europeans and Asians. Population differences are, however, best sought in the combined effects of many crown diameters, using multivariate statistics. This approach has been effective with skull morphology (Howells, 1973, 1989), which shows clear general differences between living populations, but Falk and Corruccini (1982) found dental measurements to be much less effective discriminators than skeletal measurements. This partly relates to inter-correlations between crown diameters (Harris & Rathbun, 1991), which thus duplicate one another in discriminatory power, reducing their value for identification in forensic cases, or in reconstructing the biological affinities of ancient populations (Kieser, 1990).

There is, however, some hope for crown measurements. Harris and Rathbun (1991) carried out a factor analysis of mean crown diameters from many different studies: sub-Saharan Africans, Western Europeans/Western Asians, Aboriginal Australians, Melanesians, East Asians, and Native Americans. They used a regression method to remove the effect of differences in the overall size of dentitions. Africans and Australians had shorter (mesiodistal diameter) and broader (buccolingual diameter) anterior teeth, associated with longer and narrower cheek teeth, than would be expected from their overall tooth size. Europeans similarly had narrower anterior teeth, and broader cheek teeth, than expected, whilst Melanesians, Asians and Americans formed an average group, which generally matched expectations. With discriminant analysis, it was possible to classify half the groups correctly in terms of their factor scores – Americans and Asians classifying least well, whilst Australians did best. This work, the largest multivariate study so far, suggests that general differences in the pattern of crown diameters do exist between populations, and this is echoed in the earlier work of Garn, Lewis and Walenga (1968d; 1969) using crown size profile patterns.

Crown diameters and hominid evolution

Crown diameters are reported in most descriptions of fossil dentitions, although the permanent dentition is far better known than the deciduous (Wolpoff, 1971; Frayer, 1978; White *et al.*, 1981; Semal, 1988; Calcagno & Gibson, 1991; Wood, 1991). Living and extinct hominids are distinguished from pongids (Figures 3.1 and 3.2) by several clearly defined features:

1. Anterior tooth crown diameters are small relative to molars in the same dentition for hominids. There is less difference in pongids.
2. Canine crowns are much larger than premolars in pongids. There is less difference in the hominids, and some robust forms have premolars larger than canines.
3. First premolar crown diameters are larger than second premolar diameters in pongids. The reverse is true in hominids.
4. Hominid lower dentitions increase gradually in size from mesial to distal along the dental arcade. This pattern contrasts strongly with pongids.

Within the hominids, anterior crown diameters of different taxa overlap extensively, and there are much clearer differences in the cheek teeth. The australopithecines have larger cheek teeth relative to anterior than *Homo*, particularly in *Paranthropus*, where there is a marked increase from the second incisor to distal round the dental arcade, and the premolars are larger than the canines. Within the hominines, cheek tooth diameters decrease from early *Homo*, to *Homo erectus*, to archaic *Homo sapiens*, to Neanderthals and finally anatomically modern *Homo sapiens*. Neanderthal dentitions are notable for having particularly large anterior teeth relative to cheek teeth in the same jaw (especially the buccolingual diameters). Few deciduous crown diameter measurements are available for either hominids or pongids, but patterns within the Hominidae are similar to those of permanent teeth, and *Paranthropus* thus has relatively large deciduous cheek teeth, followed by *Australopithecus* and then hominines.

Wolpoff (1976) claimed a high level of dental sexual dimorphism amongst the australopithecines on the basis of the shape of their crown diameter distributions. In anatomically modern *Homo sapiens*, if male and female canine mesiodistal diameters are plotted together they show a single, asymmetrical mode, but the distribution for australopithecines is bimodal and Wolpoff interpreted this as two, more dimorphic, sexes. Whilst this bimodality might also be due to mixture of different populations, sexual dimorphism remains an important issue in interpreting hominid fossils.

Dental reduction in Homo sapiens

Reduction in crown diameters continued within anatomically modern *Homo sapiens* (Calcagno, 1989; Kieser, 1990). The phenomenon was worldwide, but is best known in Europe (Frayer, 1978, 1980, 1984), where the most rapid reduction occurred between Early and Late Upper Palaeolithic contexts, particularly in lower anterior teeth and upper cheek teeth. Males were affected more than females and, as crown reduction appeared to be matched by reduction in body size and robustness, Frayer suggested that improvements in hunting technology had favoured less heavily built and smaller-toothed, more energy efficient males. This mechanism is unlikely to be universal. Calcagno (1989) found no evidence of accompanying sexual dimorphism reduction in Nubian dental material, and many alternative explanations have been offered. One of the most controversial is the Probable Mutation Effect (PME) proposed by Brace and colleagues (Brace, 1964; Brace & Mahler, 1971; Brace & Ryan, 1980; Brace 1984, 1987). PME suggests that increased tool use and more sophisticated food preparation removed selective pressure in favour of larger teeth and, as size became selectively neutral, it allowed reduction through the accumulated effect of random mutations. Arguments against the PME were summarized by Calcagno (1989). It is at odds with most interpretations of the synthetic theory of evolution, in which all features of an organism are subject to selective pressures (Bailit & Friedlaender, 1966). Indeed, the rapid rate of reduction suggests strong directional selection, and several directional selection models have been proposed, including Frayer's body size reduction hypothesis (above). Another theory suggests that jaw reduction was the driving force (y'Edynak, 1978; y'Edynak & Fleisch, 1983; y'Edynak, 1989, 1992) – softer food resulted in jaw reduction through decreased functional stimulation, increasing malocclusion (page 116) and dental disease, which constituted the selective pressure to reduce tooth size. Carlson and van Gerven (1977) also saw dental reduction as part of a general reduction in the masticatory apparatus due to decreased functional requirements.

Dental reduction in *Homo sapiens* is just one component of a general trend within the Hominidae. It is not a simple trend (Harris & Rathbun, 1991), and some crown diameter increases have been recorded. In the long perspective

Table 3.2. *Molar cusp terminology*

Upper molars		
Cusp location	Tribosphenic[a] system name	Dental anthropology name
Mesiolingual	Protocone	Cusp 1
Mesiobuccal	Paracone	Cusp 2
Distobuccal	Metacone	Cusp 3
Distolingual	Hypocone	Cusp 4
On distal marginal ridge	Metaconule	Cusp 5
On base of mesiolingual cusp	—	Cusp of Carabelli
On base of mesiobuccal cusp	Parastyle	Paramolar tubercle

Lower molars		
Cusp location	Tribosphenic[a] system name	Dental anthropology name
Mesiobuccal	Protoconid	Cusp 1
Mesiolingual	Metaconid	Cusp 2
Centrobuccal	Hypoconid	Cusp 3
Distolingual	Entoconid	Cusp 4
Distobuccal	Hypoconulid	Cusp 5
On distal marginal ridge	Entoconulid	Cusp 6, tuberculum sextum
Centrolingual (between mesiolingual and distolingual cusps)	Metaconulid	Cusp 7, tuberculum intermedium
Base of mesiobucal cusp	Protostylid	Paramolar tubercle

[a] The Osborn or tribosphenic system is used in palaeontology to label cusps in living and extinct forms of mammals – see Hillson (1986a).

of evolution, it seems logical to interpret changes in tooth size as part of a complex of other morphological changes, in relation to the changing functional demands placed upon the dentition and its supporting structures, and not to expect any single explanation.

Non-metrical variation in tooth form

One of the most celebrated morphological crown features is the *tuberculus anomalus* of Georg Carabelli, who was court dentist to the Austrian Emperor Franz, and first published his famous tubercle in 1842 (Hofman-Axthelm, 1981). It is a small additional cusp on the mesiolingual corner of upper molars, entirely absent in some individuals, but present in others in a variety of different forms. It is difficult to decide which to record – its absence, or its form when present – and even more difficult to define any simple measurement to describe it. Such problems apply to many similar structures normally described as non-metrical variants, or traits. Early 'odontographies' contented themselves with noting their presence or absence (Campbell, 1925; Weidenreich, 1937; Pedersen, 1949; Robinson, 1956; Moorrees, 1957a; Moorrees, 1957b). Hrdlicka (1920) was the first to define a graded scale of expression for a particular variant, but it soon became apparent that standardization was necessary and Dahlberg (1945, 1949, 1963) devised a series of plaster plaques, taken from dental impressions, showing several permanent crown variants. Hanihara (1963) developed a similar series for deciduous crowns and, more recently, Turner and colleagues at Arizona State University (1991) assembled a comprehensive series of plaques for permanent teeth. The Arizona State University (ASU) dental anthropology system (page 308) includes scoring forms as well as the plaques, and is both the most widely used scheme and the recommended standard (Buikstra & Ubelaker, 1994).

Such plaque series ensure the use of a common terminology, but must be used carefully. When present, the characters are continuous variants no less than crown diameters (page 80) and they are considered non-metrical because they are difficult to measure consistently, not because they display absolutely clear-cut, clearly divided forms. Reference plaque grades are therefore arbitrarily defined and the dividing lines are often matters of personal opinion, or school of training. Another difficulty is that frequencies are usually presented with several grades combined together (Mayhall, 1992), so that different studies may not be directly comparable because they use different grade combinations.

Commonly recorded variants

Shovelling and double-shovelling

In incisors, and sometimes canines, the marginal ridges (page 14) may be especially prominent and enclose a deep fossa in the lingual surface. Hrdlicka (1920) described such incisors as 'shovel-shaped' and defined three grades of shovelling. The ASU system (Figure 3.4) has seven grades including 'barrel-shaped' upper second incisors, in which the marginal ridges and tuberculum (Table 2.6) unite to form a high ridge enclosing a deep pit (the lingual fossa). Clear definitions are difficult because marginal ridges are normal incisor features, and are exaggerated in smaller diameter teeth. In order to clarify this, several studies have instead measured the depth of the lingual fossa with a gauge (Aas, 1979; Aas & Risnes, 1979; Aas, 1983; Mizoguchi, 1985). Shovelling is most common in permanent and deciduous upper incisors (Hanihara, 1963), but occurs in lower incisors too. Highest frequencies (> 90%) are found

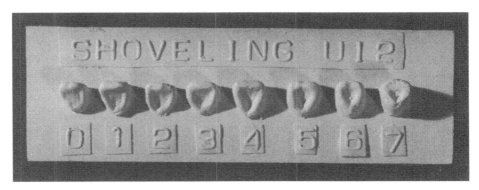

Figure 3.4 Shovelling and double-shovelling in permanent upper incisors (Arizona State University system plaques 12 and 4). Key for shovelling plaques: 0, none – flat lingual surface; 1, faint – very slight mesial and distal elevations seen and felt; 2, trace – elevations easily seen (minimum extension for most observers); 3 and 4, semi-shovel – stronger ridges tending to converge at the cingulum; 5, shovel – ridges almost contact at cingulum; 6, marked shovel – ridges sometimes coalesce at cingulum; 7, barrel (upper second incisors only). Key for double-shovelling plaque: 0, none – smooth labial surface; 1, faint – ridging seen in strong oblique illumination; 2, trace – ridging more easily seen and felt; 3, semi-double-shovel – ridging easily felt; 4, double-shovel – ridging pronounced for at least half crown height; 5, pronounced double-shovel – ridging pronounced for whole crown height; 6, extreme double-shovel (NB distal ridge may be absent).

amongst Asians and Native Americans, and lowest amongst Europeans (Carbonell, 1963). Some incisor and canine crowns have prominent marginal ridges on their labial surfaces (Figure 3.4), a condition known as double-shovelling, whether or not strong lingual ridges are also present.

Cingulum features of the anterior teeth

The tuberculum dentale of upper incisors and canines varies considerably (Table 2.6) and, in some teeth, it rises to a free cusp with a deep pit behind. In addition, tuberculum projections may extend as ridges into the concave lingual surface, and all these features are scored in the ASU system. The ASU system also recognizes the crown–root groove (interruption groove) of upper incisors (Table 2.6).

Canine ridges

In upper canines (Table 2.11, Figure 3.5), the mesial marginal ridge is always slightly larger than the distal, but Morris (1973) found that, in Africans (especially San Bushmen), the ridge was particularly prominent, with an additional buttress that ran down the lingual crown side and merged with the tuberculum dentale. Scott (1977) similarly defined a canine distal accessory ridge (DAR), which was found on both upper and lower canines but was most commonly seen on the lingual surface of upper canines, lying between the central buttress and distal marginal ridge (Table 2.11, Figure 3.5).

Uto-Aztecan upper premolar

In the permanent upper first premolar, the buccal cusp may bulge out (to buccal), with a marked fossa in its distal shoulder (Figure 3.8). The form is known only in Native Americans, with highest frequencies in Arizona, so that Morris and colleagues (1978) called it Uto-Aztecan after a regional linguistic division.

Variation in the main cusps of the upper molars

The distobuccal cusp (Table 3.2, Figure 3.9) is normally prominent in upper molars (Table 2.21) but occasionally it may be reduced, or absent, particularly in third molars (Figure 3.6). The distolingual cusp is considerably more variable, and is best developed on upper first molars, but reduced on second and particularly third molars (Figure 2.25), so that Dahlberg (1949, 1963) devised a four-point scale for upper permanent molars and deciduous second molars, ranging from a score of 4 (all four main cusps well developed), through 4– and 3+, to 3 (distolingual absent). The most common form of deciduous upper first molar has three cusps (Hanihara, 1963) – mesiobuccal, mesiolingual and

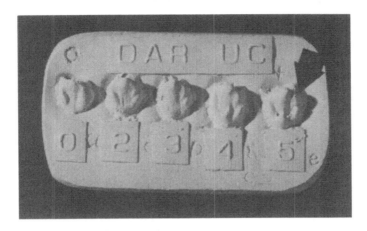

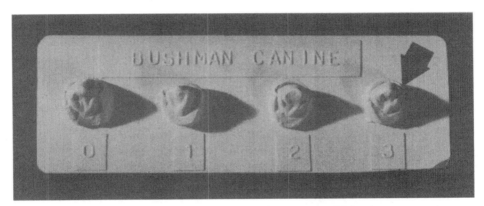

Figure 3.5 Distal Accessory Ridge and Mesial Accessory Ridge in upper canines (Arizona State University system plaques 14 and 17). Key for distal accessory ridge (DAR): 0, DAR absent; 1, DAR very faint (not illustrated); 2, DAR weakly developed; 3, DAR moderately developed; 4, DAR strongly developed; 5, DAR very pronounced. Key for canine mesial ridge: 0, mesial and distal marginal ridges similar size and neither is attached to the tuberculum; 1, mesial ridge larger than distal and weakly attached to tuberculum; 2, mesial ridge larger than distal and moderately attached to tuberculum; 3 Morris' type form – mesial ridge much larger than distal and fully incorporated into tuberculum. Both features are marked with a large black arrow.

distobuccal – although the latter is reduced in some and others have an additional distolingual cusp (Table 2.26).

The metaconule of upper molars

The metaconule is a feature that arises from the distal marginal ridge (Table 2.21) in permanent upper molars (Harris & Bailit, 1980), and ranges from a

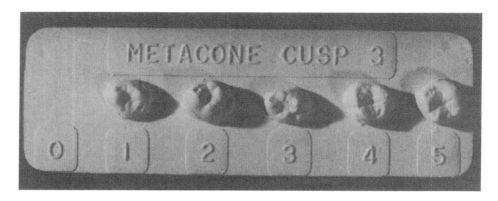

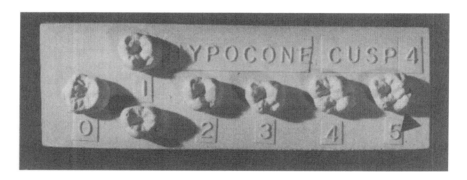

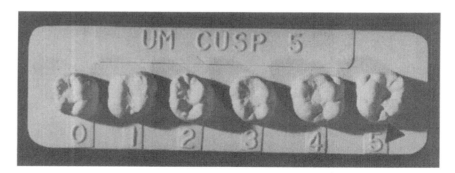

Figure 3.6 Metacone (cusp 3), hypocone (cusp 4) and metaconule (cusp 5) in upper molars (Arizona State University system plaques 23, 8 and 19). Key for metacone and hypocone: 0, cusp absent; 1, a ridge is present at the cusp site; 2, faint cuspule; 3, small cusp; 4, large cusp; 5, very large cusp. Key for metaconule: 0, metaconule absent – a single groove cuts the distal marginal ridge; 1, faint cuspule; 2, trace cuspule; 3, small cuspule; 4, small cusp; 5, medium size cusp (NB presence of double grooves may indicate metaconule in worn first and second molars). Marked by the point of a large black triangle.

tiny cuspule to a prominent cusp (Table 3.2 and Figure 3.6). It is most commonly found in first molars, followed by second, and then third (but reaches its most prominent form in this tooth).

The cusp of Carabelli and associated variants

Carabelli's cusp (Tables 2.21, 3.2) is, in fact, only one of a group of features arising from the base of the mesiolingual cusp in upper molars (Figure 3.7). The cusp may rival the main cusps in size, whereas other related forms include a small ridge, pit or furrow, and a similar structure, the lingual cingulum, is found amongst the apes and gibbons (Swindler, 1976). Dahlberg (1963), Scott (1980) and Hanihara (1963) elaborated schemes for classifying the different forms, which comprise the basis of the ASU system and encompass most variations, although some pit and groove forms may not fit completely (Reid *et al.*, 1992). Carabelli's trait is most common (and most pronounced) in permanent upper first molars and deciduous upper second molars, is less common in permanent upper second molars, and is rare in third. It is most frequent amongst Europeans (75–85% of individuals) and rarest in the Pacific Islands (35–45%), with African frequencies just slightly less than European, and Asians and Native Americans falling in between (Kolakowski *et al.*, 1980, Scott, 1980). Scott (1979c) demonstrated a relationship between Carabelli's trait and distolingual cusp size in upper molars and (1978) between Carabelli's trait in upper molars and the protostylid in lower molars.

The parastyle of upper molars

Bolk (1916) described a 'tuberculum paramolare', situated on the mesiobuccal side of upper molars, but Dahlberg (1945) felt that the cusp should more properly be called the parastyle. The feature (Figure 3.8) ranges from a pit near the buccal groove (Tables 2.21, 3.2) up to a large, well-separated cusp. It is found on all permanent upper molars, but is most common on the third and is rare on the first.

Permanent lower premolar cusps

The lingual cusps of lower premolars, especially the first, are rather variable. There is usually one small lingual cusp, but there may be two or even three, of variable size (Table 2.18, Figure 2.21). The details are rapidly obscured by even the slightest amount of wear.

Main cusps of the lower molars

Most permanent lower first molars have five cusps (Table 3.2 and Figure 3.9) – mesiobuccal, mesiolingual, centrobuccal, distolingual and distobuccal

Figure 3.7 Carabelli trait in upper molars (Arizona State University system plaque 24). Key for Carabelli's trait: 0, lingual side of mesiolingual cusp is smooth; 1, groove; 2, pit; 3, a small Y-shaped depression; 4, a large Y-shaped depression; 5, small cusp without a free apex and its distal border does not contact the lingual fissure; 6, medium size cusp – with attached apex making contact with lingual fissure; 7, large free cusp. Marked by the point of a large black triangle.

(Table 2.24) – but there may be four or three. In four-cusped forms, the distobuccal cusp is missing, whilst the distolingual is additionally missing in three-cusped forms. At the distal margin of the crown, there may be an extra cusp to lingual of the distobuccal cusp (Figure 3.10) called cusp 6, which can only be distinguished when there are five other cusps present, because a single

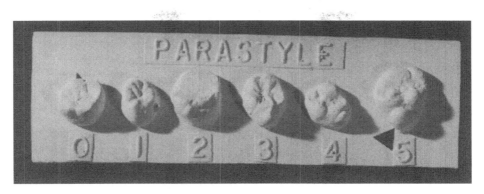

Figure 3.8 Parastyle in upper molars and Uto-Aztec premolar (Arizona State University system plaques 5 and 15). Key for parastyle: 0, buccal crown side is smooth; 1, pit in or near buccal fissure; 2, small cusp with attached apex; 3, medium size cusp with free apex; 4, large cusp with free apex; 5, very large cusp with free apex – involving mesiobuccal and distobuccal cusps; 6, a free peg-shaped crown is attached to the root of the third molar (not illustrated). Uto-Aztec premolar plaque is deliberately presented upside-down because different lighting was needed. Marked by black triangle and arrow.

distobuccal element may be either the distobuccal cusp or cusp 6, and it is conventional to assume that a single cusp of this kind *is* the distobuccal cusp (or cusp 5). A further cusp, called cusp 7, may lie on the lingual margin of the crown between the mesiolingual and distolingual cusps and may be present even when cusps 5 and 6 are not. In living humans, both cusp 6 and cusp 7 are rare, but they are common in *Paranthropus* (Robinson, 1956) and in living great apes (Swindler, 1976).

Groove and fissure patterns of the lower molars

Lower molar fissures have been studied extensively (Gregory & Hellman, 1926; Hellman, 1928; Jørgensen, 1955; Chagula, 1960; Erdbrink, 1965; Rob-

UPPER MOLAR

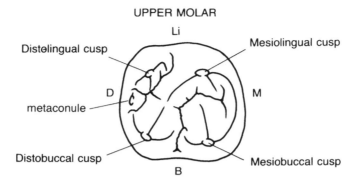

LOWER MOLAR

Figure 3.9 Upper and lower molar cusps and fissure pattern. Key: 'Y', mesiolingual and centrobuccal cusps in contact at central fossa; '+', mesiobuccal, mesiolingual, centrobuccal and distolingual cusps all in contact at central fossa; 'X', mesiobuccal and distolingual cusps in contact (Turner II *et al.*, 1991).

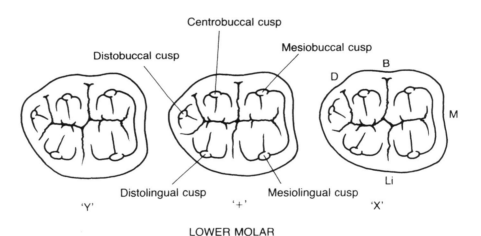

inson & Allin, 1966; Biggerstaff, 1968; Johanson, 1979), and the basic arrangement of fissures has already been described in Chapter 2 (Table 2.24). If the occlusal surface of a five-cusped molar is viewed with its lingual edge lower-most, then its central group of fissures usually takes the form of a 'Y' (Figure 3.9). The arms of the 'Y' enclose the centrobuccal cusp, whereas the tail separates the mesiolingual and mesiodistal cusps. A further fissure branches from the mesial arm, running into the mesial fossa to separate the

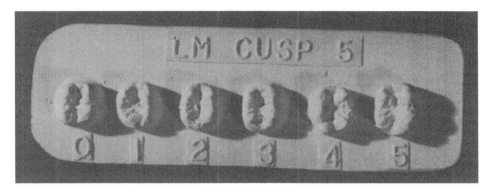

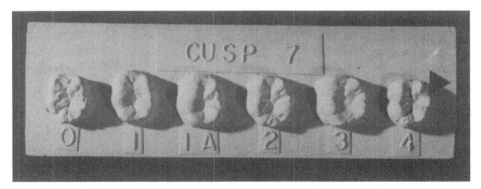

Figure 3.10 Cusps 5, 6 and 7 in lower molars (Arizona State University system plaques 16, 18 and 20). Key for cusp 5: 0, absent; 1, very small cusp; 2, small cusp; 3, medium size cusp; 4, large cusp; 5, very large cusp. Key for cusp 6: 0, absent; 1, much smaller than cusp 5; 2, smaller than cusp 5; 3, equal to cusp 5; 4, larger than cusp 5; 5, much larger than cusp 5. Key for cusp 7: 0, absent; 1, faint cusp with two lingual fissures instead of one; 1A, faint cusp forms a bulge on the lingual side of the mesiolingual cusp; 2, small cusp; 3, medium size cusp; 4, large cusp. Marked with black triangles.

mesiobuccal and mesiolingual cusps, whilst another fissure branches from the distal arm, and runs into the distal fossa to separate the distobuccal and distolingual cusps. The point at which the mesial branching fissure joins the central 'Y' varies. In the typical Y pattern, the mesiolingual cusp contacts the centrobuccal but, if the mesial fissure arises at the origin of the two main arms, then a cross-like structure (+ pattern) is formed, with all four main cusps meeting in the middle. If the mesial fissure arises from the tail of the 'Y' (X pattern), then the mesiobuccal and distolingual cusps touch, whilst the mesiolingual and centrobuccal cusps are separated. Erdbrink (1965) combined cusp number and fissure pattern in a series: X6, +6, Y6, Y5, +5, X5, X4, +4, Y4, 3, but there has been some confusion about the most suitable nomenclature (Robinson & Allin, 1966; Biggerstaff, 1968; Johanson, 1979), and the ASU system records Y, + and X as separate features to cusp numbers. The fissure patterns vary in any case, are often obscured by additional fissures and wrinkling, and may be difficult to classify (Sofaer *et al.*, 1972b; Berry, 1978).

Most first molars in living *Homo* are Y5, and second molars +4, whereas third molars are normally +5 (or +4). Australopithecines are Y5 or Y6 in most lower molars. The Y pattern is also characteristic of the fossil pongid *Dryopithecus* and, amongst living great apes, Y5 dominates all lower molars although it becomes less dominant along the series from first to second and third (Swindler, 1976; Johanson, 1979). At one time, Y5 was regarded as a primitive feature (Hellman, 1928) that showed progressive reduction through the hominid series, but it is now known that the distribution is more complex.

In some lower molars, the mesial marginal ridge is deeply indented by a groove extending from the mesial branching fissure (Table 2.24), but more commonly the marginal ridge runs continuously and encloses a well-defined mesial fossa (anterior fovea). The ASU system defines five grades, although differences are rapidly obliterated by wear.

The deflecting wrinkle and trigonid crest

First described by Weidenreich (1937), the deflecting wrinkle is a fold in the distal side of the mesiolingual cusp of permanent first molars and deciduous lower second molars (Hanihara, 1963), which gives a pronounced 'L' shape in occlusal view (Table 2.24 and Figure 3.11). Incidence in different populations varies between a few per cent to over 80% (Morris, 1970).

The distal trigonid crest is a rare variant of permanent lower molars and deciduous second lower molars. It is a high ridge that unites the mesiobuccal and mesiolingual cusps, which is not cut by the mesial fissure, and therefore isolates the mesial fossa.

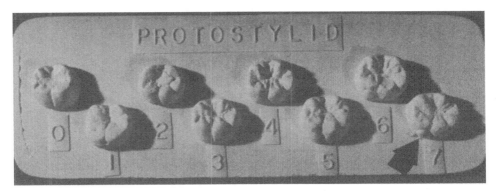

Figure 3.11 Deflecting wrinkle and protostylid in lower molars (Arizona State University system plaques 13 and 22). Key for deflecting wrinkle: 0, absent – medial ridge of mesiolingual cusp is straight; 1, medial ridge straight but constricted at its midpoint; 2, medial ridge deflected distally but does not make contact with distolingual cusp; 3, medial ridge has an 'L'-shaped form, contacting the distolingual cusp. Key for protostylid: 0, smooth buccal surface; 1, pit in buccal fissure; 2, buccal fissure curved to distal; 3, a faint secondary groove extends to mesial from the buccal fissure; 4, secondary groove more pronounced; 5, secondary groove stronger; 6, secondary groove extends across most of the buccal side of the mesiobuccal cusp (a weak or small cusp); 7, cusp with a free apex. Marked with large black arrows.

The protostylid of lower molars

The protostylid (Dahlberg, 1947) is a feature on the buccal side of the crown below the mesiobuccal cusp, ranging from a pit in the buccal groove, through a furrow to a prominent cusp (Table 2.24 and Figure 3.11). It is seen in all permanent lower molars, especially the first and third, or in deciduous lower second molars, and may be present in up to 40% of a population.

Enamel extensions and enamel pearls

In most multi-rooted teeth the cervical crown margin runs level, but it may extend down at a root furcation as a small lobe, or a long tongue of enamel running between the roots (Lasker, 1951). Enamel extensions are found in upper premolars and molars (see Figure 12.5), especially first molars, although premolars are rather less strongly affected than molars, and extensions are mainly found in teeth that bear buccal root grooves. In some cases, a separate nodule, or pearl, of enamel is present on the root surface (actually a nodule of dentine covered by an enamel cap (Risnes, 1989), varying from microscopic size to several millimetres in diameter. Pearls are often associated with enamel extensions and are particularly common in teeth with fused roots, especially on mesial or distal surfaces of permanent upper second and third molars (Pedersen, 1949). In most affected individuals only one molar bears a pearl, but some people have them on two or more teeth. The ASU system has a three-grade score for enamel extensions:

1. Extension projecting towards and along the root, and approximately 1 mm long.
2. Extension approximately 2 mm long.
3. Lengthy extension > 4 mm long.

Pearls are not recorded, and only the attached portion of an extension is scored (Turner *et al.*, 1991).

Taurodontism

The term 'taurodont' was coined by Sir Arthur Keith (1913) to describe the condition of cheek teeth in which the the root trunk is tall, encloses a high pulp chamber, and has short free roots (Lunt, 1954; Mangion, 1962; Blumberg *et al.*, 1971). Taurodontism is best measured in radiographs and, whilst it may affect all permanent or deciduous cheek teeth, is most pronounced in molars. The whole dentition may be affected, or just a few teeth. Estimates for its frequency in living people vary from < 0.1% (Pindborg, 1970) up to 5% or more in some populations (Jaspers & Witkop, 1980), and it is prominent amongst the Krapina Neanderthal specimens.

Root number in cheek teeth

Upper premolars may have one, two or three roots (Table 2.15), with multiple roots occurring most commonly in the first upper premolar of living *Homo*. There is considerable variety in the degree of separation of the roots, and the ASU system only scores roots as separate when they are one-quarter to one-third the length of the roots as a whole. For lower premolars, the ASU system records only the varying degrees of expression of Tomes' root (Table 2.18). Permanent lower molars have two main roots, but these show varying degrees of separation, and the ASU system scores a one-rooted and a two-rooted form on the same basis as the definition for upper premolars. A three-rooted form is also scored when a separate distolingual root (Table 2.24) is present.

Inter- and intra-observer error in recording non-metrical variants

Observer error is a central problem, which may limit the choice of variants to be scored (Berry, 1978), and must be affected by the type of training received by the scorer. Nichol and Turner (1986) devised a 'scorability index' varying between 0 and 100 (the highest figures representing no difference between observers), suggesting that variants with indices over 85 were acceptable. These included the parastyle, metaconule, molar cusp number, upper molar cusp development, and shovelling, whereas the anterior fovea, tuberculum dentale and DAR all scored badly. Reliability must vary between projects depending upon the observers, the prominence of variants and the conditions of study, so that independent assessments of intra- and inter-observer error are an essential part of each new study (Lukacs & Hemphill, 1991, for example).

Asymmetry and sexual dimorphism in non-metrical variants

The morphology of a tooth on one side of the dentition is not necessarily matched by its equivalent on the other side. Garn and colleagues (1966c) found a high degree of left/right concordance, but Mayhall and Saunders (1986) found that fluctuating asymmetry (page 78) of non-metrical variants increased from mesial to distal along morphological fields (page 75). Mizo-guchi (1988) criticized their statistical methods and proposed a new 'total discordance rate' for fluctuating asymmetry, although this gave similar results. There was little evidence for directional asymmetry (tested by chi-square), and this confirms that scoring may be carried out on either the left or right sides. The ASU system (Turner II *et al.*, 1991) recommends that the highest degree of expression (left or right) of a variant is used for comparative purposes, but that separate records should be kept for both sides.

Relationship of non-metrical variants with crown diameters

It is still not clear whether or not non-metrical variants are in general related to overall tooth size. Cusp number in lower molars increases with crown diameter (Dahlberg, 1961; Garn *et al.*, 1966a), whereas fissure pattern shows no such relationship. Likewise, some studies find no relationship (Garn *et al.*, 1966a) between upper molar size and Carabelli trait expression, whilst others find the opposite (Reid *et al.*, 1991, 1992).

Inheritance of non-metrical variants

In a series of classic experiments, Grüneberg (1963) found that strains of laboratory mice that consistently yielded individuals with missing third molars had both smaller and more variable teeth than other strains. He concluded that tooth size was the inherited characteristic, rather than presence or absence, and that teeth failed to form when their prospective size fell below some threshold level. He stated: 'For such discontinuous characters which arise at the extremes of continuous distributions and which are determined by multiple genes despite their discontinuous phenotype, I have proposed the name of "quasi-continuous variations" '. Sofaer (1969) further found that a molar cusp variant of laboratory mice behaved in a similar quasi-continuous way, and argued that it must be inherited through many genes acting together. A few years beforehand, Turner (1967) had suggested that non-metrical variants in human teeth might represent simple Mendelian inheritance, and this interpretation was now strongly disputed by Sofaer (1970) and Berry (1978). Most twin studies (Biggerstaff, 1970) and family studies (Saheki, 1958; Lee & Goose, 1972; Sofaer *et al.*, 1972a; Suarez & Spence, 1974; Berry, 1978) also suggested that non-metrical dental variants were inherited polygenically. Nichol (1989) has since proposed that a combination of simple and polygenic inheritance may be involved and genetics now recognizes whole spectrum from purely Mendelian, to oligogenic and polygenic (Strachan & Read, 1996).

It does seem clear that there is a strong genetic component in the distribution of at least some non-metrical features but, if they really are inherited in a quasi-continuous fashion, then all the factors that control inheritance of continuous variants (page 79) must similarly apply. Non-metrical dental variants show a higher concordance between MZ twins than DZ (Wood & Green, 1969; Biggerstaff, 1970, 1973; Berry, 1978) suggesting a proportion of genetic control, and Sharma (1992) found very high heritabilities for shovelling, upper molar and lower premolar cusp number in a family study. Jordan *et al.* (1992) illustrated dentitions of triplets, showing strikingly similar crown morphology in MZ members, with small but clear differences between DZ members.

There is little clear evidence for inheritance linked to the X and Y chromosomes, and most studies show no significant difference between males and females in non-metrical variants (Garn *et al.*, 1966b: Portin & Alvesalo, 1974; Perzigian, 1976; Aas, 1983). Some differences have, however, been shown for the metaconule (Harris & Bailit, 1980) and for third molar or second upper incisor agenesis (Davies, 1967).

Population studies of non-metrical tooth variation

Several studies have attempted to calculate distance statistics between different populations, based upon frequencies of non-metrical dental variants, which can be compared with the biological affinities between populations suggested by serological similarities, linguistic relationships, or geographical proximity. Sofaer and colleagues (1972b) compared the Pima, Papago and Zuni from Arizona and New Mexico with populations from elsewhere around the world. Closest were Inuit, Asians and other Native Americans, and furthest were Europeans, Africans and Oceanians. Scott and Dahlberg (1982) also studied a large collection of North American dentitions, and the relationships between them suggested by dental distances were very much those expected from linguistic affinities, geographical proximity, archaeological and historical evidence. Palomino *et al.* (1977) further found that it was possible to distinguish reliably between Native North and South Americans on the basis of non-metrical variant frequencies. Similar studies of ancient Nubians (Greene, 1967; Greene *et al.*, 1967; Greene, 1972; Irish & Turner II, 1990; Turner II & Markowitz, 1990) have been used to infer population stability in the Nile Valley over time.

Much work has centred on the definition of a number of dental complexes, or collections of dental features held in common by large groups of people. Hanihara (1967) defined a 'mongoloid dental complex' in the deciduous dentition of Asians and Native Americans, which included: upper incisor shovelling, the deflecting wrinkle, protostylid and cusp 7 in the lower second molar, and the metaconule in the upper second molar. The complex was extended (Hanihara, 1969; Turner II, 1987) to the permanent dentition, and included: shovelling again, with cusp 6, cusp 7, deflecting wrinkle, and protostylid in the permanent lower first molar. Hanihara also suggested a contrasting 'caucasoid dental complex', which was further defined by Mayhall and colleagues (1982) as absence of shovelling and premolar occlusal tubercles, low frequencies of cusp 6 and cusp 7, and common Carabelli trait, protostylid and counter-winging. Recent work, however, has concentrated on the mongoloid dental complex, which was divided by Turner (1987, 1989, 1990) into two patterns: Sinodonty and Sundadonty. Sinodont dentitions characterize people from north-eastern Asia (China, Japan, Siberia) and Native Americans, whereas Sundadonty characterizes the people of south-east Asia, Polynesia and Micronesia. Aboriginal Australians and Melanesians do not fit either category, but are closest to the Sundadonts. Differences between Sinodonts and Sundadonts lie in eight dental variants (Table 3.3) and further work has confirmed the division (Hanihara, 1990, 1991, 1992d, 1992b, 1992c, 1992a).

Table 3.3. *Sundadonty and sinodonty*

Variant	Sundadont Mean % frequency (range)	Sinodont Mean % frequency (range)
Upper 1st incisor shovelling (grades 3–6)	31 (0–65)	71 (53–92)
Upper 1st incisor double-shovelling (grades 2–6)	23 (0–60)	56 (24–100)
One-rooted upper 1st premolars	71 (50–90)	79 (61–97)
Upper 1st molar enamel extension (grades 2–3)	26 (0–50)	50 (18–62)
Peg/reduced/absent upper 3rd molars	16 (0–27/51)	32 (16–46)
Deflecting wrinkle in lower 1st molars (grades 2–3)	26 (0–58)	44 (0–86)
Three-rooted lower 1st molars	9 (0–19)	25 (14–41)
Four-cusped lower 2nd molars	31 (6–64)	16 (4–27)

Means rounded to nearest %.

Source: Turner II (1990).

The term 'Sundadonty' is taken from Sundaland, an area of continental shelf joining Sumatra, Java and Borneo to the Malay peninsula and Indo-China, which has been dry land for much of the past 700 000 years. Turner argued that Sundadonty evolved there between 30 000 and 17 000 BP, and spread into China/Mongolia, where it evolved into Sinodonty between 20 000 and 11 000 BP. He further suggested that Sundadonty was ancestral to the Australian and Melanesian dentition. Central Asians, people of the Indian subcontinent and Africans resemble Europeans dentally more closely than they do East Asians, and Turner has also argued (Turner II, 1987; Haeussler & Turner, 1992) not only that this demonstrated continuity across Western Eurasia from the Late Pleistocene, but that Sundadonts might also be ancestral to modern Eurasians and Africans.

These ideas sit alongside theories for the origin of modern humans based on skeletal/dental morphology, DNA analysis and new dating evidence (Smith *et al.*, 1989; Stringer, 1990). Multiregional evolution (MRE) models suggest that evolution took place locally, so that regional differences have their origin several hundred thousand years ago. Monogenesis, or recent African evolution (RAE) models emphasize instead a migration of early modern humans from Africa after 100 000 BP, replacing archaic forms. Other models fit in between, emphasizing a network of relationships.

UPPER MOLAR   LOWER MOLAR

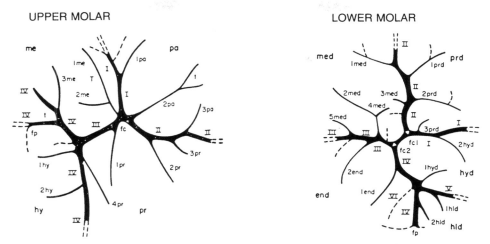

Figure 3.12 Furrow patterns in upper and lower molars. Reprinted from Zubov, A. A. & Nikityuk, B. A. (1978) Prospects for the application of dental morphology in twin type analysis, *Journal of Human Evolution, 7*, 519–524, with kind permission from Academic Press Ltd, 24–28 Oval Road, London NW1 7DX. Key: pa, paracone; pr, protocone; me, metacone; hy, hypocone; prd, protoconid; med, metaconid; hyd, hypoconid; end, entoconid; hld, hypoconulid. I, II, III and IV, intertubercular furrows; 1, 2, 3 and 4, tubercular furrows.

Odontoglyphics

Zubov (1977) considered that the system of molar furrows defined an individually distinctive pattern like that of fingerprints (dermatoglyphics), and divided them into intertubercular (first order) and tubercular (second and third order) furrows (Figure 3.12). The intertubercular furrows separate the main cusps, whilst tubercular furrows are shallower and climb the sides of cusps without dividing them. In a twin study (Zubov & Nikityuk, 1978), concordance of furrow patterns for monozygotic twins was 95%–100% in many cases, implying a very high level of heritability. The Zubov method has hardly been used outside Russia, and a system of equivalence with the ASU definitions has been worked out (Haeussler *et al.*, 1988; Haeussler & Turner, 1992) to allow utilization of published results.

Measurement of occlusal morphology

If non-metrical traits really do represent underlying continuous variation, it is then logical to find ways of measuring them. Many schemes have been devised to do this, largely in studies of fossil hominids.

Methods of measurement

Direct measurement of crowns

Molar and premolar crowns are just large enough to allow measurements within the occlusal surface, using finely pointed callipers, and Corruccini (1977a,b, 1978) defined points for measurement based upon the intersections of fissures and marginal ridge crests. Aas (1983) also directly measured shovelling as the depth of the lingual fossa (page 86), using a modified depth gauge.

Measurement using microscopy

Suwa (1986; Suwa *et al.*, 1994) traced features of the occlusal surface, including fissures, marginal ridges and cusp tips using a low-power stereo microscope fitted with a drawing tube (page 309). A digitizer was used to enter coordinates from these tracings into a computer system, which calculated areas, distances and angles. This method has the advantage that the specimen is examined directly and, with a good microscope, all the features can be seen clearly as a three-dimensional image.

Photogrammetric methods

Photogrammetry is the science of measuring photographs. The simplest method involves a single photograph, taken with the film plane parallel to the occlusal surface. Measurements are taken from the negative, or a photographic enlargement, using a planimeter to measure areas, or by entering coordinates into a digitizer/computer system. Biggerstaff (1969, 1975, 1976) and Lavelle (1984) photographed dental stone casts, inked with landmarks and fiducial marks (for calibration of measurements). Wood and colleagues (Wood & Abbott, 1983; Wood *et al.*, 1983; Wood & Uytterschaut, 1987; Wood & Engelman, 1988; Suwa *et al.*, 1994) instead photographed fossil teeth directly, adjusting the camera so that the negative image was the same size as the specimen itself. A standard enlargement could then be measured for cusp areas, or coordinates of the fissure system. Yamada and Brown (1988) used a similar photographic technique but defined a series of radii, 10° apart, about the central fossa of upper molars, and measurements along the radii were taken to the occlusal outline, so that it could be summarized by plotting angle against length of radius.

Stereo-pairs of photographs can be examined in a more sophisticated way with a photogrammetric plotter that creates a three-dimensional model of the crown, within which accurate measurements can be taken (Hartman, 1988,

1989), to give horizontal (X/Y) and vertical (Z) coordinates for cusp tips, fissures and intersections. In Moiré fringe contourography (Kanazawa *et al.*, 1984; Ozaki *et al.*, 1987; Kanazawa *et al.*, 1988; Mayhall & Kanazawa, 1989; Kanazawa *et al.*, 1990; Mayhall & Alvesalo, 1992), a grating (pitch 0.2 mm) is interposed between camera and specimen. With careful adjustment of camera, grating and lighting, this produces a series of contours at 0.2 mm height intervals, superimposed over the image of the occlusal surface of the tooth in a single photograph. Each contour outline can be digitized, to yield measurements between landmarks, height or volume of cusps, and can be used to define presence or absence of variants.

Applications of measurements of occlusal morphology

These newer forms of measurement may be the way forward for clear definition of morphology, but relatively little work has yet been done with them. Corruccini and McHenry (1980; McHenry & Corruccini, 1980) derived a cladometric analysis of the relationships between hominids and pongids, and examined the position of *Australopithecus afarensis,* with enigmatic material from Lukeino and Lothagam Hill, Kenya. Wood and colleagues (above) clarified taxonomic distinctions between *Australopithecus, Paranthropus* and early *Homo*, demonstrating the distal expansion of molar and premolar crowns in *Paranthropus*. Kanazawa and colleagues (above) demonstrated cusp height differences between populations, and questioned the more traditional definitions of variants such as the metaconule.

4

Occlusion

Definitions and methodology

Normal occlusion

For a review see Thomson (1990) or Klineberg (1991). Dental occlusion is the way in which teeth fit together, both within and between jaws. Normal occlusion (usually labelled intercuspal or centric occlusion) assumes a young adult with perfect permanent dentition – teeth all regularly and symmetrically arranged – and, although few people actually have dentitions like this, it acts as a standard against which irregularities can be measured. It is the position of the lower teeth within the upper dental arcade that affords maximum contact between the cusps and occlusal grooves of the cheek teeth (Figure 4.1). The incisal edges of upper anterior teeth overlap to labial the incisal edges of lower anterior teeth (a condition known as overbite) and the main cusp of the upper canine fits between the buccal surfaces of the lower canine and first premolar. The cusps of the upper and lower cheek teeth line up into a buccal row and a lingual row, with the buccal cusp row of the lower cheek teeth fitting into the occlusal grooves of the upper cheek teeth, and the lingual cusp row of the upper cheek teeth fitting into the occlusal grooves of the lower teeth. Even with a perfect dentition, this is not necessarily the position held normally or most comfortably, and it is not the position for chewing (which is asymmetrical – see page 244), but it still provides an important reference point.

Malocclusion

Intercuspal occlusion in a real dentition is a compromise to accommodate small irregularities. More marked departures are known as malocclusions, but the dividing line between normal and abnormal is not a sharp one. Malocclusions often do not impair dental function significantly, and are in

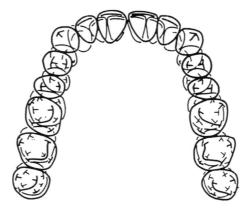

Figure 4.1 Normal occlusion. Occlusal outlines of upper teeth (heavy line) super-imposed over lower teeth (fine line). Incisal edges of incisors denoted by a line; cusps of canines, premolars and molars by 'T' shapes, and oblique ridges of upper molars by a line. Reprinted from Hillson, S. W. (1986) *Teeth,* Cambridge Manuals in Archaeology, Cambridge: Cambridge University Press.

any case modified by dental wear. Occlusal attrition (page 233) removes the cusps and grooves of cheek teeth, and lowers and broadens the incisal edges of anterior teeth, whereas approximal attrition allows the teeth to shift position and to drift in a mesial direction. All these changes tend to reduce malocclusions.

Methods of recording occlusion

The oldest method of recording occlusion is Angle's classification of 1899 (Figure 4.3), which is defined by the position of the upper first molar mesiob-uccal cusp in relation to the lower first molar. Angle Class 1 occlusion is the normal state where this cusp fits into the buccal groove of the lower molar, whereas Class 2 is when the cusp is displaced to mesial and Class 3, to distal. The Fédération Dentaire Internationale (FDI) has more recently developed a complex system of codes and measurements (Table 4.1, Figures 4.2 and 4.3) encompassing the majority of variations throughout the dentition (Baume *et al.*, 1970). Serious application requires reference to the original publication, which includes a recording form design, and is linked to the World Health Organization's international classification of diseases. Some measurements use a depth gauge which, for dental anthropologists, can be improvised from engineers' callipers. Before recording takes place, a living subject moves the lower jaw until it is in the nearest approximation to intercuspal occlusion. This position can be estimated in dental models that have been cast from impressions (page 299) by moving the upper model about on the lower until

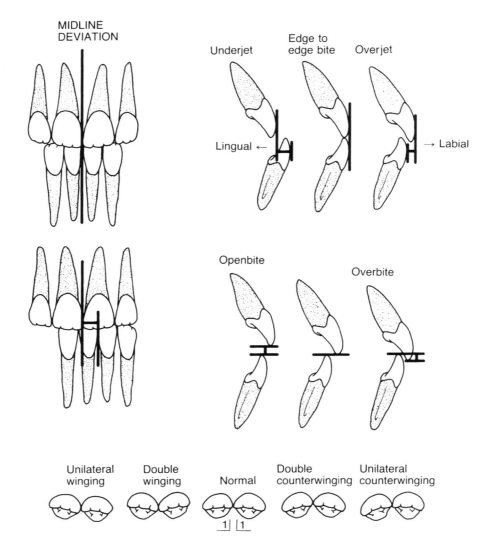

Figure 4.2 Occlusion in anterior teeth. Measurements for midline deviation, overjet and openbite (see Table 4.1). Winging and counter-winging in upper first incisors.

a best fit is achieved but, with archaeological material, mandible and skull distortions cause difficulties. The best fit achieved with dry bone specimens may misrepresent the occlusion in life and, where skulls have been reconstructed from fragmentary material, there is even less certainty. It is usually best to concentrate on occlusal variations within each jaw, rather than between jaws.

Table 4.1. *FDI system for measuring occlusal traits*

A. Dental measurements
1. Anomalies of development

a. Congenitally absent teeth. Code C. Record the FDI tooth number for the missing tooth.

b. Supernumerary teeth. Code S. Assign the FDI number for the nearest permanent tooth (the tooth to distal if equidistant). If the supernumerary is at the median sagittal plane, assign number '10'.

c. Malformed teeth. Code M. Only recorded when the mesiodistal diameter is larger or smaller than the normal range. Enter the FDI tooth number. Supernumeraries not included.

d. Impacted teeth. Code I.

e. Transposed teeth. Code T. Enter FDI numbers for each tooth involved.

f. Missing teeth due to extraction or trauma. Code X.

g. Retained primary teeth. Code R. A deciduous tooth retained for > one year beyond the normal upper limit for exfoliation.

B. Intra-arch measurements
Each arch is divided into three segments; right lateral (canines, premolars and molars), incisal (incisors only), left lateral.

1. Crowding. Code C. For each arch segment. Estimated > 2 mm shortage of space preventing correct alignment. Enter N for normal and A for non-recordable.

2. Spacing. Code S. For each arch segment. Estimated > 2 mm of space beyond that required for correct alignment. Enter N for normal and A for non-recordable.

3. Anterior irregularities. The largest irregularity for each arch. Measured between adjacent incisors/canines, to 1 mm, as distance between incisal corners at right angles to the ideal arch line.

4. Upper midline diastema. Measured to the nearest 1 mm, at the gingival margin. Enter 0 mm if no diastema exists.

C. Inter-arch measurements
1. Lateral segments

For all categories, enter A if the tooth is absent or the measurement cannot be recorded.

a. Anteroposterior – molar relation. For each lower first molar. In intercuspal occlusion:
   Code N. Upper first molar mesiobuccal cusp fits into lower first molar buccal groove.
   Code D. Upper first molar mesiobuccal cusp tip articulates with lower first molar mesiobuccal cusp tip.
   Code D+. Upper first molar distobuccal cusp fits into lower first molar buccal groove.
   Code M. Upper first molar mesiobuccal cusp tip articulates with lower first molar distobuccal cusp tip.
   Code M+. Upper first molar mesiobuccal cusp fits into lower first molar distobuccal groove, or approximal space between lower first and second molars.

Table 4.1 (*cont.*)

b. Vertical – posterior openbite. For each lower cheek tooth (except third molar). In intercuspal occlusion, viewed at right angles to the occlusal plane:

    Code N. No vertical space visible between upper and lower cheek teeth.

    Code O. Vertical space visible.

c. Transverse – posterior crossbite. For each lower cheek tooth (except third molar). In intercuspal occlusion:

    Code N. Buccal cusp of lower cheek tooth fits between cusp tips of upper tooth.

    Code B. Buccal crossbite. Buccal cusp of lower cheek tooth lies to lingual of lingual cusp tip of upper tooth.

    Code L. Lingual crossbite. Buccal cusp of lower cheek tooth lies to buccal of buccal cusp tip of upper tooth.

2. Incisal segments

For each incisor. Enter A if the measurement cannot be recorded.

a. Anteroposterior – overjet. Distance between labial surfaces, to 1 mm, measured in occlusal plane. Record as positive value for overjet and negative when upper incisors are to lingual of lower.

b. Vertical – overbite – open bite. Overlap (or open space) between incisal edges measured, to 1 mm, perpendicular to occlusal plane. Record as positive value for overbite and negative for openbite.

c. Transverse – midline deviation. Distance in transverse occlusal plane between contacts of upper first incisors and lower first incisors.

Source: Baume *et al.* (1970).

Corruccini (1991) used a number of indices to summarize occlusion. Buccal Segment Relation (BSR) is equivalent to Angle's classification for the first molar, incremented if more than one opposing pair of teeth is involved, whereas Tooth Displacement Score is the sum of rotated or displaced teeth, counting very divergent or impacted teeth twice. A range of occlusal variables was similarly summarized in an overall Treatment Priority Index (TPI).

Variations within jaws

Diastema

Gaps between teeth are known as *diastemata* (Greek; intervals), and are more common in the upper dentition than the lower. Some individuals have gaps between all upper anterior teeth, but more often there is a single gap between the two first incisors – an upper mid-line diastema.

MOLAR CUSP RELATION

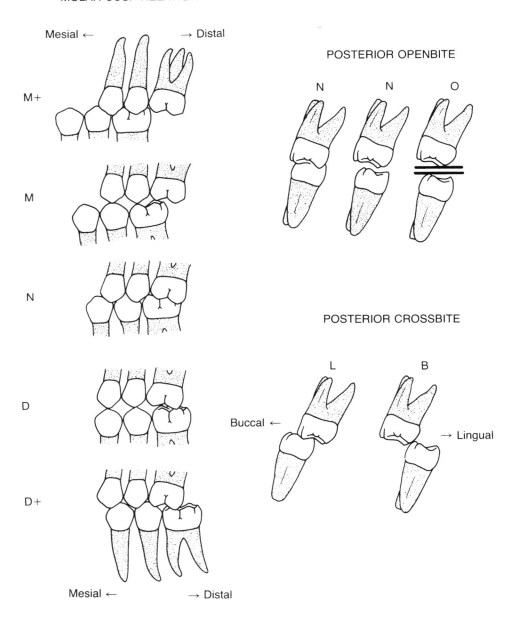

Figure 4.3 Occlusion in cheek teeth. Molar cusp relation, openbite and crossbite (see Table 4.1). Angle's Class 1 occlusion is equivalent to category N, Class 2 to D and Class 3 to M.

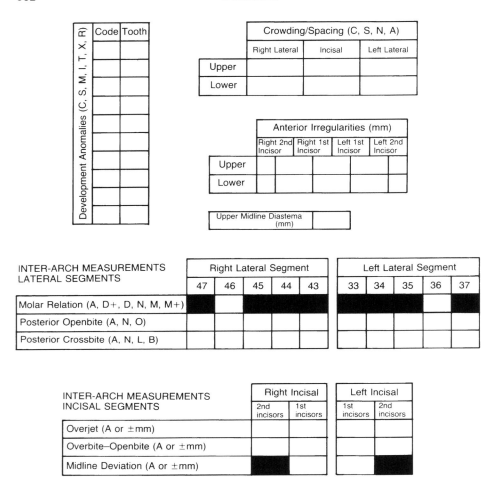

Figure 4.4 Form for recording measurements of FDI system for measuring occlusal traits. Adapted from Baume *et al.* (1970).

Irregularities and rotations of teeth

Irregularity and overlapping of anterior teeth is so common as to be almost normal. Some are merely twisted out of position, but others are wholly displaced to lingual or to labial. Tooth row irregularities frequently involve rotation of individual teeth, although this is not recorded directly by the FDI system, which merely measures the largest anterior irregularity (Table 4.1). One special case of rotation is the winging of upper first incisors (Dahlberg, 1963); a bilateral or unilateral rotation to mesial (Figure 4.2 and Table 4.2), which is particularly common amongst Native Americans (Escobar *et al.*, 1976). Counter-winging is a similar rotation to distal.

Table 4.2. *Winging and counter-winging*

1.	Bilateral winging. Rotation of both upper first incisors to mesial.
1A.	> 20° angle between incisal edges.
1B.	< 20° angle between incisal edges.
2.	Unilateral winging. Rotation of just one upper incisor.
3.	Straight. Incisal edges in a straight line, or following the curvature of the dental arcade.
4.	Counter-winging. Rotation of one or both upper first incisors to distal.

NB. Care is needed in distinguishing winging from general crowding or rotation of anterior teeth.

Source: Turner *et al.* (1991).

Transposed teeth

Neighbouring teeth may swap positions. This is generally associated with some degree of irregularity and/or rotation.

Impacted teeth

Any tooth may erupt in such a way that it fails to occlude with teeth in the opposing jaw. Properly, impaction implies that the tooth remains inside the jaw and does not emerge into the mouth at all, but there are many variations and a tooth may erupt sideways into its neighbour, presenting one of its crown sides uppermost. The most commonly impacted tooth is the third molar, especially the lower, followed by the upper canine.

Congenitally absent teeth

In human dentitions, it is quite common for teeth not to be formed at all (agenesis). Care is needed to distinguish between agenesis and impaction of teeth, or loss through injury or disease, which may leave little sign on the alveolar crest. Rarely, the entire dentition may be congenitally absent (anodontia) although, more frequently, one or only a few teeth are missing (hypodontia). The third molars are the most commonly missing teeth, followed by upper second incisors, upper or lower second premolars, lower first incisors, and upper or lower first premolars (Brekhus *et al.*, 1944; Pedersen, 1949; Brothwell *et al.*, 1963; Lavelle & Moore, 1973). The proportion of individuals with one or more third molars missing ranges from almost none, to one-third or more of a population, but absence of other teeth rarely reaches frequencies of more than a few percent.

Third molar agenesis is associated with agenesis of other teeth in the same dentition (Garn *et al.*, 1962b), with anomalies of formation and eruption in other teeth (Garn *et al.*, 1961, 1962a) and with increased general variability in dental morphology (Lavelle & Moore, 1973). Similar relationships have been observed for upper second incisor agenesis (Le Bot & Salmon, 1980). In most studies, there is little difference between left and right halves of the dentition in the proportion of teeth missing, although Meskin and Corlin (1963) recorded that upper left second incisors were twice as frequently missing than right, when agenesis was not bilateral. It has long been asserted that agenesis is inherited (Lasker, 1951), and there is support for this from studies of laboratory mice (page 100).

Supernumerary teeth

Additional, or supernumerary teeth may also be found (polydontia or polygenesis). Usually, they are teeth with a highly aberrant form, tucked-in to lingual of the normal tooth row, although they may take on the form of neighbouring teeth. Polygenesis is much less common than agenesis and it is slightly more frequent in anterior teeth than in molars (Lavelle & Moore, 1973) but, as supernumerary teeth often do not erupt, a detailed study requires careful radiography.

Retention of deciduous teeth

Occasionally, through an eruptional anomaly of the succeeding permanent tooth, a deciduous tooth is retained in the tooth row. A deciduous molar may, for example, be retained amongst the permanent premolars.

Relationship between teeth of different jaws

One individual's nearest approach to intercuspal occlusion may involve an asymmetrical fit of the lower teeth into the upper, visible as a displacement of the midline contact between the lower first incisors relative to that of the upper first incisors (Figure 4.2). This is known as a mid-line deviation, and the equivalent in cheek teeth is mesial–distal variation in the fit of the first molars (Figure 4.3). A poor fit of the upper and lower dentition may also result in some teeth failing to meet – the condition known as openbite – or to overlap in the condition of overbite, which is normal (to a limited extent) in anterior teeth but abnormal in cheek teeth. Finally, there are various crossbite conditions in which a tooth may meet its opposing tooth either to buccal or to lingual of the normal position. Anterior crossbite is usually known as overjet and, whereas a small degree of overjet is normal, it can be so large as to amount to a malocclusion. The FDI system (Table 4.1) assesses all these relationships for each tooth, without a judgement about whether or not a deviation counts as a malocclusion.

The development of malocclusion

The role of inheritance

Lundstrom (1948) studied the inheritance of occlusal features by comparing monozygotic and dizygotic twins (page 80). Some features had high heritabilities, for example overjet (75%), whereas others had lower heritabilities, but the study none the less led to an assumption that occlusion was principally related to genetic factors. More recent twin studies (Corruccini & Potter, 1980; Potter *et al.*, 1981; Corruccini *et al.*, 1986; Sharma & Corruccini, 1986; Corruccini, 1991) have found much lower heritabilities, averaging about 15%. The idea that environmental factors are more important than genetic factors was supported by a large study (Smith & Bailit, 1977; Smith *et al.*, 1978) on people from 14 villages on Bougainville Island, Papua New Guinea. No consistent differences in occlusion were found between any villages, largely because they were swamped by a large range of variation within villages.

Begg's hypothesis and the epidemiology of malocclusion

Begg (1954) argued that dental attrition in children had a controlling effect on occlusion. He suggested that human teeth, like those of most mammals, were designed to wear and that a correct occlusion could be attained only when sufficient attrition had taken place. The cusps of molars and premolars were only initial guides to occlusion, and were rapidly removed to produce a more efficient flat grinding surface. Begg's hypothesis was based upon the study of museum specimens of aboriginal Australians who had lived a hunter–gatherer existence with a tough diet, promoting faster wear than the soft diet of Westernized culture. He suggested that their dentitions showed a characteristic developmental sequence. Just after eruption, the deciduous upper incisors are always in overjet over their lower counterparts, but eruption of the remaining deciduous dentition in the aboriginal Australians was followed by rapid occlusal and approximal wear (page 242), which allowed it to drift to mesial in the jaw. Drift of the lower teeth was greater than that of the upper teeth, leading to an edge-to-edge bite in the incisors and, when the permanent first molars erupted, they were already in Angle's Class 2 position (FDI position M to M+). Continued rapid approximal attrition on deciduous molars and permanent first molars, and premolars when they erupted, maintained space in the jaws but again allowed the permanent incisors to end up in edge-to-edge bite. Begg calculated that, on average, aboriginal Australian dental arcades lost 1 cm of their length through approximal attrition before the eruption of the third molar, although later estimates (Murphy, 1964; Kaul & Corruccini, 1992) suggested less than half this figure. He maintained that this was the way in which evolution had shaped the human dentition, to produce as rapidly as possible an efficient flat grinding surface, so that the occlusal problems of the modern world were primarily related to adoption of a soft diet that did not enable the dentition to function correctly. Corruccini (1991) was critical of Begg's proposition. There are inconsistencies in its argument, and a study of dental casts of recent aboriginal Australians (eating a softer Westernized diet) failed to find the relationship between their large crown size and the malocclusion that would be anticipated from Begg's model (Corruccini & Townsend, 1990; Kaul & Corruccini, 1992). There is, however, no doubt that dental attrition modifies occlusion. Approximal wear must inevitably cause mesial drift of the dentition, and the broader attrition facet (page 242) produced by approximal wear around the contact points of teeth must also change their relationships. Similarly, progressive removal of cusps by occlusal wear must give more freedom in occlusion, and the heavily worn dentitions of many archaeological collections clearly have a different type of occlusion to the idealized intercuspal position of modern orthodontics. Attrition cannot, however, be the only factor involved and the dynamic maintenance of jaw morphology must also be important, as it constantly remodels under the influence of stresses acting through the pull of muscles and the meeting of teeth against resistance.

There is much additional evidence to suggest a relationship between occlusal anomalies and softness of diet. Some evidence comes from comparisons of older and younger generations, where the younger generation has taken on a more Westernized lifestyle involving a softer diet (Corruccini & Whitley, 1981; Corruccini *et al*., 1983a,b; *et al*., 1983b; Corruccini & Lee, 1984). Another approach is to compare living groups with those of the past (Helm & Prydso, 1979). For example, if dental casts of younger and older generations in recent aboriginal Australians are contrasted with museum specimens of pre-European contact people, the pre-contact group show fewest malocclusions, and the younger generation the most (Corruccini & Townsend, 1990). Corruccini (1991) suggested in all these cases that functional stimulation of developing jaws by a harder diet produced a larger dental arch, which could accommodate the teeth more readily.

Eating imposes heavy stresses, but it is not the only source of such stresses, and occupies a small proportion of each 24 hour period. Teeth are rarely still and jaws are often held tightly clenched when people are concentrating or are worried, whereas some of the greatest pressures are exerted during sleep (bruxism, page 242). In pre-industrial societies, heavy loads may well also have been imposed by their use of teeth as tools for craft work and food preparation (page 251). It is not clear what effect these additional factors have on either wear or jaw remodelling.

5

Sequence and timing of dental growth

Development of the dentition is important as the basis for age estimation in the remains of children. It has also been widely studied as part of a general investigation of the biology of growth.

Initiation and development of tooth germs

The surfaces lining an embryo's developing mouth are covered with a layer of tissue known as epithelium, and this is underlain by a tissue called mesenchyme, which will ultimately develop into different types of connective tissue – bone, cartilage, muscle, tendons and blood vessels, dentine and cement. At six weeks after fertilization, mesenchyme cells proliferate into an arch-shaped zone along the line of the developing jaws. Epithelium grows into this condensation to produce the so-called primary epithelial band (Figure 5.1), which itself divides into two lobes; the vestibular lamina and the dental lamina. Small swellings develop along the edge of the dental lamina so that, by the tenth week, there are ten of them for each jaw, and these are the enamel organs for the deciduous teeth, which will eventually form the enamel of their crowns. Enamel organs for the permanent dentition are initiated from around the sixteenth week after fertilization, with the latest of them appearing only after birth. This initial phase of tooth germ development is known as the bud stage, and next comes the cap stage when one side of the enamel organ bud develops a hollow, filled with mesenchyme called the dental *papilla* (Latin; nipple), and the mesenchyme outside forms a bag-like structure called the dental follicle (Latin *folliculus*; little bag). The papilla will ultimately form the dentine, and the follicle, the cement. Late in the cap stage, the enamel organ differentiates the layer which will eventually form enamel matrix – the internal enamel epithelium – and as the enamel organ grows into the

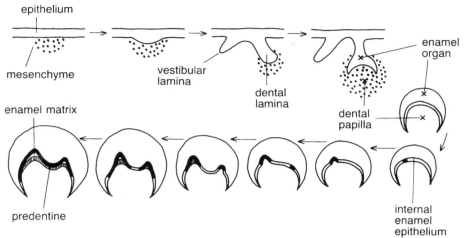

Figure 5.1 Development of tooth germ – a diagrammatic representation adapted from Ten Cate (1985). Top row, development of dental lamina; middle and lower rows, development of internal enamel epithelium and deposition of first enamel matrix and predentine.

succeeding bell stage, the hollow inside becomes deeper and takes on the pattern of folds that defines the future crown shape. Up to this stage, the internal enamel epithelium layer grows by cell division throughout but, where the cusps are to develop, small clusters of cells stop dividing and the epithelium buckles into folds, as the cells in between continue to divide. Where the epithelial cells have stopped dividing, the neighbouring dental papilla cells differentiate into odontoblasts and start to secrete predentine matrix (page 184), after which the internal enamel epithelium cells differentiate into ameloblasts and secrete enamel matrix (page 148) on top of the predentine. Bands of cells differentiate around the periphery of the first and this change spreads like a wave through the tooth germ. At the rim, where the cervical edge of the crown is to be, the enamel organ forms a cuff that extends out as a tube called Hertwig's sheath, and which dictates the shape of the roots.

The crown and root formation pattern

The enamel of the crown is laid down as a series of layers (Figure 5.2). The first formed are in the deepest infoldings of the enamel organ – tiny dome-like structures that lie at the cores of the future cusps. After completion of the first enamel matrix layers, further dome-like layers, gradually increasing in size, are deposited on top. Each cusp grows in height and width until, at its full height, the layers open out at the tip with progressively wider apertures at each successive layer, to complete the cusp sides. Ridges are formed by

MOLAR

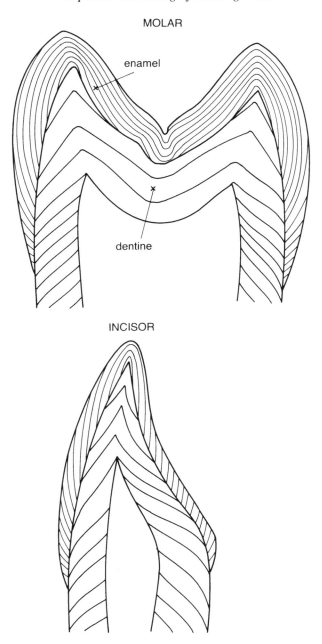

INCISOR

Figure 5.2 Enamel and dentine layering in molars and incisors. Idealized radial section in a buccolingual plane, with lines in the enamel roughly every 10 brown striae of Retzius (page 157).

additional folds in the enamel organ that indent the layers, and they grow towards one another until they coalesce. In molars, cusps and ridges link into a ring, leaving the central occlusal area open until the cusp bases spread enough to meet, and form the fossae and fissures of the crown. Once the occlusal surface is completed, the crown sides are formed by sleeve-like layers that overlap one another down towards the cervix. These layers become narrower until, eventually, enamel formation ceases. In incisors, the apically bulging cervical margin on labial and lingual sides is produced by continued apposition of layers, after completion of the mesial and distal sides (Figure 6.9).

At the same time, predentine (page 184) is also being laid down and the initial enamel matrix layer of each cusp is preceded by tiny predentine layers within the same depression of the enamel organ. More predentine layers are added on the inside as enamel matrix layers are added on the outside. Eventually their top opens out to form the diverticles of the pulp chamber, widening with each successive layer to form the roof and then the walls. The dentine layers are sharply angled, so that a growing crown is a cap-like structure with a sharp rim of dentine projecting as a narrow band below the enamel (Figures 5.3 and 5.4). When the crown is complete, predentine secretion continues with steeply angled layers and a sharp growing edge, down the root. For a single-rooted tooth, Hertwig's sheath forms a gradually tapering tube, but multi-rooted teeth have folds in the sides, which eventually meet to form the root furcation and continue as separate, tapering tubes. Predentine layers form the floor of the pulp chamber, roots and canal, and secretion finally ceases at the apex. Partially completed teeth are common archaeological finds, but may bear little resemblance to fully formed teeth. They are delicate and require careful handling, so that excavators must know where to expect them in the jaws.

Intrauterine tooth formation sequence

The accepted gestation period for human children, from fertilization to birth, is 38 weeks and most babies are born between 34 and 42 weeks. Onset of tooth formation starts with the first deciduous incisor between 14 and 16 weeks after fertilization, followed after an interval of about 2 weeks by the second incisor and, after another week has elapsed, the canine. Deposition of dentine and enamel matrix in all the anterior teeth is initiated at one centre, in the middle of the incisal edge (Kraus, 1959). Canines continue in a simple conical form, whereas incisors develop mesial and distal shoulders as the centre of deposition grows. Deciduous first molars are initiated around 15 weeks after fertilization, and second molars 3–4 weeks after that (Kraus & Jordan, 1965). The mesiobuccal cusp is always the first to start forming, followed by mesiolingual, then distobuccal and distolingual, and lower molars usually have a

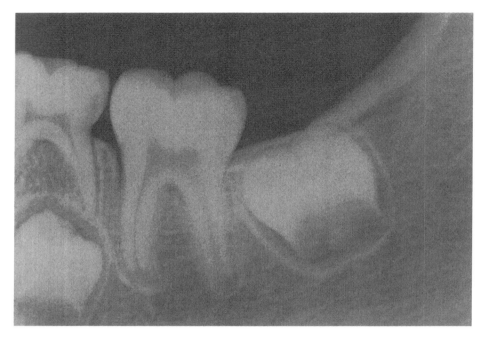

Figure 5.3 Radiograph of lower second deciduous molar (with permanent second premolar developing underneath), permanent first molar (with root apices still forming) and second molar (still being formed within its crypt). Post-medieval archaeological specimen from London.

fifth, distal cusp which is the last to start. Cusps grow initially as tiny cones, but ridges soon spread out from their sides. The ridges on mesial and buccal sides are the first to coalesce, followed by internal ridges such as the oblique ridge of the upper second molar, and then distal and lingual ridges so that, eventually, all the ridges are connected to leave openings only in the deepest part of the occlusal area. These are eventually filled in, and further enamel layers are deposited to create the final form of cusps and ridges.

The permanent first molar also starts forming *in utero*. Its sequence of development is similar to the deciduous molars. The mesiobuccal cusp is initiated 28–32 weeks after fertilization, although the average is around 30 weeks, and lower molars start to form slightly earlier than upper (Christensen & Kraus, 1965). The mesiobuccal centre is followed at around 36 weeks by initiation of the mesiolingual cusp and, in the lower permanent first molar, the distobuccal cusp may also start at about this time. Soon after this, most babies are born.

The state of development of the dentition at birth (Table 5.2 and Figure

Table 5.1. *Ages for permanent tooth formation*

	First incisor	Second incisor	Canine	First premolar	Second premolar	First molar	Second molar	Third molar
	Upper							
Ci[a]	0.25–0.3	0.8–1.0	0.3–0.4	1.5–2.0	2.0–2.5	0.0	2.5–3.0	7.0–10.0
Crc[a]	4.0–5.0	4.0–5.0	6.0–7.0	5.0–6.0	6.0–7.0	2.5–3.0	7.0–8.0	12.0–16.0
Ac[a]	9.0–10.0	10.0–11.0	12.0–15.0	12.0–13.0	12.0–14.0	9.0–10.0	14.0–16.0	18.0–25.0
Ci[d] male					3.6		3.8	9.5
female					4.0		3.7	9.2
Crc[d] male	3.7	4.0	4.9	5.8	6.3	3.8	6.7	13.3
female	3.6	3.8	4.1	5.1	5.9		6.3	12.7
Ac[d] male	10.6	11.1	13.7	13.5	13.8	10.1	14.6	18.2
female	9.3	9.7	11.9	11.8	12.6	9.2	13.6	18.8
Ci[e] male							3.7	9.0
female							3.8	9.4
Crc[e] male	3.3	4.6	4.6	6.8	7.1	3.6	7.3	13.2
female	3.3	4.4	4.5	6.3	6.6	3.5	6.9	12.8
Ac[e] male	9.8	10.8	13.6	13.3	14.0	9.8	16.2	19.5
female	9.3	9.6	12.7	12.6	13.4	9.2	15.1	19.6
	Lower							
Ci[a]	0.25–0.3	0.25–0.3	0.3–0.4	1.5–2.0	2.0–2.5	0.0	2.5–3.0	7.0–10.0
Crc[a]	4.0–5.0	4.0–5.0	6.0–7.0	5.0–6.0	6.0–7.0	2.5–3.0	7.0–8.0	12.0–16.0
Ac[a]	9.0–10.0	10.0–11.0	12.0–15.0	12.0–13.0	12.0–14.0	9.0–10.0	14.0–16.0	18.0–25.0
Ci[b] male			0.5	1.8	3.0	0.0	3.7	9.3
female			0.5	1.8	3.0	0.0	3.5	9.6
Crc[b] male			4.0	5.2	6.3	2.2	6.5	12.0
female			4.0	5.1	6.2	2.2	6.2	12.3
Ac[b] male	8.1	9.3	13.0	13.4	14.3	9.4	14.9	20.0
female	7.7	8.5	11.3	12.2	13.7	8.7	14.6	20.7
Ci[c] male			0.4–1.0	1.9–3.1	2.9–6.9	0.1–0.3	2.8–5.7	7.0–11.3
female			0.3–1.2	1.6–3.1	2.7–5.2	0.1–0.3	2.8–5.2	7.7–12.4
Crc[c] male			3.9–5.7	4.5–7.2	5.7–9.1	2.0–3.2	5.9–9.0	10.4–15.2
female			3.5–5.5	4.4–7.0	5.4–8.2	2.0–3.0	5.6–8.0	10.6–17.5
Ac[c] male			11.5–15.6	11.3–15.5	12.0–15.4	8.2–12.4	12.6–17.3	16.8–20.3
female			9.7–12.9	10.9–13.9	11.6–16.7	7.3–11.5	12.2–17.6	17.4–22.4
Ci[d] male					3.7		3.8	9.4
female					4.2		3.7	9.4
Crc[d] male	3.6	4.0	4.8	5.6	6.3	3.7	6.7	13.3
female	3.6	3.7	4.1	5.0	5.9		6.3	12.8
Ac[d] male	9.2	9.9	13.5	13.7	14.0	10.0	14.8	18.5
female	8.1	8.8	11.4	11.9	12.8	9.2	13.8	18.3
Ci[e] male							3.9	9.8
female							3.9	9.6
Crc[e] male		3.3	4.3	5.9	7.0	3.5	7.4	13.7
female			4.1	5.4	6.4	3.5	7.0	13.3
Ac[e] male	8.0	9.6	13.2	12.8	13.8	9.8	15.7	20.4
female	8.0	9.0	11.5	12.1	12.8	9.2	14.7	20.8

Sources:
[a] Schour & Massler (1940), range in years.
[b] Smith (1991b) scaling of Moorrees *et al.* (1963a), mean age in years.
[c] Fanning & Brown (1971), 3rd to 97th percentile in years.
[d] Anderson *et al.* (1976), mean age in years.
[e] Haavikko (1970), median age in years.

Table 5.2. *Deciduous tooth formation timing*

	Upper				
	Deciduous first incisor	Deciduous second incisor	Deciduous canine	Deciduous first molar	Deciduous second molar
Ci[a] At birth	13.0–16.0 Crown 80% complete	14.7–16.5 Crown 60% complete	15.0–18.0 Crown 30% complete	14.5–17.0 Occlusal surface complete	16.0–23.5 Cusps joined into 'U' by distal marginal ridge
Crc[b]	1.5	2.5	9.0	6.0	11.0
Ac[c]	1.5	2.0	3.3	2.5	3.0
	Lower				
	Deciduous first incisor	Deciduous second incisor	Deciduous canine	Deciduous first molar	Deciduous second molar
Ci[a] At birth	13.0–16.0 Crown 80% complete	14.7– Crown 60% complete	16.0– Crown 30% complete	14.5–17.0 Occlusal surface complete	17.0–19.5 Cusps joined into ring
Crc[b]	2.5	3.0	9.0	5.5	10.0
Ac[c]	1.5	1.5	3.3	2.3	3.0

[a] Age in weeks after fertilization.
[b] Age in months after birth.
[c] Age in years after birth.

Source: Lunt & Law (1974).

5.3) is an important age indicator, but depends upon the length of gestation period and timing of intrauterine tooth development. The best teeth for assessment at this stage are the deciduous molars and it is usually assumed that, at birth, deciduous first molars will have a complete occlusal cap of mineralized tissue, although this is very thin in places (it may break up in archaeological specimens). The occlusal cap of deciduous second molars is not usually complete at 36–38 weeks, and the cusps are joined by their ridges, with gaps in between. It is difficult to recognize the tiny fragments of permanent first molar development at birth.

Postnatal development sequence

The permanent incisors (with the exception of upper second incisors) are initiated around 3–4 months after birth; first incisors starting together, and lower second incisors following slightly behind. Initiation of the canine crown occurs about 1 month later, but the upper second incisor does not start until the end of the first year, and the first premolar, second premolar and second molar are initiated in sequence, running from the latter part of the second year through the third year. Completion of permanent tooth crowns is much more variable than initiation. The permanent first molar crowns are completed at approximately 3 years, followed by incisors at 4 or 5 years, canines and first premolars at roughly 6 years, second premolars and second molars at around 7 years after birth. The end of root formation is more variable still and covers a period of 2 to 4 years in most teeth. Permanent first molars and incisors usually complete their roots first, at somewhere between 9 and 12 years of age, and the rest are completed after 12 years of age. The permanent third molar is in a class of its own, in most cases not even starting formation until after the other crowns are completed. All its stages of development are also highly variable – initiation may be anywhere between 7 and 13 years, completion of the crown occurs during the mid-teens and completion of the roots is usually between 17 years and the early 20s.

There is a consistent difference between girls and boys in the timing of different stages. Garn *et al.* (1958) noted that, in common with most aspects of growth, girls were in advance of boys throughout the sequence, on average by 3%, and other studies have shown up to a year's difference between boys and girls (Table 5.1). Although few studies have been carried out, there are also clear differences between ethnic groups. Harris and McKee (1990) found that the average difference between black girls and boys was roughly twice that seen between white girls and boys, and within each of the sexes, black children attained each stage around 5% earlier than white children. Similar differences have been found in other studies (Loevy, 1983; Davis & Hägg, 1994).

Methods of study

Size and weight of tooth germs

Simple measurement of developing tooth size was the basis of early attempts to assemble a chronology (Berten, 1895). More recently Stack (Stack, 1964, 1967, 1971; Luke *et al.*, 1978) found that the dry weight of the mineralized part of deciduous tooth germs was highly correlated with gestational age in foetuses, and the same relationship holds for the height of developing deciduous tooth germs (Deutsch *et al.*, 1981, 1984, 1985; Liversidge *et al.*, 1993).

Whole jaw sections

The basis of most charts of dental formation is the classic work of Logan and Kronfeld, carried out at the Chicago College of Dental Surgery, Loyola University. They removed whole sections of the jaws during autopsies, demineralized them as a block, microtome sectioned and stained them with haematoxylin/eosin, to give a direct assessment of enamel and dentine matrix development that was considerably more accurate than radiography (Logan & Kronfeld, 1933; Kronfeld, 1935a,b,c; Logan, 1935). This work has proved fundamental, but few related studies have been carried out (Calonius *et al.*, 1970; Sunderland *et al.*, 1987).

Histology of fully developed teeth – 'tooth ring analysis'

Schour, a histologist at the College of Dentistry, University of Illinois, also in Chicago, was working at the same time as Logan and Kronfeld on growth lines in fully developed teeth. He defined a number of growth rings, each marking a stage in crown development, and the clearest of which marked the point of birth (page 159). With the exception of the last, these rings have not survived the test of time, but the work informed development of the classic crown formation chronologies. In addition, the layered enamel structure is emerging as potentially the most accurate method for determining the tooth development sequence (page 177).

Dissection of foetal tooth germs

Most of what is known about antenatal dental development is derived from an examination of tooth germs dissected from the jaws of aborted foetuses. The tooth germs are stained with alizarin red S, which colours mineralized areas a bright red (Kraus & Jordan, 1965, see frontispiece). Follicle and stellate reticulum are dissected away to reveal mineralized areas in three dimensions, under a stereo microscope (Kraus, 1959; Christensen & Kraus, 1965; Kraus & Jordan, 1965; Butler, 1967a,b, 1968, 1971).

Radiography

Each of the bony crypts within the jaws that contain the developing teeth is clearly outlined by the radio-opaque layer of the lamina dura (page 13, Figure 5.4). The sharp, growing edge of crown or root is visible and a radiolucent zone, bulging out into the base of the crypt, represents the position of the dental papilla. Enamel and dentine are only apparent when sufficiently mineralized to create a radio-opacity and, as both tissues are initially laid down as a poorly mineralized matrix, radiographic detection of each development stage must inevitably lag behind. The first stage to be detected in radiographs is the appearance of centres of mineralization at the cusps (Moorrees *et al.*, 1963a; Demirjian *et al.*, 1973), followed in cheek teeth by the coalescence of two or more centres, although this is difficult to observe as cusp outlines are superimposed. Once the occlusal outline is complete, a core of

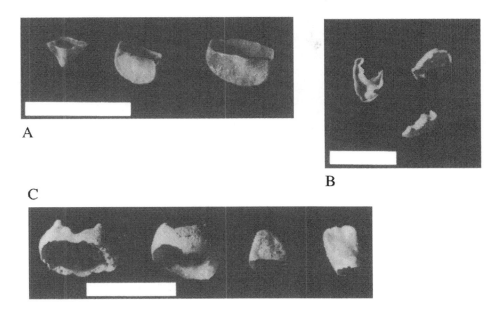

Figure 5.4 The dentition near birth. A, upper left deciduous canine and incisors in a state of development somewhat less than that expected at birth (Table 5.2). B, partly formed fragments of upper first deciduous molar from the same individual as A. C, deciduous lower second molar, first molar, canine and incisor, in a state of development slightly greater than that expected at birth. Scale bars 1 cm.

the less radio-opaque dentine becomes visible within the jacket of more opaque enamel. Completion of the crown and initiation of the root are difficult to define, because the first evidence of root formation is the projection of tiny dentine spurs below the enamel cervical margin. In a multi-rooted tooth, the next stage is the appearance of a root furcation, and the final stage is closure of the root apex, leaving a narrow gap where the root canal emerges.

Most studies of dental development in living children are based on radiography. They vary widely in methodology, and some follow the same children over a period with serial radiographs (a longitudinal growth study), whilst others are based on a single set of radiographs taken from a group of children of different ages (a cross-sectional growth study). Other differences lie in the definition of development stages, but all studies use some derivation of Gleiser and Hunt's (1955) diagram, which recognizes the essential stages of initial mineralization, completion of crown and completion of root apex. Demirjian *et al.* (1973) added a clear definition that reduced inter-observer error (Pöyry *et al.*, 1986). The difficulty with any method for dividing the

Crown

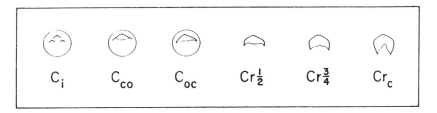

Root

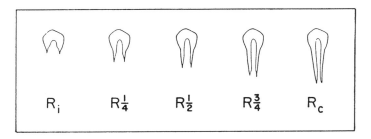

Apex

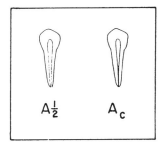

Fɪɢ. 1.—Stages of tooth formation for assessing the development of single-rooted teeth

Figure 5.5 MFH tooth formation stages. Reproduced from Moorrees, C. F. A., Fanning, E. A. & Hunt, E. E. (1963) Age variation of formation stages for ten permanent teeth, *Journal of Dental Research,* **42,** 1490–1502 with kind permission from the International Association for Dental Research, 1619 Duke Street, Alexandria, VA 22314–3406.

Crown

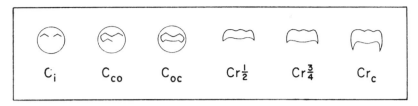

C_i C_{co} C_{oc} $Cr\frac{1}{2}$ $Cr\frac{3}{4}$ Cr_c

Root

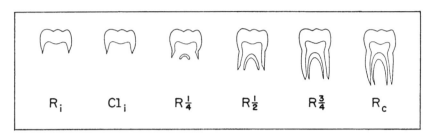

R_i Cl_i $R\frac{1}{4}$ $R\frac{1}{2}$ $R\frac{3}{4}$ R_c

Apex

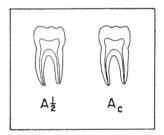

$A\frac{1}{2}$ A_c

Fig. 2.—Stages of tooth formation for assessing the development of permanent mandibular molars

TABLE 1
TOOTH-FORMATION STAGES AND THEIR CODED SYMBOLS

Stage	Coded Symbol
Initial cusp formation	C_i
Coalescence of cusps	C_{co}
Cusp outline complete	C_{oc}
Crown $\frac{1}{2}$ complete	$Cr_{.1/2}$
Crown $\frac{3}{4}$ complete	$Cr_{.3/4}$
Crown complete	$Cr_{.c}$
Initial root formation	R_i
Initial cleft formation	$Cl_{.i}$
Root length $\frac{1}{4}$	$R_{1/4}$
Root length $\frac{1}{2}$	$R_{1/2}$
Root length $\frac{3}{4}$	$R_{3/4}$
Root length complete	R_c
Apex $\frac{1}{2}$ closed	$A_{1/2}$
Apical closure complete	A_c

dynamic process of tooth formation into stages is that the majority of individuals do not fit any stage precisely at the time their radiograph was taken. Most are somewhere in between two stages, and the age at which they attained the first of these has to be interpolated. Methods for interpolation vary, but are not always clearly stated, and Smith (1991b) recognized six statistically based approaches:

A. Cumulative distribution functions. The percentages of children attaining a particular development stage by a given age are plotted as cumulative frequency graphs, from which median or mean ages of attainment are derived. This approach is considered the best basis for ageing methods.
B. Average age at first noted appearance, corrected for examination interval. In a longitudinal study, where a radiograph shows a tooth between two stages, half the expected age interval between them is subtracted from the child's age. There is little evidence to support such arbitrary rules.
C. Average age in a stage of development. Corrections cannot be applied in cross-sectional studies and the mean ages are reported, knowing that they will inevitably be over-estimates.
D. Age prediction. Leinonen *et al.* (1972) summed stage scores for each tooth in the mandible and then calculated a regression of known age on summed scores. Smith (1991b) questioned the polynomial regression model used.
E. Mean formation stage for age group. The mean is not a good measure of central tendency in tooth formation stages because they are not continuous variables.
F. Maturity scales. It is possible to produce a measure of overall dental maturity by assigning a weighted score to each development stage for each tooth and then summing the weighted scores to produce an overall score. The weightings are calculated so that there is a linear relationship between overall maturity score and chronological age. This is standard procedure in growth studies and provides a simple measure of dental maturity, which can then be compared with a similar score for, say, skeletal maturity (Wolanski, 1966; Fanning & Brown, 1971; Demirjian *et al.*, 1973; Demirjian & Goldstein, 1976), but is not suitable as a basis for general ageing methods because the overall maturity score requires all (or a particular group) of the teeth to be recorded.

It thus appears that Method A should be the basis for the standards used in age estimation, and that all the others are to various degrees unsuitable. By contrast, it matters less whether or not the standard is based on a longitudinal or cross-sectional study.

Ageing methods based on assessment of postnatal tooth formation

Moorrees, Fanning and Hunt (MFH)

Much pioneer work on dental development and morphology was carried out on a group of Ohio children in the Fels Longitudinal Study (Fanning & Brown, 1971), which started in 1929 at the Fels Research Institute, Yellow Springs and continues to this day at Wright State University (Roche, 1992). 1036 people, from a range of socio-economic groups similar to the US national distribution, have participated as subjects and were regularly examined at 6-monthly intervals. Moorrees *et al.* (1963a,b) combined radiographs of lower cheek teeth from the Fels study with incisor radiographs from Boston. They used Method A above to derive mean age at attainment for 14 dental development stages (Figure 5.5) and presented their work as a series of complex plots. It is possible to assign forensic and archaeological teeth to these stages and interpolate ages directly from the plots. Saunders *et al.* (1993) tested this methodology on a group of children, identifiable to age and sex from their coffin plates, and buried between AD 1821 and 1874 at Belleville, Ontario. Another test was carried out by Liversidge (1994) on remains which were again identifiable to age and sex from coffin plates and parish records, and buried between AD 1729 and 1856 at Christ Church, Spitalfields, London. In both studies, the estimated ages were around 6 months adrift from the true age (Table 5.3).

Smith version of MFH

Smith (1991b) reworked the MFH data to produce a series of values for predicting age (Table 5.4). Each tooth is assigned to a development stage (Figure 5.5) for which an age value is taken from the table, and a mean age calculated using the values from all available teeth. Smith tested the method on four individuals of known age and sex, and found that the estimated age was up to 0.18 years too old. In archaeological and forensic material, it is often not possible to establish the sex of juvenile material and Smith found that little extra error resulted from averaging the estimates provided by both male and female tables. Liversidge (1994) carried out an independent test on the Spitalfields material (Table 5.3), producing results that were closer to true age than those derived from the original MFH charts.

Anderson, Thompson and Popovitch (ATP)

Anderson *et al.* (1976) made a longitudinal growth study of children from Burlington, Canada. Dental development was scored according to the 14 MFH stages and Method B was used to establish mean ages at attainment for each stage. They carried out a test based on four known-age children, and the

Table 5.3. *Tests of tooth formation age estimates*

Method	Number of individuals	Mean known age minus estimated age
Schour & Massler diagram[a]	63	0.11
Gustafson & Koch table[a]	63	0.1
MFH deciduous[a]	47	0.52
MFH deciduous[b]	17	1.17
MFH permanent[a]	42	0.57
MFH permanent[b]	17	0.64
MFH permanent and deciduous[b]	17	0.47
Smith (MFH)[a]	42	0.29
ATP[b]	17	1.39
Length di–dc (birth–1 year)[a]	26	0.15
Length dm (birth–1 year)[a]	26	0.02
Weight di–dc (birth–1 year)[a]	25	0.09

Sources:
[a] Liversidge (1994) boys and girls, birth to 8 years, [b] Saunders *et al.* (1993) boys and girls, birth to 5 years.

method underestimated true age by up to 1.6 years. Saunders *et al.* (1993) tested the ATP technique on the Belleville sample (Table 5.3), finding that it consistently overestimated age in young children and underestimated age in the older children.

Haavikko

Haavikko (1970) carried out a cross-sectional study based on radiographs of Helsinki children, using the MFH scoring system and Method A to establish median age at attainment (Table 5.1). The results were broadly compatible with MFH. Haavikko (1974) selected two groups of the least variable teeth as a basis for age estimation:

A. For birth to 9 years – lower second molar, first molar, first premolar and first incisor.
B. For 10 years and older – lower second molar, first premolar and canines.

Staaf *et al.* (1991) carried out an independent test of these methods, and found that estimated ages differed from known ages by only 3–6 months.

Demirjian

Demirjian and colleagues (1973, 1976, 1980) developed methods for assessing dental maturity in a large longitudinal study of children from Montreal. They

Table 5.4. *Values for estimating age from permanent lower tooth formation stages (Smith method)*

	First incisor	Second incisor	Canine	First premolar	Second premolar	First molar	Second molar	Third molar
					Males			
Ci			0.6	2.1	3.2	0.1	3.8	9.5
Cco			1.0	2.6	3.9	0.4	4.3	10.0
Coc			1.7	3.3	4.5	0.8	4.9	10.6
Cr1/2			2.5	4.1	5.0	1.3	5.4	11.3
Cr3/4			3.4	4.9	5.8	1.9	6.1	11.8
Crc			4.4	5.6	6.6	2.5	6.8	12.4
Ri			5.2	6.4	7.3	3.2	7.6	13.2
Rcl (Cli)						4.1	8.7	14.1
R1/4		5.8	6.9	7.8	8.6	4.9	9.8	14.8
R1/2	5.6	6.6	8.8	9.3	10.1	5.5	10.6	15.6
R2/3	6.2	7.2						
R3/4	6.7	7.7	9.9	10.2	11.2	6.1	11.4	16.4
Rc	7.3	8.3	11.0	11.2	12.2	7.0	12.3	17.5
A1/2	7.9	8.9	12.4	12.7	13.5	8.5	13.9	19.1
Ac								

	First incisor	Second incisor	Canine	First premolar	Second premolar	First molar	Second molar	Third molar
					Females			
Ci			0.6	2.0	3.3	0.2	3.6	9.9
Cco			1.0	2.5	3.9	0.5	4.0	10.4
Coc			1.6	3.2	4.5	0.9	4.5	11.0
Cr1/2			2.5	4.0	5.1	1.3	5.1	11.5
Cr3/4			3.5	4.7	5.8	1.8	5.8	12.0
Crc			4.3	5.4	6.5	2.4	6.6	12.6
Ri			5.0	6.1	7.2	3.1	7.3	13.2
Rcl (Cli)						4.0	8.4	14.1
R1/4	4.8	5.0	6.2	7.4	8.2	4.8	9.5	15.2
R1/2	5.4	5.6	7.7	8.7	9.4	5.4	10.3	16.2
R2/3	5.9	6.2						
R3/4	6.4	7.0	8.6	9.6	10.3	5.8	11.0	16.9
Rc	7.0	7.9	9.4	10.5	11.3	6.5	11.8	17.7
A1/2	7.5	8.3	10.6	11.6	12.8	7.9	13.5	19.5
Ac								

Method: Stages assigned as in Figure 6.4 and age values are given for each stage in each tooth.

Source: Smith (1991b).

defined new dental formation stages and established a set of weighted scores for each so that, when summed, scores for all teeth produced an overall maturity score (0–100) that could be plotted against age. These plots were marked with percentile curves to act as standards. The Demirjian system was designed purely as a measure of dental maturity (Loevy, 1983; Nyström *et al.*, 1986), but the system has sometimes been used for age estimation (Hägg & Matsson, 1985; Staaf *et al.*, 1991; Davis & Hägg, 1994). The basis of the technique is, however, not really appropriate for age estimation in anthropology, particularly because it does not allow for missing teeth.

Gustafson and Koch

Gustafson and Koch (1974) devised a tooth development diagram (Figure 5.6) on the basis of many published studies, combining figures for boys and girls, histological and radiographic studies. Their aim was 'to construct a simple diagram for age determination', with triangular areas intended to show the age ranges for start of mineralization, crown completion and root completion. The apex of each triangle gives an indication of the average. They included an ageing test based on radiographs of 41 children, assessing the stage of development for each tooth, placing a ruler across the diagram and adjusting it until they found a best fit. Estimated age was mostly within two months of true age and other authors have found similarly low errors for the technique (Crossner & Mansfield, 1983; Hägg & Matsson, 1985). Liversidge (1994) also found only small errors in a test on the Spitalfields children, even though she noted high intra-observer variability (Table 5.3). Despite its subjectivity, therefore, the Gustafson and Koch method often works well. It is useful for anthropological material, because it is little affected by missing teeth and provides a rapid answer, with a level of precision appropriate to the uncertainties of the material. Its effectiveness lies in the range of studies that it encompasses, the averaging effect of its best fit approach, and the many years of forensic experience built into it.

Height and weight of teeth

In modern reference material, the calcified portions of teeth can be dissected out, cleaned and air-dried. The resulting specimens are weighed, and their height measured 'perpendicular to the plane of the growing end (i.e. the cervical margin in teeth which contain no root, and root end in teeth where root formation had already begun) to the incisal tip, to an accuracy of ± 0.1 mm' (Stack, 1964, 1967, 1971; Luke *et al.*, 1978; Deutsch *et al.*, 1981, 1984, 1985). These measurements are highly correlated, in deciduous incisors and canines, with age during the antenatal and early postnatal period and Liversidge (1994) found that ages estimated for the Spitalfields children from these independent data conformed well with known age (Table 5.3). Liversidge and colleagues (1993) also developed their own formulae (Table 5.5) for both deciduous and permanent teeth, based upon the Spitalfields children themselves, aged from birth to 5.4 years. Israel and Lewis (1971) similarly found high correlations

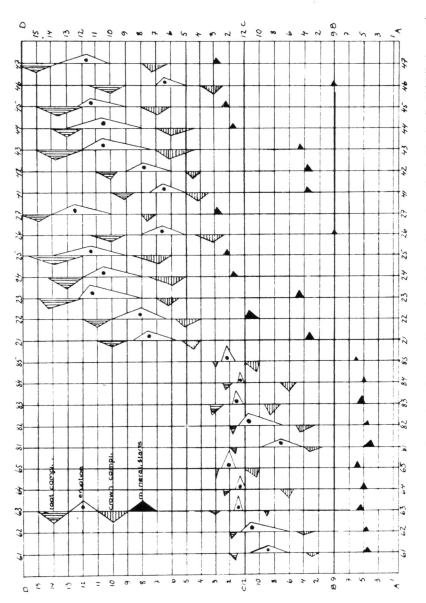

Figure 5.6 Gustafson and Koch tooth formation diagram. Reproduced from Gustafson G. & Koch, G. (1974) Age estimation up to 16 years of age based on dental development, *Odontologisk Revy*, **25**, 297–306, with kind permission from Professor Göran Koch and Swedish Dental Journal, Odontologiska Institutionen, Box 10030, S-551 11 Jönköping, Sweden. Key: A–B, intrauterine life; B–C, first year of life; C–D, 2–16 years. Base of triangles represent range and peak, mean age.

Table 5.5. *Regression formulae for estimation of age from crown/root length*

Tooth	Regression formula for age in years after birth
Deciduous first incisor	−0.653 + 0.144 length
Deciduous second incisor	−0.581 + 0.153 length
Deciduous canine	−0.656 + 0.210 length
Deciduous first molar	−0.814 + 0.222 length
Deciduous second molar	−0.904 + 0.292 length
Permanent first incisor	0.052 − 0.06 length + 0.035 length
Permanent upper second incisor	−0.166 + 0.533 length + 0.003 length
Permanent lower second incisor	0.411 − 0.035 length + 0.05 length
Permanent canine	−0.163 + 0.294 length + 0.028 length
Permanent molar	−0.942 + 0.441 length + 0.01 length

Method: Use only during active tooth formation (birth to 5 years). Tooth 'length' is measured with sliding calipers to 0.1 mm, 'parallel to the long axis of the tooth from the central mammelon of incisor, or cusp tip of canine, and molar (mesiobuccal cusp of mandibular molars, palatal cusp tip of maxillary molars), to the developing mineralizing front of the crown or (incomplete) root margin'. Inter-observer error < 1%.

Source: Liversidge *et al.* (1993).

($r = 0.87$–0.93) between age and permanent tooth height measurements in dental radiographs.

Third molar ageing methods

The third molar is the most variable tooth in size, shape, presence or absence, formation and eruption timing. At first sight, it seems unpromising for age estimation but its development covers the period between 15 years and the early 20s, for which there are few alternatives. Garn *et al.* (1962a) pointed out that third molar variability was no more than would be expected as an extension of the gradual increase in variability of dental development with age, and the third molar is unusual in showing little or no sexual dimorphism in its development timing (Levesque *et al.*, 1981; Mincer *et al.*, 1993). So, in spite of the difficulties, there are several third molar development standards for ageing (Johanson, 1971; Nortjé, 1983). The largest study was by the American Board of Forensic Odontology (Mincer *et al.*, 1993), who employed Demirjian stage definitions and a large group of examiners to establish means

Table 5.6. *Ageing from third molar development*

	Probability (%) of an individual being ≥ 18 years of age			
Grade of formation	Upper M3 Males	Upper M3 Females	Lower M3 Males	Lower M3 Females
D	15.9	9.7	6.1	11.3
E	27.8	28.4	69.4	27.4
F	44.0	50.4	40.5	43.2
G	46.8	63.3	56.0	69.8
H (assuming < 25 years)	85.3	89.6	90.1	92.2

	Regression formulae for individuals under 25 years of age		
Tooth	Formula to calculate age in years	Mean difference between known and predicted age	Standard deviation of difference
Upper right M3	0.33 + 0.98 Grade	1.64	1.25
Upper left M3	−0.39 + 1.02 Grade	1.6	1.22
Lower left M3	−0.18 + 1.01 Grade	1.55	1.13
Lower right M3	0.25 + 0.99 Grade	1.57	1.19

Grades of formation

D – crown complete
E – root furcation present and root length less than crown height
F – root length as great as crown height, and roots have funnel-shaped openings
G – root walls parallel, but apices remain open
H – apical ends of roots completely closed

Source: Mincer *et al.* (1993).

and percentiles for five root development stages in White Americans (Table 5.6), but conceded that the wide variation made it difficult to use them in forensic age estimates.

The development of tooth supporting structures

Whilst tooth germs are developing, bone forms in the jaws around them. Plates of bone extend on their buccal and lingual sides, and eventually thin walls grow up between them to complete the crypts. After birth, the roots of the deciduous teeth start to form and their crowns move into the overlying tissues, which remodel to accommodate them, until they erupt through the gingivae and into the mouth. The roots continue to grow, while periodontal ligaments and alveolar bone are remodelled to adapt to changing forces. All permanent teeth except second and third molars develop underneath already erupted deciduous teeth, which are lost (exfoliated) as their roots are resorbed.

Dental eruption

Eruption is the process by which teeth, in their bony crypts, migrate through the jaws and emerge into the mouth. It continues as each tooth moves into occlusion and beyond, to compensate for the effects of wear, so that eruption is a continuous process that never completely ceases. Clear-cut stages are therefore difficult to define:

1. Alveolar emergence. The appearance of the tooth through the crest of the alveolar process is not a sudden event. In a dry bone specimen, it is first seen as a small aperture, which gradually widens as the tooth crown rises higher until it has opened out to the full crown diameter. In a radiograph the same process is seen as a gradually decreasing rim of lamina dura overlying the tooth. Anthropologists studying dry specimens define alveolar emergence as the first appearance of tooth cusps above the alveolar crest whereas, in radiographs, it is the stage 'at which the alveolar bone has been completely resorbed on the occlusal side of the tooth' (Haavikko, 1970).
2. Gingival emergence (clinical eruption). The appearance of teeth through the gingivae is again a gradual process, as anyone with young children knows only too well. Cusp tips appear in tiny pinpricks some weeks before the bulk of the occlusal surface follows. Haavikko (1970) defined a tooth as 'clinically erupted when the crown of the tooth or part of it has roentgen-ologically observed to have penetrated the mucous membrane'. Saleemi *et al.* (1994) recorded teeth as emerged 'if any part of the crown was visible'.
3. Entry of the crown into occlusion. With the dentition *in situ* each crown may be judged with reference to its neighbours, or by first signs of wear, a definition that is difficult to use clinically but may be useful for anthropo-logical purposes.
4. Exfoliation of deciduous teeth. The process of root resorption can be observed both in dry specimens (Figure 5.7) and in radiographs. Resorption timing has been established in several studies (page 142).

Depending upon the definitions used, alveolar emergence may be in advance of gingival emergence by a few months to a year or more (Haavikko, 1970).

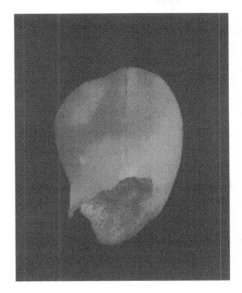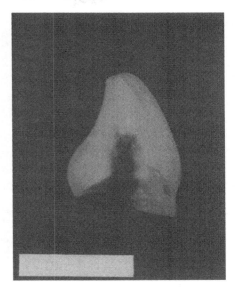

Figure 5.7 Root resorption in a deciduous lower canine. Left, appearance from labial surface. Right, appearance in fractured section. Scale bar 5 mm.

In addition, the sequence with which teeth emerge at the alveolar crest may be different from the gingival emergence sequence (Garn & Lewis, 1963).

Timing and sequence of dental eruption

Almost all studies are based upon gingival emergence, and show that eruption is arranged into three periods during childhood and early adulthood:

1. Period of deciduous dentition.
2. Period of mixed dentition, when the permanent first molars emerge at the distal end of the deciduous tooth row, and deciduous incisors are replaced by permanent incisors.
3. Period of permanent dentition, when all the deciduous teeth have been replaced.

The deciduous dentition usually erupts within the first two-and-a-half-years after birth (Robinow *et al.*, 1942; Lunt & Law, 1974), but the permanent dentition has been studied in much greater detail (Hurme, 1949, 1951; Dahlberg & Menegaz-Bock, 1958; Hurme & van Wagenen, 1961; Jaswal, 1983; Smith & Garn, 1987). There is considerable variation in both sequence and timing, although it is possible to state a normal sequence that applies to many populations around the world (Table 5.7); noted as $M^1I^1I^2P^1CP^2M^2$ for

Table 5.7. *Initiation of resorption in deciduous tooth roots*

	Upper				
	Deciduous first incisor	Deciduous second incisor	Deciduous canine	Deciduous first molar	Deciduous second molar
Boys, mean age in years (standard deviation)	4.4 (1.4)	5.9 (1.1)	7.2 (1.5)	4.8 (0.7)	6.0 (1.1)
Girls, mean age in years (standard deviation)	4.4 (0.9)	5.7 (1.1)	6.9 (0.9)	4.7 (0.9)	5.9 (0.7)
	Lower				
	Deciduous first incisor	Deciduous second incisor	Deciduous canine	Deciduous first molar	Deciduous second molar
Boys, mean age in years (standard deviation)	4.8 (0.9)	5.4 (1.0)	6.6 (1.3)	4.4 (0.5)	6.2 (1.2)
Girls, mean age in years (standard deviation)	4.5 (0.8)	5.0 (1.0)	6.6 (1.1)	4.2 (1.4)	5.8 (1.7)

Source: Haavikko (1973).

the upper dentition and $M_1I_1I_2CP_1P_2M_2$ for the lower. Emergence times are grouped into three phases:

Phase 1. Emergence of permanent first molars and incisors (5–8 years of age).

Phase 2. Emergence of canines, premolars and second molars (9.5–12.5 years).

Phase 3. Emergence of third molars (late teens–early twenties).

Gingival emergence (Table 5.8) is strongly correlated between antimeres (left and right equivalents) and isomeres (equivalent teeth in upper and lower jaws), whereas neighbouring teeth are only moderately correlated (Garn & Smith, 1980). Lower teeth emerge earlier than their equivalents in the upper jaw, especially the anterior teeth. The permanent dentition (especially canines) in girls usually emerges before that of boys in the same population, but the difference varies between populations (Garn *et al.*, 1973b), and Europeans erupt their permanent teeth (particularly molars) later than other populations, whereas children from poorer families show slightly later tooth emergence than the more well off (Garn *et al.*, 1973a).

Table 5.8. *Order of gingival emergence*

Deciduous dentition

First incisors (lower then upper)
Second incisors (upper then lower)
First molars
Canines
Second molars (lower then upper)

Permanent dentition

First molars
First incisors
Second incisors
Upper first premolars, or lower canines
Upper canines, or lower first premolars
Second premolars
Second molars
Third molars

Some pairs of teeth are particularly close in eruption timing and in these tooth pairs, the normal order is frequently reversed (Smith & Garn, 1987). Within Phase 1, the most common variation (especially in the lower jaw) is a reversal of first incisors and first molars – I_1M_1 instead of M_1I_1 During Phase 2, there is considerably more variation, but the most stable sequences are P_2P_1 in the lower jaw, and P^2M^2 in the upper jaw, whereas the most common deviations are P^2C, CP^1, P_1C and M_2P_2. Within one population the most common sequence may be slightly different in boys and girls, and there are differences between populations, especially relating to the late emergence of molars in Europeans. The following sequences therefore encompass the most likely pattern of variation (the brackets indicate that the order is commonly reversed):

Upper jaw $M^1I^1I^2(P^1CP^2)M^2$

Lower jaw $(M_1I_1)I_2(CP_1)(P_2M_2)$.

Eruption sequence in primates has been the subject of some discussion (Garn *et al.*, 1957; Koski & Garn, 1957; Garn & Lewis, 1963; Smith & Garn, 1987), since Weidenreich (1937) speculated that the jaws of young fossil hominids showed ape-like eruption sequences (page 178), but such sequences are still common in modern humans.

Gingival and alveolar emergence and anthropology

Saunders (1837) established gingival emergence of molars as an indicator of children's age. Exploitation of young children in factories became widespread following the Industrial Revolution in Britain, and led to a series of legislative measures to apply age limits, which were enforced by factory inspectors who used Saunders' method to verify the children's ages. Modern dental anthropology or forensic odontology rarely relies on gingival or alveolar emergence for age estimates, because it is much better to use the whole development of the dentition as seen in radiographs or dissections. Gingival emergence has, however, found a place as an index in growth studies of children who cannot be routinely radiographed (Filipsson, 1975; Moorrees & Kent, 1978).

Deciduous tooth root resorption

Timing of root resorption in deciduous teeth (Table 5.7) is known from several radiographic studies (Fanning, 1961; Moorrees *et al.*, 1963b; Haavikko, 1973). All stages show considerable variation and would not be the best basis for age estimation, but resorption may have a place in the study of isolated deciduous teeth.

The Schour and Massler combined formation and eruption chart

Schour and Massler (Schour & Massler, 1941) published a well-known diagram summarizing both formation and eruption (Figure 5.8), dividing the sequence into 22 stages, each associated with a particular age. The American Dental Association published a larger version of this chart, which had just 21 stages and included ± figures after each age to express variation, but in neither case were sources of information quoted. They stated that the figures were modified from the work of Logan and Kronfeld (1933), although there are differences and Smith (1991b) suggested that it should more properly be attributed to Kronfeld (1935a) and Kronfeld and Schour (1939). In any case, their work was based upon a small number of terminally ill children, most of whom were under 2 years old when they died. Nevertheless, the chart has been tested on known-age children (Miles, 1958; Gray & Lamons, 1959) and it generally performs well in comparison with alternatives (Table 5.3). A major revision (Figure 5.9) was carried out by Ubelaker (1978), who removed one prenatal stage, added a new stage at 18 months, and applied ages drawn from many studies. The Ubelaker chart was developed for studies of Native Americans, but is a recognized standard throughout the world (Ferembach *et al.*, 1980; Buikstra & Ubelaker, 1994).

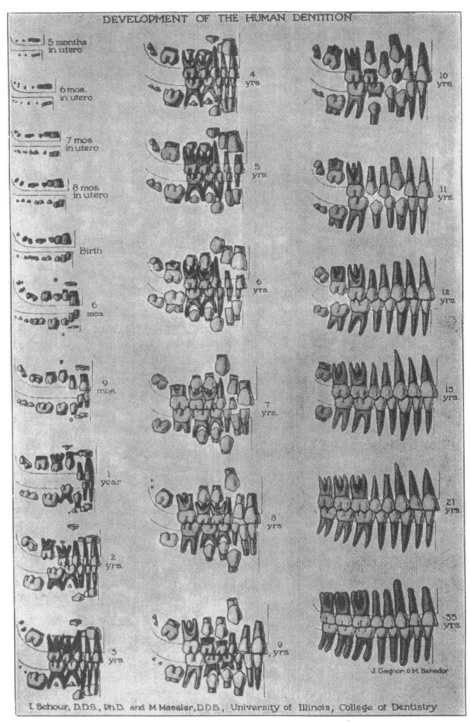

Figure 5.8 Schour and Massler's original dental development diagram. Reproduced from Schour, I. & Massler, M. (1941) The development of the human dentition, *Journal of the American Dental Association,* **28,** 1153–1160, with the kind permission of the American Dental Association, 211E Chicago Avenue, Chicago 60611.

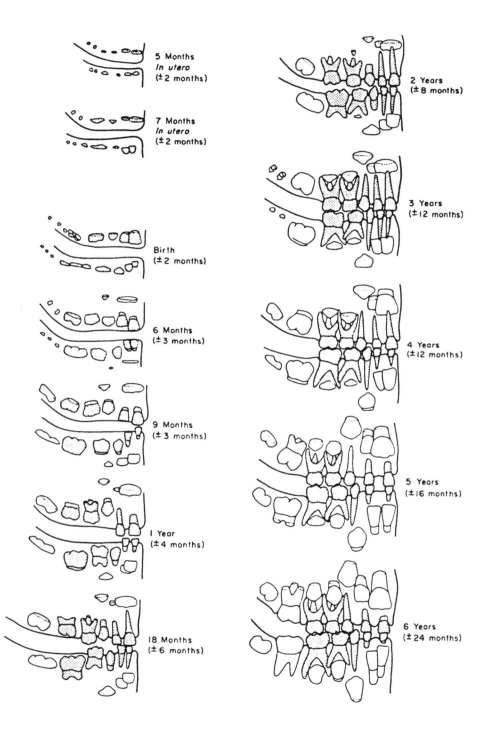

5 Months
In utero
(±2 months)

7 Months
In utero
(±2 months)

Birth
(±2 months)

6 Months
(±3 months)

9 Months
(±3 months)

1 Year
(±4 months)

18 Months
(±6 months)

2 Years
(±8 months)

3 Years
(±12 months)

4 Years
(±12 months)

5 Years
(±16 months)

6 Years
(±24 months)

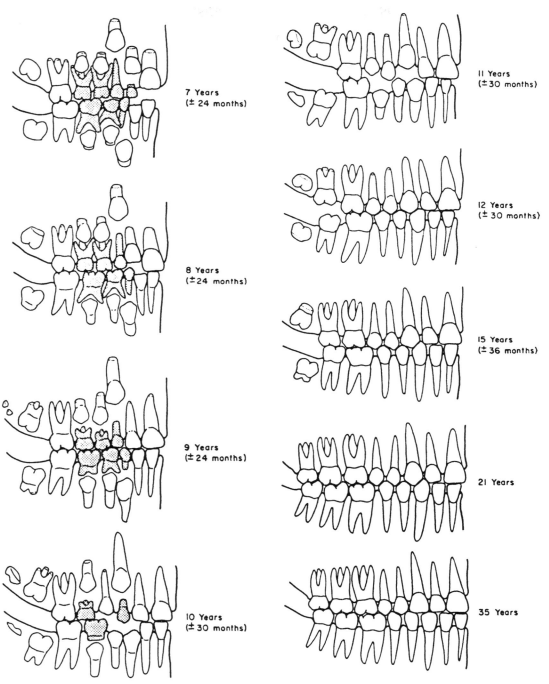

7 Years
(± 24 months)

8 Years
(± 24 months)

9 Years
(± 24 months)

10 Years
(± 30 months)

11 Years
(± 30 months)

12 Years
(± 30 months)

15 Years
(± 36 months)

21 Years

35 Years

Figure 5.9 A version of Ubelaker's revised dental development diagram. Reproduced from Ferembach, D., Schwidetzky, I. & Stloukal, M. (1980) Recommendations for age and sex diagnoses of skeletons, *Journal of Human Evolution,* **9,** 517–549, with kind permission from Academic Press Ltd, 24–28 Oval Road, London NW1 7DX.

Conclusions – estimation of age from dental development

Estimation of age is one of the main tasks required of physical anthropology, and most anthropologists recognize that dental development provides the best evidence for age at death in children (Saunders, 1992). Such is the reliance placed on dental development that it is particularly important to establish well-supported procedures and standards. The most widely used standard in anthropology is the Schour and Massler/Ubelaker chart, which is easily used and taught, although there is a subjective element in assigning individual dentitions to a particular stage. In forensic studies, the Gustafson and Koch diagram is as widely used (Cottone & Standish, 1981), but once again has a subjective element. In both cases, the basis of the chart is not clearly stated, and the best argument in their support is that they work reasonably well when tested on known-age material. The MFH graph (page 131) and Smith table (page 131) are the most commonly used methods based on single large growth studies, although Haavikko (page 132) is used in Scandinavia. Of the three, Smith's method is easiest to use and can probably be regarded as the standard of the future.

Different methods may produce different age estimates from the same material (Ubelaker, 1987). The nature of the reference population, on which each developmental standard was based, must be an important factor and the size of any mismatch between study collection and reference population depends upon the degree of biological affinity, similarities of nutritional plane and health. One reason for the success of the Schour and Massler chart is that it divides dental development into broad stages, to which ages are only loosely attached, so that the act of assigning a dentition to one of these stages averages out the variation. There has been some discussion (Smith, 1991b; Hillson, 1992d; Liversidge, 1994) of the group of terminally ill children used by Logan and Kronfeld (page 125), on which the Schour and Massler chart is probably based. Their illness must have affected their dental development. The MFH graph was based on healthy children which, on the face of it, would provide a better standard of reference but the children whose remains are found on archaeological sites were, by definition, *not* healthy children. All died young, and it is most unlikely that their nutrition or health-care approximated to that of modern middle-income Americans. It is thus important to maintain a distinction between the dental development stage and assigned age, equivalent to maintaining a normal scientific distinction between observation and interpretation. As a basic minimum, the state of development should be recorded separately for each tooth, using the MFH scores (Buikstra & Ubelaker, 1994), even if an overall development stage is assigned as well.

Seriation of dentitions within a collection can help to assign stages and acts as an independent check on variation. The only alternative for an independent chronology of dental development for forensic or archaeological material is to count incremental structures within the enamel or dentine (pages 177 and 187). This is difficult to relate to the pattern of development seen in radiographs, because it records the deposition of enamel matrix before it would be sufficiently mineralized to show as a radio-opacity.

6

Dental enamel

Dental enamel coats the crown with a heavily mineralized layer that can be over 2 mm thick over unworn cusps of permanent molars, decreasing to 1 mm down the crown sides and a fraction of that at the cervix. Deciduous teeth typically have much thinner enamel than permanent (1 mm or less). For a review of enamel structure and physiology, see Boyde (1989).

Enamel formation

Enamel is itself non-cellular, but is formed by the internal enamel epithelium (page 118), a sheet of closely and regularly packed cells called ameloblasts. Formation proceeds in two stages:

1. Matrix secretion. Enamel matrix is one-third organic and one-third mineral (page 227, Table 10.1). Filament-like apatite crystallites, each 30 nm in diameter, are seeded into the matrix and grow with orientations that depend upon the shape of each ameloblast and the packing of the cell sheet as a whole. During the bulk of matrix secretion, the ameloblasts bear a protuberance on their ends called Tomes' process. If the ameloblast layer is pulled away from the developing matrix surface, the Tomes' processes leave a hexagonal network of pits. Crystallites seeded in the pit floors grow with a different orientation to those in the walls, to create discontinuities that extend through the matrix.
2. Maturation. On completing their quota of matrix secretion, ameloblasts metamorphose and start instead to break down the organic component of the matrix. At the same time, the crystallites grow to reach 50–100 nm in diameter, so that fully mature enamel is almost entirely mineral

(Table 10.1). Despite the removal of organic material, the discontinuities in the enamel structure are preserved.

In a developing tooth crown, the forming edge is therefore characterized by a zone of enamel matrix. Higher up, there is a transitional zone where loss of the enamel matrix organic component starts, followed by a maturation zone through which the mineral content rises sharply, passing into the fully mature part of the crown (Robinson & Kirkham, 1982).

Prismatic structure of enamel

The bulk of the mature enamel is characterized by the discontinuities that were established around each Tomes' process during matrix secretion, although they are absent from thin zones next to the EDJ and at the crown surface. If a ground section is examined in transmitted light microscopy, the discontinuities are visible as regular lines (Figure 6.2) whereas, in SEM images of fractured specimens, the enamel divides up along them into bundles of crystallites, called prisms or rods (Figure 6.1). The thin zones without discontinuities at the EDJ and crown surface are distinguished as prism-free enamel but, elsewhere, three patterns of prisms are recognized (Figure 6.3):

1. Pattern 1 enamel. Separate prisms, with a tubular discontinuity right the way around, fit into a honeycomb mesh of inter-prismatic enamel to give a spacing between prism centres of c. 5.5 μm. If they are sectioned longitudinally, the appearance is one of alternating broad and narrow bands.
2. Pattern 2 enamel. The prisms are packed one above another in vertical rows, with sheets of inter-prismatic enamel between them. If they are sectioned transversely, the rows of cut prisms alternate with clear inter-prismatic sheets and the distance between prism centres is c. 4.5 μm. When sectioned longitudinally, the structure produces a variety of layering.
3. Pattern 3 enamel. Each prism has a flange of inter-prismatic enamel, attached to its cervical side, which interlocks between the two prisms directly below. This gives a keyhole-like appearance when prisms are sectioned transversely, with the cut prism called the head and the flange, the tail. The head centres are spaced c. 6 μm apart. When the prisms are sectioned longitudinally, a regular pattern of lines represents the discontinuities, with a spacing that varies depending upon the plane of section.

All three patterns may be found in any human or ape tooth crown, but the dominant form is Pattern 3. Irregularities and intermediate forms are often seen and Gantt (1982) suggested that there were two variants of Pattern 3,

Figure 6.1 Enamel prisms in a human molar from Medieval York. The prisms run diagonally from top left to lower right, and the cross striations are represented by alternate swellings and constrictions along their length. Fractured, etched preparation, examined in ET mode in the SEM. Scale bar 20 μm.

although this seems unlikely (Vrba & Grine, 1978). Martin and colleagues (Boyde & Martin, 1984; Martin & Boyde, 1984; Martin, 1985, 1986; Martin *et al.*, 1988) found that most higher primates had a layer of prism-free enamel at the crown surface, which was underlain in turn by a layer of Pattern 1 and then the Pattern 3, which characterized the bulk of the enamel thickness. In the hominids, gibbons and *Sivapithecus* the Pattern 1 and prism-free layers together occupied less than 10% of total enamel thickness, whereas in the orang-utan the Pattern 1 layer occupied 20%, increasing to 40% or more in the gorilla and chimpanzee.

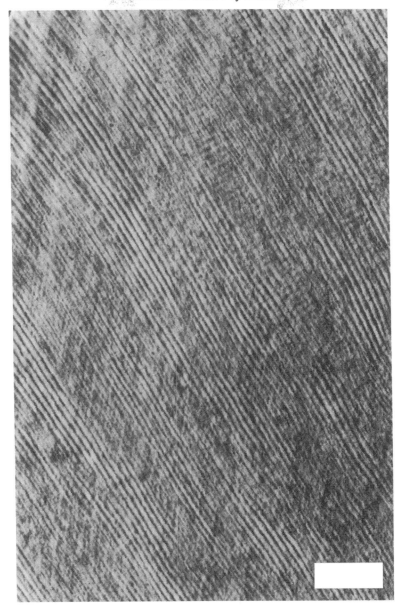

Figure 6.2 Enamel prisms and cross striations, as seen in polarizing microscopy of a ground radial tooth section. The prism boundaries are the lines running from lower right to top left, and the cross striations are seen in patches cutting perpendicularly across them. Scale bar 50 μm.

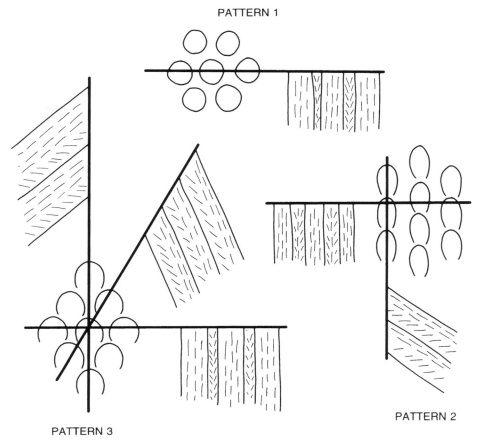

Figure 6.3 Enamel patterns 1, 2 and 3. Adapted from Boyde (1989). The prism boundaries are seen end on (ie. with the prisms transversely sectioned), and in various other planes of longitudinal section, which result in a pattern of variously spaced lines.

Prism boundaries and crystallite angles

Polarizing microscopy allows the average orientation of enamel crystallites to be established through the phenomenon of extinction (page 312). In Pattern 3, there is always a difference in orientation between the head and the tail so that, in polarizing microscopy, the prisms are exaggerated by an alternation of light and dark bands. The angle of extinction relative to the long axis of the head increases from 0° at the cusp tips (parallel to the head long axis), to 17° midway down the crown side and 28° at the cervix. In theory this could be used to identify the part of the crown from which a small fragment of enamel came, but other clues are easier to use (pages 157 and 162).

The course of prisms through the enamel

The prisms do not run straight through the enamel. In radial tooth sections (Figure 2.3), they generally form an angle of about 120° with the crown surface on their occlusal side and only at the cusps or cervix do they run perpendicular. In transverse sections, the prisms are perpendicular to the surface in the middle of each crown side but, elsewhere, they approach the surface obliquely. In addition, each prism undulates from side to side in a sinusoidal way. Osborn (1970) likened this to a waveform (Figure 6.4), with the waveform of each prism slightly out of phase with its occlusal and cervical neighbours to produce an additional cyclic variation passing down the crown side. The phenomenon is known as decussation (Latin *decussare*; to cross at an acute angle), and produces features that can be seen with the naked eye in radially fractured enamel. It gives a fibrous appearance described by several workers in the 1600s, but is today most commonly named after Hunter (1771) and Schreger (1800). Hunter–Schreger bands are clearly seen as alternating bright and dark zones (Figure 6.5) in radial tooth sections under the polarizing microscope, and represent groups of prisms, sectioned at different points in their undulations. In alternate bands (labelled parazones), the prisms are sectioned longitudinally so that their crystallite long axes are parallel to the section plane, and they show up bright, whereas the intervening bands (diazones) contain more obliquely sectioned prisms that show darker (page 312). These bands appear to extend out from the EDJ for about two-thirds the enamel thickness, each with a slight occlusal curve, but they are also shown in the SEM using deep-etching techniques (Boyde, 1989), which demonstrate continued decussation closer to the crown surface than is apparent in polarized light microscopy (Boyde, 1976b).

Incremental structures in enamel

Prism cross striations

If enamel is fractured so that the occlusal surfaces of prisms are exposed, a series of regular swellings and constrictions, 2–5 μm apart, can be imaged in the SEM (Figure 6.1). In compositional BSE imaging (page 315), an alternating light and dark banding of prisms implies a similarly spaced variation in mineralization and, in transmitted light microscopy of ground sections, structures known as prism cross-striations are seen at the same spacing (Figure 6.2). These are alternating bright and dark lines that are emphasized by acid treatment, or under the polarizing microscope (Schmidt & Keil, 1971; Boyde, 1989), and sometimes give an impression of swelling and constriction.

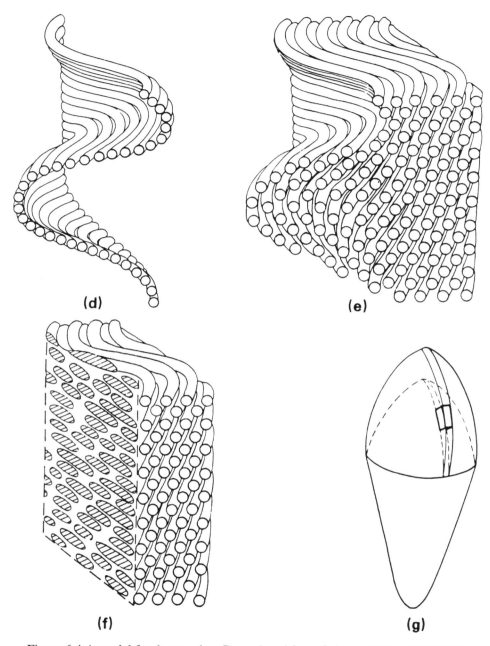

Figure 6.4 A model for decussation. Reproduced from Osborn, J. W. (ed.) (1981) *Dental anatomy and embryology, A Companion to Dental Studies*, Oxford: Blackwell Scientific Publications, with the kind permission of Blackwell Science Ltd, Osney Mead, Oxford, OX2 0EL. (d), a single vertical row of prisms; (e), several such rows together; (f), the effect of sectioning a rectangular block of prisms; (g), the location of the block.

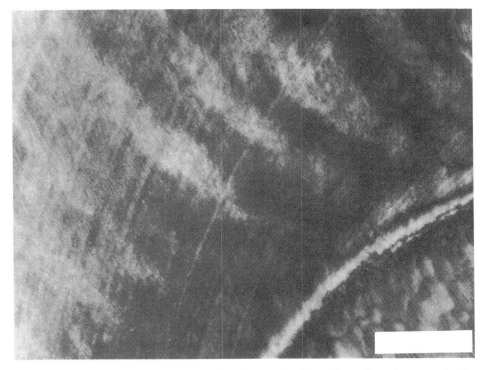

Figure 6.5 Hunter–Schreger bands in a human incisor. Ground section examined in polarized light microscopy. The EDJ crosses the lower right corner and brown striae of Retzius curve up from lower left to upper right. The Hunter–Schreger bands are the diffuse, broad bands, alternating dark and bright, which spread out from the EDJ. Scale bar 200 μm.

It seems clear that these are all manifestations of the same phenomenon, although it is not clear why cross-striations should be visible at all in transmitted light, because the prism boundaries are so narrow (page 312). Some workers (Weber & Glick, 1975) have suggested that they are an artefact of the section plane and thickness, but their existence is confirmed by TSM examination (page 311), phase contrast microscopy and transmission electron microscopy (Gwinnett, 1966; Hinrichsen & Engel, 1966). The repeat interval of cross-striations varies through the crown (Beynon *et al.*, 1991b) – closer together near the EDJ and further apart towards the crown surface, and also closer in the cervical region of the crown than in occlusal parts (Table 6.1).

It seems likely that cross-striations are caused by a cyclical variation in enamel matrix secretion rate (Boyde, 1976a; 1989) in which the bulges of prism heads probably represent faster secretion, and the constrictions, slower. This may also lead to fluctuations in the ratio between carbonate and phosphate in the

Table 6.1. *Cross-striations*

Mean spacing of cross-striations in different parts of human tooth crowns

	Premolars (μm)	Molars (μm)
Cuspal (appositional enamel zone)		
Outer	4.9 ± 0.4	5.1 ± 0.7
Middle	4.2 ± 0.5	4.3 ± 0.7
Inner	2.7 ± 0.4	2.7 ± 0.4
Lateral (occlusal half of imbricational zone)		
Outer	4.9 ± 0.4	5.0 ± 0.5
Middle	4.1 ± 0.6	4.0 ± 0.4
Inner	2.8 ± 0.5	2.6 ± 0.5
Cervical (cervical half of imbricational zone)		
Outer	3.0 ± 0.5	2.8 ± 0.3
Inner	2.3 ± 0.3	2.3 ± 0.3

Source: Beynon *et al.* (1991b).

Counts of cross-striations between brown striae of Retzius
Asper (1916) – 5, 7, 8, 10.
Gysi (1931) – commonly 7.
Bromage & Dean (1985) – range 6–9, mean 7.75, standard deviation 0.69, number of teeth 20.
Bullion (1987) – range 7–10, mean 8.15, standard deviation 0.69, number of teeth 40.
Beynon (1992) – range 6–10, mean 7.7, standard deviation 0.83, number of teeth 100.
Fitzgerald *et al.* (1996) – range 7.8–12.3, mean 9.7, standard deviation 1.0, number of teeth 96.

mineral phase (page 217), which would explain acid etching effects and the BSE appearance of cross-striations. The cross-striation spacing matches well with the estimated enamel matrix secretion rate in humans of 4–4.5 μm per day (Schour & Hoffman, 1939b; Massler & Schour, 1946), and cross-striation counts through a crown correspond well with independent estimates for formation time (Asper, 1916; Gysi, 1931; Boyde, 1963, 1990). In experiments with laboratory animals using periodic marker injections, counts of cross-striations between markers also correspond well with the number of days between injections (Mimura, 1939; Bromage, 1991). There is thus now strong evidence that cross striations represent a 24 hourly, or circadian rhythm.

Brown striae of Retzius

A routine 60–150 μm thick ground enamel section, examined in transmitted light at low magnification, shows a series of lines (cutting across the prism boundaries) following the pattern of crown growth layering (Figure 5.2). These lines are poorly defined in deeper and more occlusal enamel (Figure 6.6) but more sharply defined near the surface, especially near the cervix, often with a 'staircase pattern' in which each step is an exaggerated prism cross-striation (Figure 6.7). The lines vary from light brown to almost black, and Retzius (1836; 1837) described them as *bräunlicher Parrallel-Striche* (German; brown parallel lines), so that they have been known ever since as the brown striae of Retzius. They mark a discontinuity in enamel structure that causes light to be scattered (page 311) as it passes through the specimen and, in reflected light, they are brighter than the surrounding enamel (Boyde, 1976a; Beynon & Wood, 1986). The sharp brown striae at the crown surface are regularly spaced, at 30–40 μm apart in the occlusal half of the crown and 15–20 μm in the cervical half, but those in deeper enamel are less obviously regular. In radial tooth sections, the prism boundaries run through the brown striae at an angle of 90° under the cusp tips, decreasing to about 40° (on the cervical side) at half crown height. The brown striae also run up to the crown surface at an oblique angle of 10–15° (again on their cervical side) just below the cusps, increasing to 60° at the cervix.

Enamel can be made to fracture along the planes of the brown striae (Figure 6.8). In the SEM, these planes are pock-marked with Tomes' process pits (Boyde *et al.*, 1988), clearly indicating that the ameloblasts were interrupted in the midst of matrix secretion. If a radial section is polished and then acid etched, the line of some brown striae can also be made out in the SEM (Risnes, 1985b), sometimes with a discontinuity (Frank, 1978), or a zig-zag translocation of prism boundaries as they cross the line (Osborn, 1973; Weber & Ashrafi, 1979).

Brown striae vary in prominence throughout the crown (Figures 6.5 & 6.6), in a pattern that is identical in left/right pairs of teeth, and in any crowns of one dentition that were formed at the same age, so that teeth from the same person can be matched (Fujita, 1939; Gustafson, 1955; Gustafson & Gustafson, 1967). Most sequences also show less distinct, finer layering (Figure 6.6) in between the main striae, which is sometimes strong enough to dominate over the main pattern. Counts of cross-striations between brown striae within the sequence provide estimates of their periodicity but, in deeper enamel, cross-striations are difficult to count between the less sharply developed striae. The sharp surface zone striae however usually provide a constant count for each individual, varying between individuals from 7 to 10 with an overall average of almost 8 (Table 6.1). The cause of this circaseptan rhythm is not clear, although human growth and physiology display many different rhythms (Dean, 1987b).

Dental enamel

Figure 6.6 Brown striae of Retzius in deep enamel, showing pronounced prism cross-striations and other finer patterns of layering. The prism boundaries are running diagonally from lower right to upper left, and two prominent brown striae cut across them. Other brown striae are less well defined. Polarizing microscopy of a thin ground section (too thin to show the brown striae well). Scale bar 50 μm.

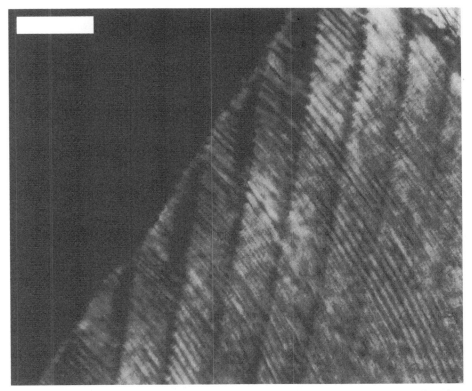

Figure 6.7 Brown striae of Retzius at the crown surface. The striae are sharply defined and run almost vertically, whilst the prism boundaries cut across them and the dark area at the top left is beyond the surface of the crown. Ground section examined in polarized light microscopy. Scale bar 50 μm.

The neonatal line in the enamel

Following Rushton (1933), Schour (1936) described a prominent line in the enamel of deciduous teeth and permanent first molars marking, in each tooth, the expected state of development at birth (page 122). This neonatal line is essentially the first, accentuated brown stria of Retzius, as prenatal enamel does not normally contain brown striae (Scott & Symons, 1974). In light microscopy of thick (150 μm) sections the neonatal line appears as a broad, diffuse band, but in thin sections it is sharper and has a staircase appearance (Weber & Eisenmann, 1971) and, in the SEM, the prisms are constricted and interrupted (Whittaker & Richards, 1978). The neonatal line varies in prominence, presumably reflecting the circumstances of the birth (Schour & Kronfeld, 1938).

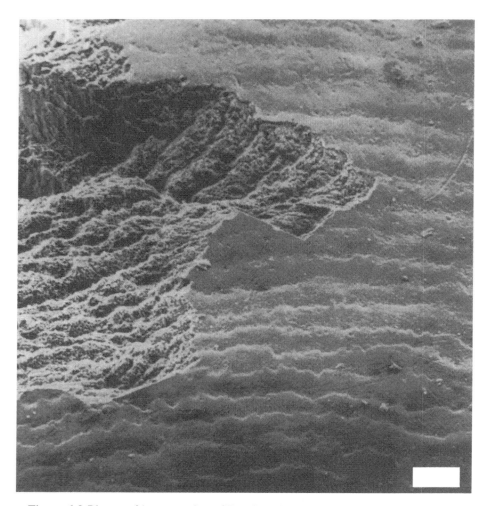

Figure 6.8 Planes of brown striae of Retzius, showing in a fractured preparation of enamel in a molar. The surface of the crown (right half of the picture) is marked with perikymata, and the brown stria planes each join with a single pkg. ET mode in the SEM, operated by Sandra Bond. Scale bar 100 μm.

Pathological bands and Wilson bands

The idea of 'pathological' brown striae, representing a larger disruption of matrix secretion than 'normal' striae, was first suggested by Gustafson (1959; Gustafson & Gustafson, 1967). It was taken further by Wilson and Shroff (1970) who defined a pathological band as 'an accentuated striae of Retzius . . . seen as a relatively broad band in enamel, where there was a sudden change in the prism direction associated with atypical rod forms'. One particular feature of Wilson and Shroff's work was a microscope adjustment 'with the high power condenser arranged so that oblique lighting of the section occurred, producing a three dimensional image', which allowed light to scatter through the thick sections, blurring subsurface details to leave a clearer image of the surface. A more conventional alternative is dark-field microscopy (page 310). Rose termed the features 'Wilson bands' to avoid the implication that they were necessarily pathological, and tested a variety of techniques in order to establish standards for recording them (Rose, 1977, 1979; Goodman & Rose, 1990). Condon (1981) defined Wilson bands as sharp deviations of prism boundaries that produced an optical illusion of a ridge or trough, running continuously from the EDJ to the crown surface, but Goodman and Rose (1990) felt that a band need be continuous only for three-quarters its length. They also defined 'cluster bands'; sequences of closely spaced Wilson bands extending from the surface for only half the enamel thickness.

The causes of variation in the brown striae are still not clearly established, but it is assumed that they relate to the same growth disrupting factors as hypoplasia (page 165). Condon (1981) found at least one Wilson band underneath all hypoplastic defects, even though Goodman and Rose (1990) found them in only 80% of defects. Many factors could cause a mismatch, as both Wilson bands and hypoplasia have a strong element of personal preference in their diagnosis.

Incremental lines on the crown surface

The average human permanent incisor contains about 150 regular brown striae or their equivalent in cross-striations, counting from the first enamel deposited underneath the incisal edge to the last increment at the cervical margin (Figure 5.2). Canines usually contain over 180 striae, whereas molars contain 120 to 150 (Bullion, 1987). For incisors and canines, the first 30–40 brown striae (10–20% of the total) are hidden under the incisal edge and are distinguished as appositional striae. For molars, the first 50–70 striae are appositional (30–50% of the total), presumably reflecting the thicker enamel of molar cusps. Outside the zone of appositional increments, down the sides of the crown, the brown striae angle up to the surface and are called imbricational striae, each marked where it meets the surface by a small groove (Figures 6.8 and 6.9) around the circumference of the crown (Pantke, 1957; Newman & Poole, 1974; Risnes, 1985a,b). In cheek teeth, the grooves around the cusps are 200 μm or more apart (visible with the naked eye), decreasing to 50 μm half way down the crown side and 20–30 μm in the cervical region (Hillson, 1992a). The spacing on anterior teeth is over 130 μm near the incisal edge, decreasing to about 50 μm one-third the way down the crown side. In the occlusal one-third of all crowns the grooves are shallow and merge smoothly with the intervening

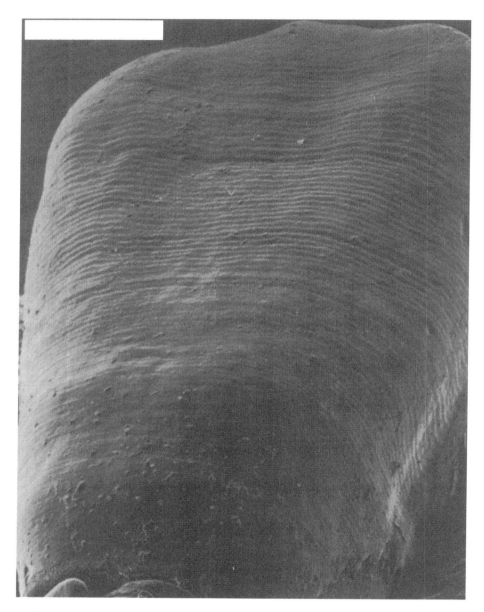

Figure 6.9 Crown surface in an unworn lower incisor from post-medieval London. A progression of perikymata can be seen, running down from the mamelons at the incisal edge. Epoxy resin replica, examined in ET mode in the SEM. Scale bar 2 mm.

ridges in a gentle wave-like pattern (Boyde, 1971). In the middle they are deeper and more sharply defined, becoming sharper still in the cervical region, and sometimes taking on an overlapping sheet structure (Figure 6.10). The wave-like form was termed perikymata (Greek *peri*; around and *kymata*; waves) by Preiswerk (1895), whereas the tile-like pattern of the cervical region was christened imbrication lines (Latin *imbricare-atum*; to tile) by Pickerill (1912). 'Perikymata' is commonly used to describe the pattern as a whole, distinguishing between perikyma grooves (hereafter pkg) and perikyma ridges (Risnes, 1984). Perikymata vary in form between individuals and populations (Pedersen & Scott, 1951), and this can be very marked in the imbrication line zone. On the labial surfaces of some incisors and canines, pkg run through a series of regular bulges and depressions that may be related to the presence of diazones and parazones near the surface (page 153). Furthermore, on anterior teeth, pkg are difficult to follow through the folds of the lingual surface, and are interrupted by the deep mesial or distal curve of the cervical crown margin. Deciduous teeth do not bear perikymata, although a shallow pattern of ridges is sometimes found (Boyde, 1975).

The crown surface is also decorated by a pattern of Tomes' process pits (Boyde, 1971). They are poorly marked and shallow on the perikyma ridges, which are coated with prism-free enamel, but are strongly developed in the pkg where the prism-free layer is missing (Figure 6.10). In some fractured specimens, it is possible to follow a spread of Tomes' process pits continuously from brown striae planes into each pkg, which thus represent the exposed edges of the planes (Figure 6.8 and page 157). Abrasion (page 231) removes the shallower Tomes' process pits on the perikyma ridges first, whilst those in pkg are much more persistent (Scott & Wyckoff, 1949).

Other features of the enamel surface

In addition to perikymata and Tomes' process pits, there may be other microscopic crown surface features. These include 10–15 µm diameter Isolated Deep Pits, sometimes with Surface Overlapping Projections covering them (Fejerskov *et al.*, 1984; Boyde, 1989). There are also fields of brochs, which are bulges 30–50 µm in diameter and 10–15 µm high, most with the outer layer of enamel missing from their tops. Found in all tooth classes, but especially in molars, they are usually scattered in bands around the cervical region. Lamellae are longitudinal fault planes radiating out from the EDJ through the whole thickness of the enamel layer and forming lines of weakness along which crowns may fracture in archaeological material. They are present before the tooth is erupted, as thin irregular layers of poorly mineralized material.

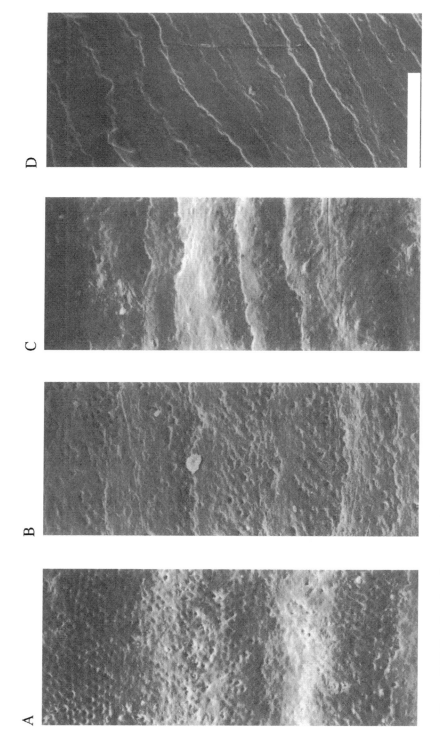

Figure 6.10 Details of perikymata, of occlusal (A), mid-crown (B and C) and cervical type (D). Epoxy resin replica, examined in ET mode in the SEM. Scale bar 100 μm.

Enamel defects

Viewed with the naked eye, the crown surface is normally smooth, white, and translucent, but it may show three types of defect:

1. Hypoplasia. A deficiency of enamel thickness, disrupting the contour of the crown surface, initiated during enamel matrix secretion.
2. Opacities (also called hypocalcifications). A disruption to mineralization at the maturation stage.
3. Discolourations. Deposits of pigment resulting from metabolic disorders (Pindborg, 1982) or later staining of deficiencies in mineralization.

Amelogenesis imperfecta (AI)

Dental defects may be inherited, either on their own, or as part of a much broader syndrome (Pindborg, 1982). Such inherited defects are often called amelogenesis imperfecta, or AI for short, and are very rare – various estimates range from 1 in 14 000 people affected (Winter & Brook, 1975) to 1 in 1000 for a Swedish population (Bäckmann, 1989). AI may involve both opacities and hypoplasia, and is usually defined clinically as an enamel defect that is not attributable to disease or toxicity in a particular period. It affects the whole of both permanent and deciduous dentitions in many cases, but some forms are apparently more localized and this must make diagnosis difficult. Only one archaeological case has been published (Cook, 1980).

Defects relating to local disturbances

Trauma, electrical burns, ionizing radiation, or localized osteitis may all disrupt formation of just one part of the dentition. They result in hypoplasia of one or two teeth on their own, a rare condition, for which only Goodman and Rose (1990) have proposed an archaeological example.

Developmental defects related to generalized disturbances

Hypoplasia

Most hypoplastic defects are arranged in a band around the circumference of the crown, following the trend of the perikymata, and representing episodic disruptions to matrix secretion throughout the growing dentition. Bunon (1746) found them on the unerupted teeth of children with rickets, scurvy, measles and smallpox, thus proving that they were caused during crown formation. Zsigmondy (1893) introduced the term *hypoplastische emaildefecte* or 'hypoplastic enamel defects' (Greek *hypo*, under; *plastikos*, mould), and Berten (1895) showed that the defects paralleled the layering of crown development and could be matched between teeth from one dentition. Much research has involved laboratory animal experiments, in which Mellanby (1929; 1930; 1934) and Klein (1945) related vitamin A or D deficiency to hypoplastic defects, whilst Kreshover and colleagues concentrated on non-dietary factors. Experimentally induced diabetes (Kreshover *et al.*, 1953) and fever (Kreshover & Clough, 1953) in pregnant rats interrupted enamel formation in their offspring, and innoculation with pathogenic viruses (Kreshover *et al.*, 1954; Kreshover & Hancock, 1956) or bacteria (Kreshover, 1944) also disrupted amelogenesis. In a review paper, Kreshover (1960) concluded that 'abnormal tooth formation is a generally nonspecific phenomenon and can be related to a variety of local and systemic disturbances'.

Clinical studies have concentrated on the deciduous dentition. Schour and colleagues (Schour & Kronfeld, 1938; Kronfeld & Schour, 1939) described linear hypoplastic defects associated with the neonatal line (page 159) in children with birth injury, notably on the labial surface of deciduous incisors. Sweeney *et al.* (1971) found similar defects in 73% of Guatemalan children recovering from third degree malnutrition, as compared with 43% with second degree malnutrition and 22% in the population at large. Infante and Gillespie (1977) further found that members of the same family had similar frequencies of defects, and noted a pattern relating to different seasons of birth. In these studies, economic factors that varied through the year apparently controlled an aetiology that was related to malnutrition. Other pre- and neonatal conditions that have been related to hypoplasia in deciduous teeth include allergies, congenital defects, neonatal haemolytic anaemia, maternal rubella, diabetes or syphilis (Pindborg, 1970, 1982), but it is more difficult to demonstrate a relationship with later childhood diseases. Eliot *et al.* (1934) found higher frequencies of defects in children with progressively more severe rickets, and malnutrition has been linked with permanent tooth hypoplasia in Mexican children (Goodman *et al.*, 1991). In some children, it is possible to match episodes of diarrhoea with particular lines of hypoplasia (Pindborg, 1982). Sarnat and Schour (1941; 1942) attempted to match such lines with the medical histories of 60 children, which included rickets, chickenpox, convulsions, diarrhoea, diptheria, measles, pneumonia, scarlet fever, vomiting and whooping cough. Only half the children showed good matches, but much depends upon the level of defect that was recorded as hypoplasia.

The defects, in fact, vary widely in form, and Berten (1895) described three types that remain the basic classification: *Grübchen* (German; little pits), *Fürchen* (furrows) and *flächenformig* (plane-form) defects.

1. Furrow-type defects. Furrows (Figure 6.11) are much the most common type, and are often referred to as Linear Enamel Hypoplasia or LEH

(Goodman & Rose, 1990). They can only occur in the imbricational zone of crown formation (page 161) and represent an exaggeration of the perikymata (page 163), with a greater than normal pkg spacing down their occlusal wall (Hillson, 1992a). A continuous range of furrows is found, from large examples involving 20 pkg or more down to just one pkg. The normal pkg spacing varies, in any case, down the crown side, and this has an effect on the prominence of the defects, independently from the severity or duration of the growth disruption causing it. Furrows are particularly prominent on the sides of anterior tooth crowns, although they are also seen in the cervical half of cheek tooth crowns. Sometimes, several are found one above another, 'washboard' style.

2. Pit-type defects. The pits vary in size; some representing an interruption to hundreds of ameloblasts, and some just a small group (Figure 6.12). Cusps of cheek teeth or incisive edges of anterior teeth may be covered with a spread of pits, or they may be arranged in bands around the crown sides, sometimes following furrow-type defects. When they occur along a broad furrow, pits are themselves large and widely separated, but in a narrower furrow defect they are small and close together, like stitching. The pit floors represent the exposed plane of a brown stria, and are often marked by Tomes' process pits, showing that matrix secretion was abruptly and permanently interrupted. Other pit floors may show some irregular continued matrix secretion.

3. Plane-type defects. These defects (Figure 6.13) expose large areas of brown stria planes, marked with Tomes' process pits. Sometimes the edge of the defect follows the circumferential line of perikymata, producing a regular step with normal enamel below but, around the cusp tips of cheek teeth and incisal edges of anterior teeth, they may instead create irregular sloping facets. In the floor of a prominent furrow defect, very widely spaced pkg occasionally expose broad brown stria planes (Boyde, 1970).

Furrow defects are produced by a broader than normal band of ameloblasts ceasing matrix secretion at each pkg, whereas pits result from compact clusters of ameloblasts ceasing matrix secretion, at some distance from the pkg currently being formed, but the reason for this difference is still obscure. It is important to note that one growth disruption may produce a different type or prominence of defect on different teeth in the dentition because of the different geometry of crown formation (page 161).

Opacities or hypocalcifications

Opaque white patches in the normally translucent enamel are the surface expressions of poorly mineralized zones beneath the surface, which reflect

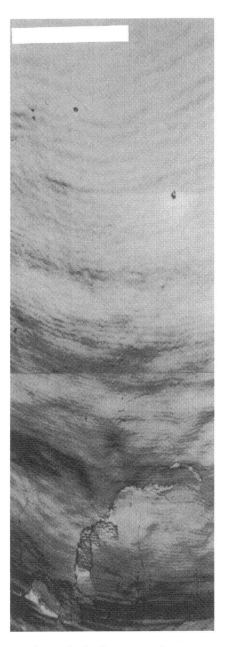

Figure 6.11 Furrow-type hypoplasia. Permanent lower second molar crown from the Predynastic site of Badari, Egypt. The cervical crown margin is at the lower edge of the picture, and the main furrow shaped defect runs across just above. Other, smaller defects can be seen as accentuations of the pkg above. Epoxy resin replica, examined in ET mode in the SEM. Scale bar 1 mm.

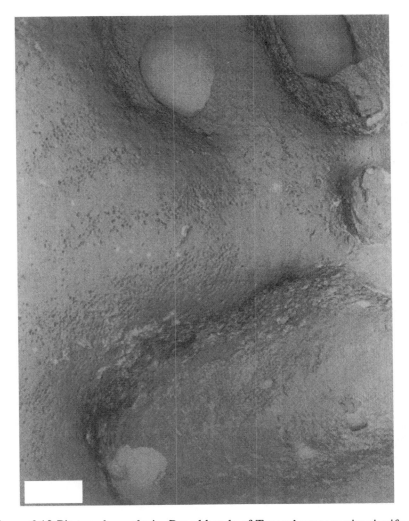

Figure 6.12 Pit-type hypoplasia. Broad bands of Tomes' process pits signify the presence of pkg running across the picture. A cluster of large pit-type defects is shown, with exposed Tomes' process pits in their floors. Most have air bubbles trapped in them by the impression process (difficult to avoid). Epoxy resin replica, examined in ET mode in the SEM. Scale bar 100 μm.

more light back to the observer (page 311). Two types are recognized (Suckling *et al.*, 1989): demarcated opacities and diffuse opacities. The demarcated variety vary in size, position and depth of enamel affected, but all have a clear boundary, whereas the diffuse form affects only the surface 150 μm of enamel, and has no clear boundary. Five per cent of teeth or more may have opacities in living people, but they are rarely reported in archaeological

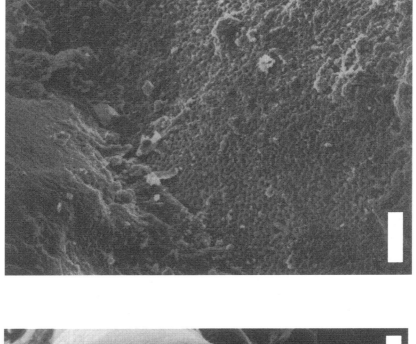

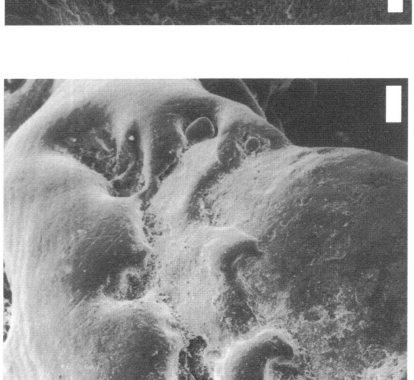

Figure 6.13 Plane-form hypoplasia. A, lower first molar from post-medieval London, seen upside down. The tip of the cusp is below the lower edge of the picture. Two broad exposed planes are visible, with a low step between them and an irregular, pitted and bulging boundary. Unaffected pkg can be seen towards the top. Scale bar 250 μm. B, enlarged surface of exposed plane showing Tomes' process pits. Scale bar 50 μm. Epoxy resin replica, examined in ET mode in the SEM, operated by Sandra Bond.

material. This could be due to diagenetic changes or altered water content, and they may be better viewed under ultraviolet illumination, which is more affected than visible light by scattering phenomena. Poorly mineralized zones have also been monitored by hardness testing (Suckling *et al.*, 1989) or microradiography (Suga, 1989) of ground sections. Some zones occupy the whole enamel thickness, implying that the maturation was interrupted permanently, whereas others do not reach the surface, and the affected ameloblasts must have returned to normal maturation. Opacities are often associated with hypoplasia, implying disruption to both matrix secretion and maturation.

Dental fluorosis

Fluoride in drinking water (page 220) is responsible for a range of dental defects (Møller, 1982; Pindborg, 1982; Fejerskov *et al.*, 1988). Low levels of fluorosis produce opacities (Figure 6.14), that may be stained brown through taking up colours from plaque or food, but higher levels produce pitted hypoplastic defects. Cheek teeth are more severely affected than anterior teeth, in both deciduous and permanent dentitions, and defects match between left/right tooth pairs. Lukacs and colleagues (1985) reported opacities, mottled yellow-brown stains and pitting (distinct from the usual pattern of hypoplasia) in permanent teeth from the Neolithic/Chalcolithic site of Mehrgarh, Pakistan. They considered the role of diagenetic effects, but found high levels of fluoride in the enamel and in local drinking water, although they recognized that fluorine concentrations varied greatly through the enamel (page 220). It would be very difficult to confirm fluorosis by these means in archaeological material.

Hypoplasia and syphilis

Dental defects are one of the classical stigmata of congenital syphilis (Jacobi *et al.*, 1992). The disease disrupts crown formation between birth and one year of age and affects only those parts of the dentition being formed at that time. Hutchinson's (1857) Incisors are found in 10–60% of patients, and are permanent upper first incisors with a pumpkin-seed-shaped labial outline, tapering to a narrow incisal edge that bears a pronounced notch. Lower incisors may also be somewhat peg-like and, although the upper second incisors are not affected, canines may show a corresponding furrow defect around the main cusp tip. The permanent first molars are affected in up to 60% of patients, but the form of defect varies. Moon's (1877) Molars have a narrow occlusal area, giving a domed or bud-like, bulbous crown form. Mulberry Molars (Karnosh, 1926) have prominent plane-form defects around the cusps, with a marked step at their base below which the crown side is normal. The overlap between these specific malformations and more general hypoplastic defects makes diagnosis difficult (Condon *et al.*, 1994).

Methods for scoring hypoplasia and hypomineralization

Views on recording systems divide on the issue of fluorosis (Clarkson, 1989). Some maintain that a clear group of diagnostic criteria can be established (Horowitz, 1989; Richards *et al.*, 1989), as defined by the Community Fluorosis Index (F_{ci}) (Dean, 1934), or by the Tooth Surface Index of Fluorosis (Horowitz *et al.*, 1984) and the TF index (Figure 6.14) (Thylstrup & Fejerskov, 1978; Fejerskov *et al.*, 1988). Others (Clarkson, 1989) feel that it is better to describe defects without making a diagnosis, as fluorosis is just one possible factor. Mellanby (1927; 1934) devised a scale of hypoplasia severity defined by photographs, whilst Sarnat and Schour (1941; 1942) evolved a descriptive technique. A Working Party of the Fédération Dentaire Internationale Commission on Oral Health, Research and Epidemiology was set up in 1977 to propose 'a system of classification of developmental defects of enamel suitable as an international epidemiological index'. This (Table 6.2) became the Developmental Defects of Enamel (DDE) Index (Commission on Oral Health, 1982), which has since been widely used and recommended as a standard in anthropology (Hillson, 1986a; Goodman & Rose, 1990), although its classification is still not ideal. Some have also found it cumbersome, proposing a simpler Modified DDE Index (Clarkson & O'Mullane, 1989), but this does not provide sufficient detail for anthropology.

Methods for estimating the timing of enamel defects

Massler and colleagues (1941) divided the dentition into phases by the definition of 'growth rings' (page 126): neonatal (birth to 2 weeks), infancy (10 months of age), early childhood (2.5 years) and later childhood (5 years). They produced a chart, which was used directly by Sarnat and Schour (1942) and Schultz and McHenry (1975) to estimate the age at which hypoplastic defects were initiated. Swärdstedt (1966) instead measured the position of each defect on the crown surface as a distance from the CEJ, converting these to ages by applying mean crown heights to Massler and colleagues' chart and producing a table (Goodman *et al.*, 1980) that has been widely used in anthropology. Further developments include regression equations (Goodman & Rose, 1990) and a computer program (Murray & Murray, 1989) for converting measurements to ages. The table is not reproduced here, because there are drawbacks with the method, which fall under five main headings (Goodman & Rose, 1990; Hillson, 1992d):

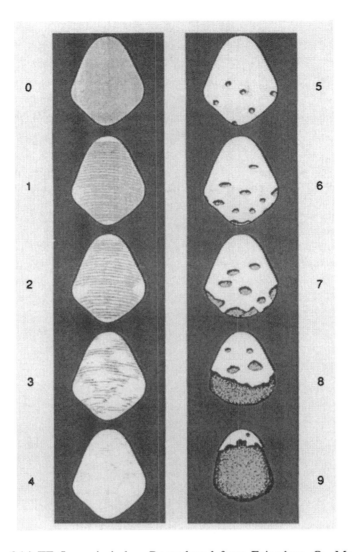

Figure 6.14 TF fluorosis index. Reproduced from Fejerskov, O., Manji, F. & Baelum, V. (1988) *Dental fluorosis – a handbook for health workers,* Copenhagen: Munksgaard, with kind permission from Munksgaard, 35 Nørre Søgade, PO Box 2148, DK-1016 Copenhagen K. TF0 is normal. TF1 to TF14 show fine white opaque lines following the trend of the perikymata, together with increasing diffuse opacities. TF5 to TF9 show pitted- and plane-type hypoplastic defects.

1. The table assumes a constant growth rate for formation of the crown surface that is not supported by the known geometry of crown formation.
2. It assumes that there is no appositional zone of crown formation, an error that is particularly large in cheek teeth (page 161).

Table 6.2. *Index of developmental defects of dental enamel (DDE Index)*

Type of defect	Code for permanent teeth	Code for deciduous teeth
Type		
Normal	0	A
Opacity (white/cream)	1	B
Opacity (yellow/brown)	2	C
Hypoplasia (pits)	3	D
Hypoplasia (grooves: horizontal)	4	E
Hypoplasia (grooves: vertical)[a]	5	F
Hypoplasia (missing enamel)	6	G
Discoloured enamel (not associated with opacity)	7	H
Other defects	8	J

Number and demarcation of defects

	Code for permanent teeth	Code for deciduous teeth
Single (one well-demarcated defect)	1	A
Multiple (> one well-demarcated defect)	2	B
Diffuse (fine white lines – following pattern of pkg)	3	C
Diffuse (patchy, irregular, lacking well-defined margins)	4	D

Location

	Code
No defect	0
Gingival one-half	1
Incisal one-half	2
Gingival and incisal halves	3
Occlusal	4
Cuspal[b]	5
Whole surface	6
Other combinations	7

Each tooth (labelled by the FDI system) is recorded separately. Defect combinations are recorded as multiple scores for each tooth affected.

[a] 'Grooves: vertical' is not well defined and its relationship to the development of hypoplasia is obscure.
[b] 'Cuspal' is reserved for the tip of the cusp.

Adapted from Commission on Oral Health (1982).

Table 6.3. *Matches for defects of dental enamel*

	First molar	Upper and lower first incisors	Lower second incisor	Canines	Upper second incisor	First and second premolars	Second molars	Third molars
A	O	O						
B	I	I	O					
C	C	I	I	O				
D		C	C	I	O	O	O	
E				I	C	I	I	
F				C		C	C	
G								OIC

Key: O, occlusal part of crown side; C, cervical part of crown side; I, intermediate.

Method: The table shows the point at which hypoplastic defects normally match between teeth in the same dentition. Defects can therefore be assigned to a category by these matches, with due allowance for attrition. Defects should not be assigned if a match cannot be made.

Source: Hillson (1992d).

3. A single mean crown height for each class of tooth is used as a basis for the table, when crown height is known to vary (page 161).
4. The table was from a single set of standards (devised in 1941) for dental development, when many different studies since then have shown considerable variation.
5. The position of a defect on the crown surface is not simply related to the timing or duration of the growth disruption that caused it.

Buikstra and Ubelaker (1994) never the less suggested recording measurements for the position of enamel defects, as a basic minimum, but it would be better to take impressions (page 299) of the buccal surfaces of all tooth crowns. Measurements could be taken from these if required, but they would also allow the alternative approach (Hillson, 1992a) of odontochronology (page 177), which can also be used to refine tables of crown surface formation timing (Hillson, 1979; Bullion, 1987). For rapid surveys it may instead be best to return to the ideas of Berten (1895) and Logan and Kronfeld (1933), who devised tables of defect matches between teeth (Table 6.3), which can be used to assign defects to broad development categories, without making assumptions about age.

Condon (1981) and Rudney (1983) also measured the position of Wilson bands at the EDJ, with reference to the most occlusal tip of the EDJ under the cusp. Ages for the growth disruptions were then calculated by applying the MFH (page 131) timings for start and finish of crown formation. This method also makes assumptions about the linear growth of tooth crowns, but these are easier to accept when the measurement is along the EDJ rather than the crown surface.

Anthropological studies of dental defects

Summaries include Goodman and Rose (1990; 1991), Skinner and Goodman (1992). Hypoplastic defects are common in most living monkeys and apes, especially in their tall anterior tooth crowns (Schuman & Sognnaes, 1956; Vitzthum & Wikander, 1988; Miles & Grigson, 1990). They were also common in *Australopithecus* and *Paranthropus* (Robinson, 1956; White, 1978), and Neanderthalers showed a similar pattern of occurrence to modern *Homo* (Molnar & Molnar, 1985; Ogilvie *et al.*, 1989). The varying methodologies used make it difficult to quote normal levels for defects, or to summarize the results of the many studies carried out, but there are recurring themes in interpretation:

1. Transition from hunter–gatherer subsistence to agriculture. It is suggested that increased reliance on one food resource carries risks of seasonal deficiencies, and increased sedentism might enhance transmission of childhood diseases. Examples include the increase in hypoplastic defects from Mesolithic to early agricultural contexts in the Levant (Smith *et al.*, 1984), with adoption of maize agriculture in North America (Sciulli, 1977, 1978; Cook, 1984; Larsen, 1995), and with more intensive agriculture in Ecuador (Ubelaker, 1984). The most detailed study was carried out on the Dickson Mounds site, Illinois, where the change from Late Woodland to Mississippian cultural contexts was not only associated with increased defect frequencies (Goodman *et al.*, 1980), but also increasingly annual spacing between growth disruptions (Goodman *et al.*, 1984b) and earlier average age at disruption (Goodman *et al.*, 1984a).

2. Modal age of growth disruption associated with weaning. Breast-feeding ensures a supply of maternal antibodies, buffers the impact of nutritional deficiencies, and so has a protective effect that is lost at weaning which might be reflected in higher defect rates. Examples of this interpretation include ancient Egyptians and Nubians (Hillson, 1979), Dickson Mounds (Goodman *et al.*, 1984a), West Indian slaves (Corruccini *et al.*, 1985; Rathbun, 1987) aboriginal Australians (Webb, 1995) and nineteenth century Florentines (Moggi-Cecchi *et al.*, 1994), whereas similar suggestions have

been made for Wilson bands (Rudney, 1983). This interpretation is, however, not without problems (Skinner & Goodman, 1992).

3. The arrival of Europeans in the New World. Introduction of new pathogens to indigenous populations might increase growth disruption in children, and newly arrived Europeans may in turn have encountered difficulties in food supply. In Georgia (Hutchinson & Larsen, 1990), the contact period Spanish mission of Santa Catalina de Guale had more growth disruptions than pre-contact agriculturalists. In Ecuador, however, the opposite occurred (Ubelaker, 1994).

Most studies have interpreted both hypoplastic defects and Wilson bands as indicators of Selyean stress (Goodman *et al.*, 1988; Goodman & Rose, 1990). Stressors are seen as non-specific environmental factors that induce a disturbance to normal physiological balance and dental defects, as good indicators of unspecified growth disruptions, are thus ideal evidence for such models.

Odontochronology

Matching tooth crowns by their growth sequences

Fujita (1939) and Gustafson (1955) showed clearly that variation in brown striae (page 157) can be used to match up different tooth crown sections belonging to one individual. A less time-consuming or destructive technique is to match pkg (page 163) and hypoplasia sequences at the crown surface, and the small furrow defects are often easier to match than major hypoplastic episodes (Hillson, 1992a).

Age at death estimation from counts of incremental structures in the crown

Boyde (1963, 1990) suggested that age could be estimated from prism cross striation counts. In an Anglo-Saxon child's dentition, excavated from the site of Breedon-on-the-Hill, England, it was possible to identify the neonatal line in the lower first molar, match prominent brown striae between the first molar and a first incisor, and make a combined cross striation count from neonatal line to completion of the first incisor, just before the death of the child. Dean and Beynon (1991) similarly used brown stria/pkg counts to estimate crown formation timing and age at death in a child from the 18th/19th century AD crypt of Christ Church, Spitalfields, London. Bromage and Dean (1985) developed an ageing method based on pkg counts alone, suggesting that lower incisor crown formation started at 3 months of age, and that around 6 months was taken up by appositional enamel growth so that the first pkg would appear at 9 months of age. Applying pkg counts down the side of uncompleted fossil hominid crowns to an assumed 7-day repeat interval, they derived ages at death (Table 6.4) but the underlying assumptions have been challenged (Mann *et al.*, 1990a, 1991; Huda & Bowman, 1994):

1. Cross-striations might not represent a 24 hour rhythm.
2. The count of cross-striations between brown striae varies, at least between individuals.
3. The assumption of an age at first pkg may not be applicable to all hominid taxa.

Some of these objections can be answered directly by further research, but the main argument in favour of the method is that, where it can be tested, it works (Table 6.5). A similar methodology can be used to calibrate the sequence of dental development in past populations, or to estimate timing of dental enamel defects (Hillson, 1992a, 1992d).

Dental development in fossil hominids

There have been several reviews of this subject (Beynon & Dean, 1988; Mann *et al.*, 1991; Smith, 1991a). Much discussion has centred upon the Taung child, discovered in 1924 near Kimberley, South Africa; the first juvenile early hominid skull to be found, and the type specimen for *Australopithecus*. Dart (1925) saw it as a man-ape, with human-like teeth, and used living *H. sapiens* standards to arrive at an age at death estimate of 6 years. Mann (1975) took this idea further in a group of australopithecine jaws, concluding again that the sequence of dental development was human-like. Ape dental development is less well-known than human, but is clearly different (Beynon *et al.*, 1991a; Smith *et al.*, 1994):

1. The ape dental development sequence takes place over 10–12 years, compared with 18 or more in living humans.
2. Ape first permanent molars erupt at around 3–4 years of age, compared with 5–7 in humans.
3. Ape incisor and canine crowns are still forming after the first molar roots are completed. Human incisor and canine crowns are fully developed long before this.

Smith (1986a) developed a graphical method for comparing the development of fossil hominid dentitions with human and great ape standards. She interpreted these to indicate that most early hominids showed the ape dental development pattern but that *Paranthropus* differed from both apes and humans.

Table 6.4. *Age estimates from counts of brown striae and perikyma grooves*

Site	Specimen	Taxon	Teeth used for counts	Age estimate (years)	Age estimated from counts of pkg or bsR
Unspecified[a]	Unspecified	Homo sapiens	Lower 1st molar	0.9	bsR
Breedon-on-the-Hill, United Kingdom[a]	Unspecified	Homo sapiens	Upper 1st incisor, lower 1st molar	4.6	bsR
Laetoli, Tanzania[b,c]	LH2	Australopithecus afarensis	Lower 1st incisor	3.25 (3.2–4)	pkg
Sterkfontein, South Africa[b]	Sts24a	Australopithecus africanus	Upper 1st incisor	3.3 (3.2–4)	pkg
Swartkrans, South Africa[b,c]	SK62	Paranthropus robustus	Lower 1st incisor, Lower 2nd incisor	3.35, 3.48 (3.4–3.75)	pkg
Swartkrans, South Africa[b,c]	SK63	Paranthropus robustus	Lower 1st incisor	3.15 (3.2–3.9)	pkg
Koobi Fora, Kenya[b,c]	KNM-ER820	Early Homo	Lower 2nd incisor	5.3 (5.3–6)	pkg
Devil's Tower, Gibraltar[d]	Gibraltar child	Neanderthal	Upper 1st incisor	3.1	pkg
Koobi Fora, Kenya[e]	KNM-ER1477	Paranthropus boisei	Lower 1st incisor	2.5–3.0	pkg
Koobi Fora, Kenya[e]	KNM-ER812	Paranthropus boisei	Lower 1st incisor	2.5–3.0	pkg
Koobi Fora, Kenya[e]	KNM-ER1820	Paranthropus boisei	Lower 1st incisor	2.5–3.1	pkg
Olduvai, Tanzania[e]	OH30	Paranthropus boisei	Lower 1st incisor	2.7–3.2	pkg
Spitalfields, London, United Kingdom[f]	2179	Homo sapiens	Upper 1st molar, Lower 1st incisor	5.25	bsR, pkg
Swartkrans, South Africa[g]	SK63	Paranthropus robustus	Lower canine	3.18–4.23	bsR

Key: pkg, perikyma grooves; bsR, brown striae of Retzius.

Sources: [a] Boyde (1990), [b] Bromage & Dean (1985), [c] Aiello & Dean (1990), [d] Dean et al. (1986), [e] Dean (1987a), [f] Dean & Beynon (1991), [g] Dean et al. (1993b).

Table 6.5. *Tests of age estimation from enamel layering using known-age archaeological specimens*

Specimen	Tooth/teeth	Known age (years)	Counts of pkg or bsR	Assumed cross-striation count between bsR	Estimated age (years)
724[a]	Lower first incisor	1.1	17 pkg	8	1.4
282[a]	Upper and lower first incisor	1.3	8/15 pkg	8	1.4/1.3
440[a]	Lower first incisor	1.4	25 pkg	8	1.6
505[a]	Upper and lower first incisor	1.7	15/28 pkg	8	1.6
455[a]	Upper first incisor	2.5	61 pkg	8	2.6
420[a]	Upper and lower first incisor	2.5	70/75 pkg	8	2.8/2.6
520[a]	Lower first incisor	2.8	84 pkg	8	2.8
562[a]	Upper and lower first incisor	3	75/88 pkg	8	2.9
365[a]	Upper and lower first incisor	3.3	90/105 pkg	8	3.2/3.4
302[a]	Upper and lower first incisor	4.5	140/144 pkg	8	4.3/4.2
SB 30/125 A, Emily L.[b]	Lower canine	0.6	28 bsR	5.8	0.4–0.6
SB 31/125 B, William B. B.[b]	Lower first molar	0.7	36 bsR	7	0.7
SB 38/125 H, George B. H.[b]	Lower first molar	3.42	125 bsR	6–8	2–2.7

Key: pkg, perikyma grooves; bsR, brown striae of Retzius.

Sources: [a] Stringer *et al.* (1990), [b] Huda & Bowman (1995).

This view was strongly disputed by Mann and colleagues (Mann *et al.*, 1987), and a lively correspondence ensued in the pages of *Nature* (Conroy & Vannier, 1987; Smith, 1987; Mann, 1988; Wolpoff *et al.*, 1988; Mann *et al.*, 1990b; Beynon & Dean, 1991; Smith, 1991a). Dean and colleagues (1993b) sectioned a lower canine from one of the main *Paranthropus* jaws at issue, and cross-striation counts provided an age at death estimate that fitted best with the ape timing of dental development.

The molar development sequence seems to have been short for all australo-pithecines, and this rapid growth was achieved by a fast extension rate down the crown sides, with wide pkg spacing. Hominid cheek tooth enamel is considerably thicker over the cusps than it is in apes, and the greatest thickness is found in the large cheek teeth of *Paranthropus*, although the pkg counts show that they took no longer to form than those of more gracile hominid forms. The rapid growth rate required to achieve this was both by faster enamel matrix secretion and more rapid extension of the growing crown edge (Beynon & Wood, 1987; Beynon *et al.*, 1991a). The newly defined hominid species *Ardipithecus ramidus* (page 4) instead has thin enamel, and this is one of the features used to characterize it as close to the common ancestor of hominids and pongids (White *et al.*, 1994).

Archaeological preservation of enamel

Enamel structure is usually little affected in archaeological specimens, under most conditions of burial. On occasion, the enamel caps of the crowns may be the only part of a body to survive (Beeley & Lunt, 1980), but enamel is lost in very acid soils, when some organic remnants of dentine may survive (Stead *et al.*, 1986). In a cremation, the enamel of erupted teeth rapidly flakes away from the dentine, but unerupted teeth receive a measure of protection from their bony crypts and their enamel may be retained (McKinley, 1994). In this situation the enamel surface is cracked and no longer glossy, and exper-imental studies (Shipman *et al.*, 1984) have shown that the crystallites sinter to produce rounded masses although the prism structure can still be seen after heating to 800 °C or more.

7

Dentine

For summaries of dentine histology see Jones and Boyde (1984), Bradford (1967), Frank and Nalbandian (1989), and Schmidt and Keil (1971). The main bulk of tissue in the crown and roots is called circumpulpal dentine. A thin peripheral layer, under the EDJ, is differentiated as mantle dentine, although it may be so thin as to be indistinguishable even under the microscope. Other distinct layers underlie the CDJ in the roots. Kuttler (1959) further divided dentine development into primary, secondary and tertiary phases. Primary dentine comprised the main period of tooth formation, secondary dentine (page 193) was a slow, regular continued formation, which lined the pulp chamber in adult life, and tertiary dentine patches were laid down to repair damage.

Cells and tubules

Odontoblasts

The cells of dentine are called odontoblasts. Each is cylindrical and 4–7 μm in diameter, with a long, fine process arising from the end, which tapers as it penetrates the full thickness of dentine. The process bears many tiny off-shoots along its length, some of which interconnect with the offshoots of other processes. Odontoblasts secrete the initial predentine matrix and mineralize it to produce mature dentine. Once fully differentiated, they do not divide again and each remains viable for the whole life of the tooth.

Dentinal tubules

The odontoblast processes occupy dentinal tubules (Figure 7.1) that radiate out from the pulp chamber, tapering along their length and increasing in spacing (Table 7.1). In the bulk of the dentine, especially the cervical region, the tubules run a gentle S-shaped course that curves down to apical and then up

Figure 7.1 Dentinal tubules in a Medieval tooth from York. Fractured preparation, examined in ET mode in the SEM. Scale bar 20 µm.

Table 7.1. *Dimensions of dentinal tubules*

	Near the pulp chamber	Mid-way between pulp and EDJ	Near the EDJ
Tubule diameter (µm)	2–3	1–2	0.5–0.9
Spacing (µm)	4–5	5–6	7–8

Source: Frank & Nalbandian (1989).

to occlusal, and many also show a secondary corkscrew curvature. The odontoblast process offshoots are accommodated in tiny canals running out from the sides of the tubules. They are particularly dense towards the EDJ, where the processes end in Y-shaped branches.

The dentinal tubules can be seen clearly in ET images of fractured prep-

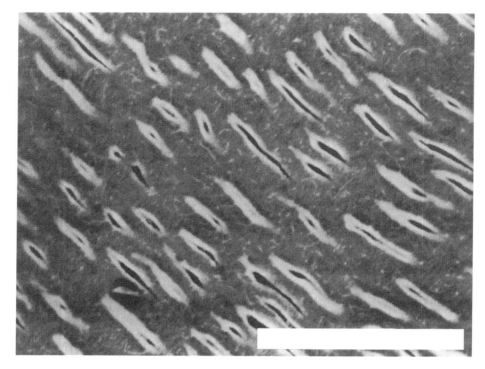

Figure 7.2 Peritubular dentine. The lumen of the dentinal tubules shows black where it is exposed, and the highly mineralized peritubular dentine as white. The intertubular dentine is dark, with bright spots representing calcospherite centres. Polished preparation, examined in compositional BSE mode in the SEM. Scale bar 50 μm.

arations (Figure 7.1) and in compositional BSE imaging (page 315), in which the embedding medium occupying the tubule lumen contrasts (Figure 7.2) with the heavily mineralised peritubular layer (page 185). In transmitted light microscopy of ground sections, the appearance of the tubules depends upon whether they are filled with mounting medium or air – in mounting medium they are translucent, whereas in air they are opaque (page 311).

Dentine matrix

Dentine formation occurs in two phases: organic matrix secretion, and mineralization. The actively forming surface that underlies the odontoblast layer is thus coated by a 10–40 μm thick layer of organic matrix, or predentine. Dentinal tubules, containing the odontoblast processes, run through the predentine and into the mineralized dentine underneath. Three types of dentine matrix are secreted: mantle, intertubular, and peritubular. They are all a mixture of collagen fibres (page 226) and a non-fibrous component called ground substance which can be differentiated by stains in demineralized preparations (page 304).

Intertubular dentine matrix

Intertubular matrix characterizes the bulk of dentine in both crown and root. Its largest component is fine collagen fibrils each about 100 nm in diameter, interspersed with ground substance. The weave of the fibrils within the matrix varies in complexity. Most form a random felt-work in a plane parallel to the developing predentine surface but, where tubules are dense, the felt-work is less random and the fibrils are gathered into bundles. In addition, a few fibrils cut across this plane, to run parallel with the dentinal tubules.

Mantle dentine matrix

Mantle dentine matrix has a denser accumulation of coarser collagen fibres than intertubular matrix. These coarse fibres are classically referred to as the fibres of von Korff, but there is some confusion about application of this term. In the occlusal crown, the fibres are arranged with their long axes parallel to the dentinal tubules. Under the sides of the crown and in the root they are more angled to apical and, in places, they are parallel to the EDJ.

Peritubular dentine matrix

Peritubular matrix is composed largely of ground substance, with very few collagen fibrils. A better name for it might be intratubular dentine matrix, because it lines the dentinal tubules rather than surrounding them. It is not present at the surface of the predentine layer, although it is secreted before any mineralization takes place. The thickness of peritubular matrix varies along the length of the tubules, but the junction between peritubular and intertubular matrix is always abrupt.

Mineralization of dentine

The odontoblasts control mineralization in a way that is not fully understood. Matrix vesicles bud-off from their cell walls to act as sites for the initiation of crystallites, and it is believed that they are involved in the production of calcospherites. These are 1–50 μm diameter spherical or paraboloid agglomerations of crystallites arranged with their long axes radiating out from the centre. Most of them coalesce to form a continuous structure, but some do not grow sufficiently to do this and leave a poorly mineralized patch of matrix

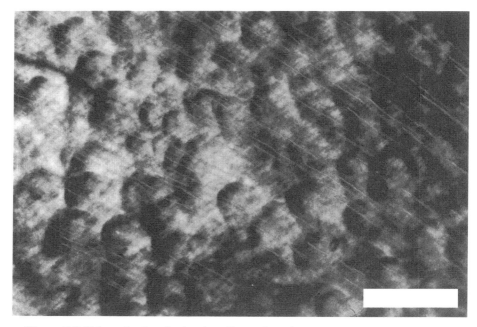

Figure 7.3 Calcospherites in dentine. Ground section examined in polarized light microscopy. The pattern of extinction outlines each calcospherite as a dark rim around the tip of the paraboloid mass of crystallites. Dentinal tubules cross diagonally from top left to lower right. Scale bar 100 μm.

in between. This calcospheritic structure develops independently from the features of the original dentine matrix, and the radial crystallites are thought to be located in the ground substance. A small proportion of crystallites is, however, deposited in and aligned with the collagen fibrils of the intertubular matrix, and the peritubular matrix is also mineralized independently.

Calcospherites of varying size, shape and packing thus dominate the intertubular dentine. Near the EDJ and pulp chamber they are small and spherical but, in between, through the bulk of the dentine thickness, they are large and paraboloid. These calcospherites are like cones, with the apex pointing towards the EDJ and the open base pointing towards the pulp, and the crystallites radiating out from a centre near the apex of the cone. Calcospherites are best seen in polarizing microscopy of ground sections, where the pattern of extinction (page 312) is controlled by the radial organization of the crystallites, and they appear as nebulous, cloud-like bodies, through which the tubules pass (Figure 7.3). They are also seen in compositional BSE images, with the centre of mineralization of each calcospherite visible as a dense spot (Figure 7.2) and, where calcospherites fail to meet, they are clearly outlined by patches of low density, poorly mineralized matrix.

Peritubular dentine matrix is not involved in the main calcospheritic mineralization pattern, and its individual crystallites are randomly orientated. It is very heavily mineralized (40% more than the surrounding intertubular dentine) and clearly visible in compositional BSE images (Figure 7.2). In ET images of fractured specimens, it is seen as a smooth lining to the tubules, penetrated by tiny canals.

The large collagen fibres of mantle dentine matrix dominate its mineralization pattern. The matrix is lightly mineralized and, in compositional BSE images, it is identified as a thin low density layer. Polarizing microscopy of ground sections distinguishes it by the pattern of extinction, which contrasts with the spheritic structure of intertubular dentine.

Peripheral dentine of the root and the cement–dentine junction

A thin layer of dentine (the granular layer of Tomes) underlies the CDJ in the root, with a granular appearance that is due to incomplete mineralization. The tiny calcospherites fail to meet and are left as isolated spheres in a poorly mineralized matrix. Tomes' granular layer may lie directly against the CDJ, but is sometimes separated from the cement by another layer. This may be mantle dentine, but it is often a heavily mineralized, homogeneous layer (15–30 μm thick) with a glass-like appearance, and called the hyaline layer of Hopewell Smith. The heavy mineralization shows up as a strong compositional contrast in BSE images but, with the prominent layered structure of cement (page 204), it can be difficult to locate the CDJ and label the neighbouring layers.

Incremental structures in the dentine

Dentine lines

The course of dentine formation is marked out by a variety of lines that can be seen by transmitted light microscopy. Sir Richard Owen (1845) described lines that were defined by parallel deviations of neighbouring dentinal tubules. These contour lines of Owen are irregular and appear to represent intermittent disturbances rather than a regular formation rhythm. Andresen (1898) did, however, note a regular pattern of prominent dark lines, together with a fainter, finer layering that was further described by von Ebner (1906). More recent work has found that the most prominent dentine layering has a spacing of around 20 μm (range 15–30 μm), and has identified it with Andresen's lines

Figure 7.4 Incremental lines in dentine. Coarser spacing lines (Andresen's lines) run diagonally from top left to lower right (some more prominent than others). The dentinal tubules cross them obliquely. Ground section examined in polarized light microscopy. Scale bar 100 μm.

(Kawasaki *et al.*, 1980; Dean, 1993; Dean *et al.*, 1993a). At higher magnification it is often possible to make out faint lines in between the main layering, with a spacing of 2–5 μm, which may be the incremental lines of von Ebner (Dean *et al.*, 1993a). The correct usage of these terms is still disputed and the lines (Figure 7.4) are only rarely clear, varying greatly in strength throughout each tooth. They are best demonstrated by polarized light microscopy of ground sections, or in microtome sections of demineralized teeth stained with haematoxylin, and they are not usually apparent in SEM preparations. Schmidt and Keil (1971) suggested that collagen fibril orientation was involved, but this implies movement of odontoblasts whilst secreting matrix, which seems unlikely because they are firmly anchored by their processes. It is much more likely that the lines represent a cyclic variation in the proportion of collagen to ground substance in the intertubular dentine matrix.

Dentine layering has been related to matrix apposition rate, through experiments based upon the injection of marker dyes during life (Table 7.2). Markers include sodium fluoride, which interrupts tooth formation and produces lines of defective dentine; alizarin, which stains calcium salts during mineralization; and tetracycline, which binds both to the mineral and to the organic matrix

Table 7.2. *Dentine apposition rates*

Animal studied	Marker(s) used	Daily rate in µm/day	Authors
Rhesus macaques	Alizarin/sodium fluoride	3.97–4.1	(Schour & Hoffman, 1939b; Schour & Hoffman, 1939a)
Rhesus macaques	Lines matched with medical/ behavioural history	6.2–8.3 deciduous 2.2–6.2 permanent	Bowman (quoted in Dean, 1993)
Rhesus macaques	Tetracycline	3–4.03	(Dean, 1993)
Human	Sodium fluoride	3.16–4.42 deciduous	(Schour & Poncher, 1937)

components. Bromage (1991) determined that fine (2.7 µm apart) lines in a macaque permanent molar represented a circadian (24 hourly) rhythm, and Yilmaz *et al.* (1977) found a circadian rhythm for fine layering (4.4 µm apart in their photographs) in pig teeth. Molnar *et al.* (1981) showed that coarser layers in macaques had an approximately 4–6 day rhythm whereas, in human teeth, Okada (Dean, 1993) established a 7–10 day interval like that of the brown striae of Retzius in enamel (page 157). Tetracycline antibiotics are frequently prescribed in humans and often leave useful marker lines, even when their relative timings are unknown. Kawasaki *et al.* (1980) used prism cross-striation counts in the neighbouring enamel to determine intervals between such dentine markers, and found by this means that the coarser 20 µm layering represented a 5-day rhythm. Dean *et al.* (1993a) used a similar approach to arrive at an average of 8 days between these lines.

Dentine layering has great potential for unravelling the chronology of tooth formation because it covers the whole process, from crown initiation to root completion. There are, however, difficulties. Although the fine lines seem to represent a 24-hourly rhythm, they are faint and patchy in distribution, and very difficult to count in many teeth. The more prominent, wider-spaced lines have not yet yielded a consistent rhythm and there are particular problems with archaeological specimens, where dentine is rarely well enough preserved. But whatever their spacing, dentine lines vary in prominence in a pattern that can be matched between teeth that were formed at the same time (Gustafson, 1955). Gustafson (1947) carried out a blind test on isolated teeth from several individuals and was able to identify them all correctly by this method, finding that the dentine matches were clearer than those based on brown striae of Retzius in the enamel (page 157). Such matches confirm that systemic factors control the prominence of layering. Schour (1936) identified a neonatal line in dentine as well as in enamel (page 159), seen in both transmitted light examination of ground sections and haematoxylin/eosin stained preparations of decalcified teeth. There was a staining contrast between prenatal and postnatal dentine, suggesting a difference in the proportions of ground substance to collagen.

Interglobular spaces

In some parts of most dentine sections, calcospherites fail to meet and leave patches of poorly mineralized matrix (Figure 7.5). These were first noted by Czermak (Hofman-Axthelm, 1981), who coined the phrase *interglobu-larräume* (German; interglobular spaces), but microradiographic evidence shows that the 'spaces' are not completely unmineralized (Blake, 1958). Both collagen fibrils and dentinal tubules pass through unaltered, although peritubular matrix is not normally found. Most human teeth show a few small interglobular spaces somewhere in a radial section, in addition to those of Tomes' granular layer (page 187). Sometimes, there are spreads of scattered spaces, giving a granular texture, but the spaces may also line up parallel to the incremental lines. In human deciduous teeth, Mellanby (1927) found interglobular spaces in 85% of sections (64% showing pronounced spaces), whereas all permanent teeth were affected (92% pronounced). Archaeological human teeth (Soggnaes, 1956) frequently show interglobular spaces, but they are somewhat less common in apes and gibbons (Schuman & Sognnaes, 1956).

Mellanby (1927) further demonstrated a strong relationship between enamel hypoplasia (page 165) and interglobular spaces. Patients attending public dental clinics had more frequent and prominent spaces than patients attending for private treatment, suggesting that social and economic factors were important. Experiments with beagle dogs (Mellanby, 1929, 1930, 1934) demonstrated that vitamin D or A deficient diets increased frequency and prominence of both interglobular spaces and hypoplasia, and rickets has been associated with SEM evidence of interglobular spaces in humans (Seeto & Seow, 1991). Prominent spaces can also be matched with accentuated incremental lines in the enamel (Soggnaes, 1956). Bermann and colleagues (1939) found that periods of artificially induced fever in a laboratory rabbit produced lines of interglobular spaces, but little work has been done to relate them to medical history, because enamel hypoplasia is so much easier to record. In addition, enamel hypoplasia is clearly related to the sequence of enamel matrix secretion, whereas interglobular spaces represent deficient mineralization of dentine matrix, and are thus not strictly comparable. There is a similar difficulty in correlating lines of interglobular spaces with the other dentine lines (page

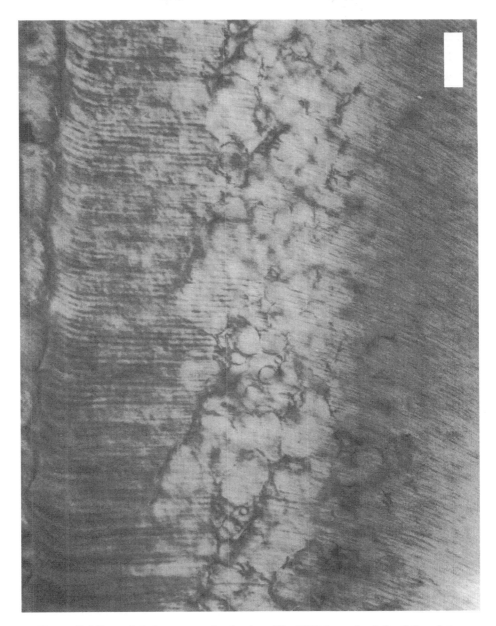

Figure 7.5 Interglobular spaces in dentine. The EDJ is at the left of the picture. Dentinal tubules run a curving, but roughly horizontal path, and a band of interglobular spaces runs down the centre. Ground section examined in polarized light microscopy. Scale bar 100 μm.

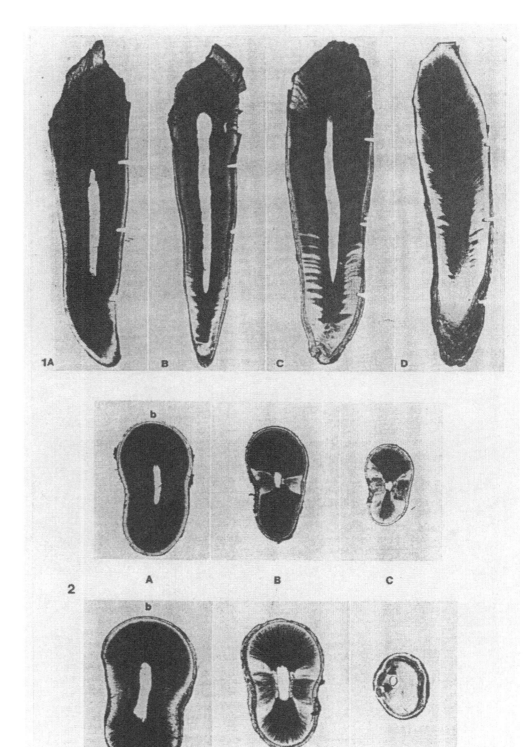

1A   B   C   D

b

2   A   B   C

b

187), which are features of the original matrix. The patterns of spaces are however repeated in teeth from one individual, so that they can help to match up isolated teeth (Gustafson, 1947, 1955).

Dentine sclerosis

Dentine sclerosis (Greek *sklerosis*; hardening) is the infilling of tubules with mineralized material. Sclerosis of dentine in the tooth crown is related to the exposure of tubules through attrition and caries, when the tubules are progressively filled with large crystals of apatite, whitlockite, octacalcium phosphate, or tricalcium phosphate (page 217) in cuboid or rhombohedral form. Mendis and Darling (1979) found that the extent of crown dentine sclerosis was not age-related but that, by contrast, dentine sclerosis in the root was strongly correlated with age. The material filling the root dentine tubules is indistinguishable from peritubular dentine, but there is a clear dividing line separating it from the peritubular dentine layer (Vasiliadis *et al.*, 1983b). Root dentine sclerosis starts in the late teen ages near the apex of the root at the CDJ and the zone of sclerosis spreads gradually up the root, particularly at the centres of the mesial and distal sides, to give a complex three-dimensional form (Figure 7.6). The boundary between the sclerotic zone and unaffected dentine has a serrated or feathered appearance, as groups of unaffected tubules alternate with infilled tubules. The zone of sclerosis is best monitored in thick (250 μm) ground sections, examined in a dry state (page 311) so that, in transmitted light, the sclerotic areas appear transparent and the normal areas opaque. This gives rise to the alternative name for sclerosis of root dentine transparency. Vasiliadis *et al.* (1983a,b) used serial sections to establish the relative volume of sclerotic to unaffected dentine in human canine tooth roots and found that, in the apical and middle one-thirds of the root, there was a strong correlation between the percentage of sclerotic dentine and age at extraction of the tooth (page 214).

Figure 7.6 Root dentine sclerosis. Reproduced from Vasiliadis, L., Darling, A. I. & Levers, B. G. H. (1983) The amount and distribution of sclerotic human root dentine, *Archives of Oral Biology,* **28,** 645–649, with kind permission from Elsevier Science Ltd, The Boulevard, Langford Lane, Kidlington, OX5 1GB. 1A to 1D show radially sectioned teeth with increasing translucent areas of root dentine (from individuals aged 25, 29, 57 and 77 years respectively). 2A to 2C show a series of transverse sections of two canines, from a 35-year-old in the top row, and a 60-year-old in the lower row (positions of sections correspond to notches cut in radial sections 1A to 1D).

Secondary dentine and dead tracts

With age and attrition, the pulp chamber and root canal are progressively infilled by a combination of regular and irregular secondary dentine. The process is broadly age related, but shows considerable variation (page 215).

Regular (physiological or senile) secondary dentine

Even after the tooth has been completed, regular secondary dentine slowly continues to form around the pulp chamber walls. Where this involves odontoblasts occupying the same dentinal tubules, there is little difference between primary and secondary dentine, and the boundary may be no more than a sharp curve in the tubules. Sometimes, a distinct line shows in stained preparations or in microradiography, but the clearest difference is seen in polarising microscopy. The boundary is sharpest at the sides and floor of the pulp chamber, and it is much harder to recognize a clear secondary dentine layer lining the root canal.

Irregular (reparative) secondary dentine

Irregular secondary dentine (Kuttler's tertiary dentine, page 182) generally occupies a larger volume of pulp and root canal than the regular form. Often it contains no tubules at all, or just a few irregular tubules that are not continuous with the tubules of primary or regular secondary dentine. Mineralization does not follow a calcospheritic pattern, and the uneven appearance in polarizing microscopy suggests great variability in collagen fibre orientation. The deposition of irregular secondary dentine is a response to exposure of the overlying primary dentine by penetration of the enamel or cement covering (Figure 7.7). The odontoblast processes in the exposed dentine tubules die, leaving a 'dead tract' that is sealed at the pulp by secondary dentine patches and demarcated from the primary dentine by a narrow sclerotic zone. Under transmitted light microscopy, the secondary dentine patch prevents the mounting medium from entering the open tubules of the dead tract, so that they remain filled with air, making the dead tract appear dark (page 312).

Pulp stones

Pulp stones or denticles are irregular, roughly spherical mineralized masses, which may be found in any part of the pulp and occur in most permanent tooth classes, especially in older adults. Mineralization follows a concentric, layered pattern, rather than a calcospheritic one, and so-called 'true pulp stones' have odontoblasts on their surfaces and contain tubules (which are lacking in 'false pulp stones'). All start as free bodies within the pulp but

Figure 7.7 Dead tract in dentine, with associated secondary dentine patch. The side of the crown is seen in the top left corner, with the layer of enamel. The pulp chamber is the dark triangular area extending out of the lower left corner, and this has the dead tract, with its secondary dentine patch at its tip. Ground section examined in polarized light microscopy. Scale bar 1 mm.

they may later become attached to one another, or to the secondary dentine of the pulp chamber wall. Most are of the false variety and lie free in the pulp, and many remain small although some grow to occupy most of the pulp chamber. In dry specimens where the pulp has decomposed, pulp stones may rattle about inside the tooth.

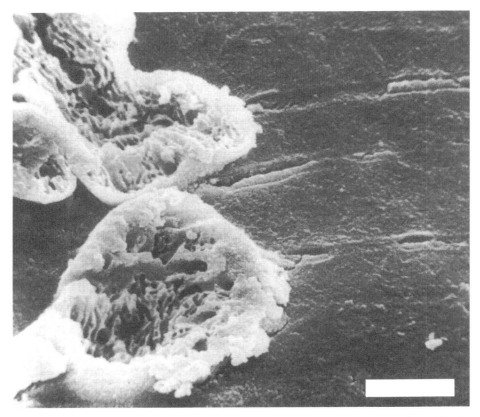

Figure 7.8 Diagenetic foci in dentine belonging to a Neolithic tooth from North Africa. The dentinal tubules are filled with sclerotic material, and can be seen to pass through the large, spongy foci. Polished, etched preparation, examined in ET mode in the SEM. Scale bar 40 μm.

Archaeological preservation of dentine

Archaeological dentine is often softened, or brittle (Beeley & Lunt, 1980) and fracturing of underlying dentine causes the break-up of otherwise intact enamel. Viewed in tooth sections with the naked eye, affected dentine is chalky or opaque and is found especially near the root surface or pulp cavity. In light microscopy of ground sections (Soggnaes, 1950, 1956; Werelds, 1961, 1962; Poole & Tratman, 1978) the damage appears to be due to irregular canals meandering through the dentine, some the diameter of dentinal tubules but most larger, and it is usually assumed that they are made by invading fungal mycelia. The SEM image of affected dentine is, however, often different from the light microscope appearance, and BSE imaging demonstrates marked changes in mineralization (Bell, 1990; Bell *et al.*, 1991). Clear canals are visible in some specimens, although the dominant form of disruption is usually scattered foci of diagenetic change (Figure 7.8), some with low densities, but many with higher densities than the intervening unaffected dentine. They tend to follow the incremental lines of intertubular dentine, whereas peritubular dentine is often left intact. Poor structural integrity of ancient dentine appears related to collagen loss, but not all specimens are affected and some even show organic matrix stain reactions (Falin, 1961; Poole & Tratman, 1978). Preservation is difficult to predict and there is no relationship with date. Fully fossilized teeth often have better dentine preservation than much younger archaeological material (Doberenz & Wyckoff, 1967) – it is even possible that diagenetic foci are repaired by remineralization. Dentine is preserved in most cremations, but is strongly affected by the heat (Shipman *et al.*, 1984). At lower temperatures, the peritubular dentine splits away from the intertubular dentine but, at higher temperatures, these boundaries are obliterated, the dentine has a granular texture and the tubules are distorted.

8

Dental cement

Introduction

A review of cement anatomy and physiology is given by Jones (1981). In primate teeth, cement coats the root surface only, mostly covering the underlying dentine, but frequently overlapping the edge of the enamel at the cervical crown margin. The thickness of the cement layer is highly variable – in an adult it may be 100–200 μm half-way up the root from the apex, and 500–600 μm at the apex itself. Thickness increases with age, and the unerupted teeth of a young individual may have a cement layer of only 10–20 μm.

The function of the cement is to attach the periodontal ligament (page 260) to the surface of the root. The ligament contains a complex of collagen fibres, embedded in the alveolar bone on one side, and embedded in the cement on the other. The cement has no blood or nervous supply of its own, but the periodontal ligament is richly supplied and the cement-forming cells (the cementoblasts) lie within it, on the cement surface. Collagen fibres in the periodontal ligament turn over rapidly, in response to dynamic changes in mechanical forces at the joint, so that the attachment of the fibres to the tooth root also changes constantly and cement is laid down throughout life to provide new attachments. The cement does not continuously turn over like this (as bone does), but it can be remodelled by the combined activities of cement-destroying cells (odontoclasts) and cementoblasts, which are able to repopulate the damaged areas (unlike odontoblasts in dentine).

During the formation of the roots, the CDJ is mapped out by the epithelial root sheath of Hertwig (page 119), within which dental papilla cells differentiate into odontoblasts and initiate predentine formation. Cells of the neighbouring dental follicle just outside the root sheath then differentiate into cementoblasts. The root sheath disintegrates and its epithelial cells exchange position with cementoblasts which, once fully differentiated and in position, deposit precement on the template of the predentine, so that cement formation is initiated right at the start of root formation. When the predentine is mineralized, this mineralization spreads across into the precement, so that the CDJ is never very sharply defined (page 187).

Components and related elements of cement

Besides the cement itself, it is necessary to consider the neighbouring periodontal ligament, because a large proportion of the collagen fibres in cement have their origin there, and the cementoblasts and odontoclasts are housed there.

The organic matrix of cement

Fresh cement contains about 24–26% organic matrix (roughly the same as bone), which is almost entirely made up of collagen fibres (page 226) of two types:

1. Extrinsic fibres (Figure 8.1) are incorporated from the periodontal ligament. They are variable in size and shape, but are typically round to oval in cross-section with a diameter of 6–12 µm or so, and lie in orientations within the cement that depend upon the organization of the periodontal ligament. As the periodontal ligament remodels, new cement layers build up, with varying extrinsic fibre orientations, but typically the extrinsic fibres penetrate into the cement at a relatively large angle to the surface.
2. Intrinsic fibres are produced by the cementoblasts themselves and are laid down in mats parallel to the developing surface of the precement. They are relatively small (1–2 µm in diameter), especially where they are mixed with large numbers of extrinsic fibres.

The non-collagenous component of cement matrix is called ground substance, as in dentine and bone. It is amorphous under the microscope, but can be differentiated by a number of stains.

The mineral component of cement

Within mature cement the degree of mineralization (Figure 8.2) is highly variable. Both ground substance and collagen fibres are mineralized, and the apatite crystallites (page 217) are probably similar to those of bone, which are needle-like and up to 100 nm or so long. The collagen fibres are intimately mineralized, with the crystallite long axes aligned so that the orientation of fibres can clearly be traced from the mineral shadows left behind, even in fossil material where much collagen has been lost. The surface of actively forming cement is coated by a thin layer of less-well-mineralized precement.

Figure 8.1 Extrinsic fibres in cement. Two extrinsic fibres run up the right-hand side of the picture. Fractured preparation, examined in ET mode in the SEM, operated by Sandra Bond. Scale bar 10 μm.

Cells of the cement and periodontal ligament

Cementoblasts

Cementoblasts are positioned within the periodontal ligament, near the developing precement surface, and in between the fibres of the ligament where they enter the cement. They vary in size, shape and orientation, but all have small, thread-like cell processes, most of which extend towards the cement.

Cementocytes

Cementocytes are cementoblasts that have been engulfed by the developing precement matrix, and are contained within spaces in the cement known as cementocyte lacunae. The pattern of incorporation is irregular and the lacunae

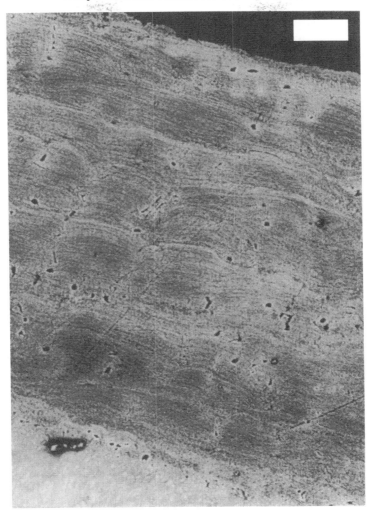

Figure 8.2 Cement layering. The surface of the root is seen at the top of the picture, with the CDJ (dentine whiter) at the lower edge. A pattern of fine, irregular layers runs across the image, with black spots representing the cementocyte lacunae. The surface is marked by two or three scratches, which are artefacts of preparation. Polished preparation, examined in compositional BSE mode in the SEM by Sandra Bond. Scale bar 50 μm.

are highly variable in size and shape, although most are about 7–20 μm across. Where extrinsic fibres are widely spaced, the lacunae are oval and flattened parallel to the intrinsic fibre mats but, where the fibres are more densely packed, lacunae are more irregular. Cementocytes have fine processes extending out on all sides, mostly towards the cement surface from which the cells draw their nutrition. The processes run within tiny channels known as canaliculi, which spread irregularly through the cement, and sometimes interconnect with those of neighbouring lacunae. Cementocytes are smaller than the lacunae that they occupy and the gap is filled by ground substance, which sometimes also lines the canaliculi. This ground substance may become mineralized, to create a thin amorphous lining. As the nourishment of cementocytes is purely derived from the periodontal ligament at the cement surface, those more deeply buried are inevitably less active, and the deepest lacunae may not contain viable cells.

Fibroblasts

The collagen fibres of the periodontal ligament are both secreted and resorbed by the fibroblasts. Most fibroblasts are highly active, maintaining rapid tissue turnover, unlike other ligaments in the body where the low turnover means that some cells enter a resting state, and become fibrocytes.

Odontoclasts

Odontoclasts resorb cement and dentine from the root surface – a similar job to the osteoclasts of bone. They are large cells, 50 μm or more across, which are associated with irregular depressions cut into the cement or dentine surface, known as resorption hollows or Howship's lacunae (Figure 8.3). These are highly variable in size, from a few to several hundred micrometers across, and their scalloped edge is readily recognizable in section – a clear sign of odontoclast activity. Resorption of this type happens rarely in adults but, in some specimens, it is possible to see a reversal line where the scalloped edge is visible but the hollow has been filled with new cement.

Types of cement

Cement is classified according to its contents. It may incorporate cementocytes, but some does not. Similarly, some cement contains no collagen fibres whereas some includes intrinsic fibres on their own, some has extrinsic fibres instead, and some contains a mixture. There is considerable variation within one section and, in practice, it is rarely easy to classify the tissue.

Afibrillar cement

This has no fibres and no cells. It is simply mineralized ground substance, produced early in the development of a cement layer as a thin skin covering parts of the dentine and, sometimes, the enamel at the cervical margin. Thickness and continuity of afibrillar cement deposits are highly variable.

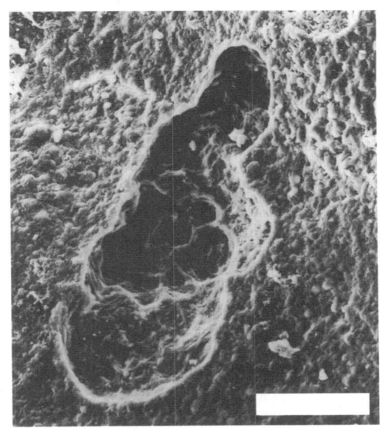

Figure 8.3 Resorption hollow at the root surface, near the apex of a post-medieval tooth from London. In the unresorbed area around the hollow, the irregular bulges represent the mineralized ends of extrinsic fibres. Surface examined in ET mode in the SEM, operated by Sandra Bond. Scale bar 50 μm.

Extrinsic fibre cement

This has no cells and no intrinsic fibres, with an organic matrix containing only tightly packed extrinsic fibres and intervening ground substance. Extrinsic fibre cement is common early in the development of the cement, before the tooth is erupted. Its level of mineralization is high, sometimes approaching that of dentine, and the extrinsic fibres are usually mineralized right the way through.

Mixed fibre cement

This form contains extrinsic and intrinsic fibres, ground substance, and may or may not incorporate cementocytes. The density of extrinsic fibres varies.

Where they are densely packed, intrinsic fibres run randomly in between but, when they are widely spaced, the intrinsic fibre mats are organized into small patches or domains of similar fibre orientation. Successive domains overlap one another to produce a composite layered structure. Mineralization varies and the central core of extrinsic fibres may remain unmineralized. Mixed fibre cement in its various forms constitutes the bulk of the cement around the root, and the variation in its components probably represents the rate at which the tissue was secreted. At a slow rate, small quantities of intrinsic fibres are secreted and cementoblasts are rarely engulfed by the growing tissue. Rapid growth is by fast production of intrinsic fibres, which then entrap more extrinsic fibres and incorporate more cementocytes. Cellular forms are therefore common at the apex of the root, and the fork of the roots in multi-rooted teeth, where cement is thickest and grows fastest.

Intrinsic fibre cement

Patches of cement, without extrinsic fibres but with cementocytes, may be found near the root apex.

Cement layering

In demineralized preparations of cement, microtome sectioned and stained with haematoxylin, alternating heavily and less heavily stained bands can clearly be seen. Cement also gives an impression of layering in ordinary light microscopy of ground sections – both in transmitted and reflected light. Alternations of light and dark layers may be visible in polarizing microscopy and these presumably relate to the changing orientations of different layers of extrinsic fibres. The layers are very thin in human teeth, however, and it is difficult to prepare a ground section thin enough to resolve them. With compositional BSE imaging (page 315), a pattern of bright and dark banding suggests an alternation in the degree of mineralization, although the effect is masked by a patchy variation in mineralization (Figure 8.2). The layers are also visible in topographic BSE images and both fibre orientation and degree of mineralization may produce relief in the polished specimen surface (page 301). These layered alternations may thus represent differences in the density of cells, in the density or orientation of extrinsic fibres, in the balance between collagen fibres and ground substance, or in the degree of mineralization. The layering is always patchy and difficult to follow over more than a small area and, in any one tooth section, there are always areas without any clear layers. Several studies (page 207) have found correlations between the number of layers in human cement and the number of years elapsed since root formation. In view of the patchy growth of cement, which is a response to localized changes in the periodontal ligament, it is not clear what features are actually being counted in these studies. It is, however, possible that a seasonal variation in growth rate causes differences in the density of cells, proportion of intrinsic to extrinsic fibres, and degree of mineralization in mixed fibre cement.

Resorption and remodelling

Deciduous teeth are resorbed before exfoliation (page 138). This proceeds irregularly, and periods of resorption are interspersed with periods of repair, when new cement is laid down over the resorbed surface and a connection is re-established with the periodontal ligament. After loss of the deciduous dentition, expensive resorption of the root surface is uncommon, but cycles of resorption and repair are found in the apical region of older adults. The edges of resorption hollows are marked by reversal lines and the first reparative cement is a thin highly mineralized layer, followed by cellular and fibrillar forms.

Appearance of the root surface

The most commonly visible features in SEM images of the surface (Jones & Boyde, 1972; Jones, 1987) are mineralized outlines of intrinsic fibres, running parallel to the forming surface, and the mineralized ends of the extrinsic fibres. In slowly forming cement, well mineralized extrinsic fibres bulge up out of the surface. Rapidly forming cement has larger proportions of intrinsic fibres and the poorly mineralized extrinsic fibre cores are seen as pits in the surface. The different types of surface texture occur as patches, demonstrating that cement formation does not take place as even layers over large areas of the root. Resorption pits may be seen (Figure 8.3), and sometimes rapidly forming cement within a resorption pit indicates a process of repair.

Hypercementosis and cementicles

Some teeth, in some individuals, display a massive overproduction of cement known as hypercementosis, which results in a bulbous, irregularly bulging root. If this occurs in several roots of a multi-rooted tooth, the whole can be united into one mass. Hypercementosis of this type can readily be detected in dental radiographs. The cause is unknown, but is unlikely to be related to excessive wear or malocclusion, because it also occurs in unerupted teeth. One special case of hypercementosis is associated with Paget's disease, where a very irregularly bulging mass of cement occurs, with associated bony changes in the jaws.

Cementicles are small nodular bodies of cement that may develop within the periodontal ligament or the alveolar bone. They are rare, and their cause is unclear. Sometimes they become incorporated into the main cement of the root surface.

Archaeological preservation of cement

Cement is often poorly preserved in archaeological specimens. It cracks and may become detached from the dentine. Often, it shows similar diagenetic foci to those seen in dentine (page 196) and, exposed at the root surface as a very thin layer, is greatly disrupted. Histology of archaeological cement is difficult, and it is best impregnated with resin before sectioning, using the simplest possible preparation techniques. The least disruptive technique for archaeological cement is thus BSE imaging (page 315) or confocal light microscopy (page 311).

9

Histological methods of age determination

Alternative ageing methods – how accurate is accurate?

Dental histology provides a variety of methods for estimating age at death in adult remains, but they compete with techniques based upon the macroscopic examination of skeleton and dentition. The most commonly used methods are based upon changes to the bony parts of joints in the pelvis (the pubic symphysis and the auricular area), and dental attrition (page 239). Other methods include changes to the sternal end of the fourth rib, and cranial sutures. The closeness of the relationship between estimated ages and known ages is usually expressed by a correlation coefficient, the mean difference between estimated age and known age, or a standard error of estimate. Table 9.1 shows that no adult ageing method is highly accurate and, to be as good as the competition, dental histological methods need only have correlation coefficients of 0.7, or average differences between estimated and known age of less than 7 years. Several of the dental histological techniques emerge as relatively good methods and their limited application outside forensic odontology is probably due to the degree of training required, and the time-consuming, destructive sectioning.

Counts of layers in cement

In marine mammals, hibernating land mammals, migratory ungulates and their dependent carnivores, cement layering is routinely used as an ageing method (Morris, 1978; Perrin & Myrick, 1980; Hillson, 1986a, 1992; Lieberman, 1993). Several studies have suggested that cement layering can can also be used for ageing monkeys (Wada *et al.*, 1978; Stott *et al.*, 1980; Yoneda, 1982; Kay *et al.*, 1984), but the first study to suggest cement layer counts as an ageing method in

Table 9.1. *Tests of age estimation methods in adults*

Method	Number of individuals in study	Correlation (Pearson's r) between known ages and estimated ages	Mean difference between known ages and estimated ages (years)	Standard error of difference for age estimates
Acsadi/Nemeskeri pubic symphysis[1]	64		10.8	7.9
Todd/Brooks pubic symphysis[1]	64		18.2	12.5
McKern/Stewart pubic symphysis[1]	65		22.5	15.0
Modified summary age[2]	80	0.85	5.2	
Revised Todd pubic symphysis[2]	109	0.78	6.5	
Revised auricular surface[2]	108	0.71	7.3	
Proximal femur[2]	97	0.53	9.3	
Revised sutures[2]	117	0.53	9.9	
Dental wear[2]	117	0.71	7.9	
Miles dental wear[3]	127		0.1–2.35	
Kerley lamellar bone histological age[1]	20	0.82	12.5	8.3
Kerley average bone histological age[1]	20	0.88	10.1	8.9
Cement thickness measurements[4]	1000	0.31–0.72[a]		
Gustafson cement thickness score[4]	1000	0.07–0.4[a]		
Johanson cement thickness score[4]	1000	0.22–0.48[a]		
Cement layer counts[5]	73	0.78[a]	6.0	9.7[a]
Root dentine translucency score[6]	46	0.86[a]		7.1[a]
Root dentine translucency measurements[7]	24		6.5	
Root dentine translucency volume measurements[8]	69	0.87[a]	3.5[a]	
Root dentine translucency measurements[9]	454	0.67[a]		
Johanson secondary dentine score[10]	1000	0.59–0.74[a]		
Cervical pulp width/tooth width (i.e. infilling by secondary dentine)[10]	1000	−0.46–0.77[a]		
Recalculated Gustafson method[11]	51			11.28
Gustafson method, ST scores, weighted for tooth[12]	66			9.1
Bang and Ramm root dentine sclerosis method[13]	33		5.15	
Johanson method[13]	33		4.52	
Gustafson method, Maples & Rice modification[13]	33		5.03	
Johanson method, Bayesian prediction[14]	71		7.0	

[a] No test on independent sample.

Sources: [1] Aiello & Molleson (1993), [2] Lovejoy *et al.* (1985) Test **II**, [3] Kieser *et al.* (1983), [4] Solheim (1990), [5] Condon *et al.* (1986), [6] Johanson (1971), [7] Bang & Ramm (1970), [8] Vasiliadis *et al.* (1983a), [9] Miles (1978), [10] Solheim (1992), [11] Maples & Rice (1979), [12] Maples (1978), [13] Lucy *et al.* (1994) [14] Lucy *et al.* (1996).

humans was that of Stott and colleagues (1982). Several teeth were selected from each of three cadavers and sectioned serially from root apex to cervix. The 100–150 μm thick transverse root sections were stained with alizarin (page 126), cement layers counted by transmitted light microscopy and the counts simply added to published eruption ages, to produce age estimates very close to the known age at death. These remarkable results were tested by Miller *et al.* (1988) on 100 extracted teeth, using one section per tooth. Only 71% of the sections demonstrated cement layering suitable for counting and, even in these, the relationship between counts and age at extraction was at best moderate (and weak for individuals over 35 years old).

Cement layers in human teeth are only a few micrometres thick. It seems inherently unlikely that > 100 μm thick sections would allow them to be resolved under the light microscope, so Miller and colleagues' results are perhaps less surprising than those of Stott and colleagues. Lipsinic *et al.* (1986) prepared 5 μm thick microtome sections from the decalcified roots of 31 extracted teeth, and stained them with haematoxylin and eosin. Two sections of each tooth were counted by three different observers, and the average layer count was added to age at eruption to give a predicted age. The predicted age was highly correlated with true age ($r = 0.85$) and the relationship was particularly strong for individuals younger than 30 years of age, although there was not a one-to-one relationship. In another study (Charles *et al.*, 1986; Condon *et al.*, 1986; Charles *et al.*, 1989) lower canines and first premolars from 42 individuals were demineralized, sectioned and stained with haemotoxylin. The clarity and continuity of the layers varied considerably between and within sections, so the two clearest were selected for each tooth, and independently scored twice by two observers. Mean counts from the premolars were added to standard eruption ages to produce an 'adjusted count'. The relationship between age and adjusted count was only moderately close, although it was stronger in females than in males, and as good as the best alternative methods (Table 9.1). This work suggests that cement layer counts could act as the basis of an age-estimation method, although several questions still remain:

1. The best methodology involves microtome sectioned, decalcified, haematoxylin stained preparations, but these are generally unsuccessful for archaeological material. Alternatives need to be found.
2. Cement layering varies from place to place within one tooth root, and several sections per tooth are needed in order to arrive at a consistent count. Some teeth produce highly anomalous counts.
3. There will always be a subjective element in counting, although interobserver variation is less than might be expected and can be minimized by

increasing the number of observers and the number of sections per tooth. Image analysis (Lieberman *et al.*, 1990) may provide more consistent results.

4. The relationship of layer counts to age is not straightforward. Most workers use the average age of eruption to represent the first cement layer, but this is not necessarily appropriate.

5. The biological basis of cement layering is not clear – no one knows exactly what is being counted (page 204).

The Gustafson technique and its derivatives

Gustafson's original technique

In a blind test to match isolated teeth, Gustafson (1947) estimated age from a general impression of secondary dentine deposition, cement thickness and periodontosis (level of periodontal attachment, page 260). This was so successful that Gustafson (1950) devised a system based on six age related factors:

1. Dental attrition
2. Periodontosis
3. Secondary dentine deposition
4. Cement apposition
5. Root resorption
6. Transparency of root (root dentine sclerosis).

In each ground section, the features were scored (Figure 9.1) and the points summed to give an overall score (Table 9.2). These scores were highly correlated ($r = 0.98$) with known age for 41 individuals, and a regression formula was derived to estimate age from points score with a standard error of ± 3.63 years. Nalbandian and Sognnaes (1960) independently scored a group of 40 teeth and plotted estimated age against known age. The slope and intercept of the regression line, and the correlation, were very similar to those of Gustafson, with a standard error of ± 7.9 years. Maples and Rice (1979) found statistical errors in the original Gustafson paper, and Burns and Maples (1976) revised the regression formula, finding a much larger standard error of ± 11.28 years when it was tested on an independent sample. Their revised formula was also tested by Lucy and colleagues (1994, 1995), who found still more statistical problems, but obtained results comparing favourably with the alternatives (Table 9.1).

Multiple regression derivatives

The original Gustafson method is based on simple linear regression of all six scores summed together, giving each factor the same weighting, and implying that they all have the same age estimating value. This is demonstrably not the

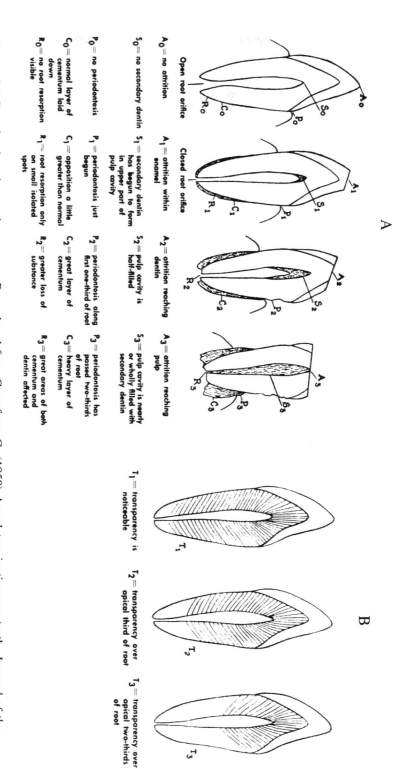

A

Open root orifice

$A_0 =$ no attrition

$S_0 =$ no secondary dentin

$P_0 =$ no periodontosis

$C_0 =$ normal layer of cementum laid down

$R_0 =$ no root resorption visible

Closed root orifice

$A_1 =$ attrition within enamel

$S_1 =$ secondary dentin has begun to form in upper part of pulp cavity

$P_1 =$ periodontosis just begun

$C_1 =$ apposition a little greater than normal

$R_1 =$ root resorption only on small isolated spots

$A_2 =$ attrition reaching dentin

$S_2 =$ pulp cavity is half-filled

$P_2 =$ periodontosis along first one-third of root

$C_2 =$ great layer of cementum

$R_2 =$ greater loss of substance

$A_3 =$ attrition reaching pulp

$S_3 =$ pulp cavity is nearly or wholly filled with secondary dentin

$P_3 =$ periodontosis has passed two-thirds of root

$C_3 =$ heavy layer of cementum

$R_3 =$ great areas of both cementum and dentin affected

B

$T_1 =$ transparency is noticeable

$T_2 =$ transparency over apical third of root

$T_3 =$ transparency over apical two-thirds of root

Figure 9.1 Gustafson's point scoring system. Reproduced from Gustafson, G. (1950) Age determinations on teeth, *Journal of the American Dental Association*, **41**, 45–54, with the kind permission of the American Dental Association, 211E Chicago Avenue, Chicago 60611. A, scores for attrition, secondary dentine, periodontosis, cement, root resorption. B, scores for root dentine transparency.

Table 9.2. *Regression formulae for Gustafson method and derivatives*

Gustafson (1950) original:

Age (years) = 11.43 + 4.56X ± 3.63 Error of Estimation

X = A + P + S + C + R + T

Maples & Rice (1979) corrected Gustafson formula:

Age (years) = 13.45 + 4.26X ± 7.03 standard error

Johanson (1971) method:

Age (years) = 11.02 + 5.14A + 2.3S + 4.14P + 3.71C + 5.57R + 8.98T ± 5.16 standard deviation

Maples (1978) method:

Age = 6.54S + 10.88T + 16.08 + position value ± 9.1 standard error

Position of tooth	Value
1	0.00
2	11.24
3	13.18
4	4.39
5	5.21
6	−5.37
7	3.73
8	8.04

The scores are: A, attrition; P, periodontosis; S, secondary dentine; C, calculus; R, root resorption; T, root dentine transparency.

case and several authors have developed multiple linear regression formulae, assigning different weights to each of the factors. Johanson (1971) revised the scoring system (Figure 9.2), and found that root dentine transparency (T) showed the highest correlation with known age, followed by secondary dentine deposition (S), attrition (A) and cement apposition (C). Periodontosis (P) and root resorption (R) were poorly correlated with age. Multiple regression formulae were defined for the combinations T, ST, PST, RPST, CPRST, and CPASRT (Table 9.2). Johanson's method was tested by Lucy and colleagues (1994, 1996), who suggested an alternative statistical approach to regression (page 216) (Table 9.1).

Maples and colleagues (Maples, 1978; Rice & Maples, 1979) independently developed multiple regression formulae for different tooth classes in a large

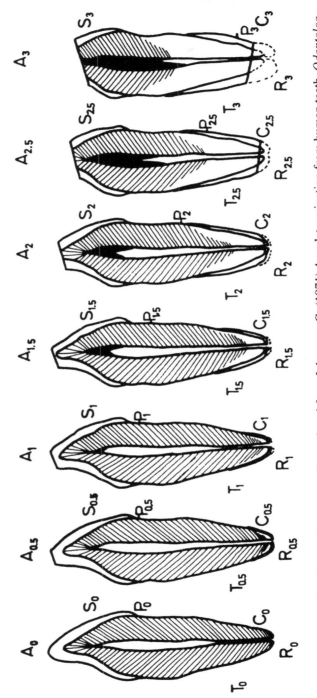

Figure 9.2 Johanson's point scoring system. Reproduced from Johanson, G. (1971) Age determination from human teeth, *Odontologisk Revy*, **22**, 1–126, with the kind permission of Professor Göran Koch and Swedish Dental Journal, Odontologiska Institutionen, Box 1030, S-551 11 Jönköping, Sweden.

working sample, and determined standard errors of estimate independently from a control sample. Second molars gave the best results, and the best combinations of factors were ST and SCT (Table 9.2). This is useful for archaeological material, as it uses the features that are best preserved.

Other variants

Metzger and colleagues (1980) developed closely defined methods for assessment of Johanson stages in a large forensic study. They scored attrition and apical resorption before sectioning, and measured the position of periodontal attachment from the CEJ. The tooth was then sectioned to make a 1 mm slice that was radiographed to provide an outline of the pulp chamber, and photographed over a striped background to record the translucent dentine zone. Section thickness was reduced to 250 μm by grinding, and photographed in colour between crossed polarizing filters to record secondary dentine and cement distribution. Photographs and radiographs were then projected and traced onto a sheet of paper, and the periodontal attachment scaled on, before scoring. Kilian and Vlcek (1989) redefined Johanson's stages and derived a new table for age prediction, whilst Matsikidis and Schulz (1982) tested the practicality of assessing Gustafson's factors from radiographs of whole teeth.

Applications in archaeology and anthropology

In archaeology, Gustafson methods have rarely been applied (Hillson, 1986b; Kilian & Vlcek, 1989), but they are likely to be limited by preservation of tissue. Whittaker (1992) used the technique to estimate an age of 72 years for an individual from the 18th century crypt of Christ Church, Spitalfields, in London, when the true age from parish records turned out to be 78 years.

It is always sensible to use several different factors to arrive at age estimates (Lovejoy *et al.*, 1985), but not all of Gustafson's original factors add much. Root dentine transparency, cement apposition and attrition are all used as ageing methods in their own right, with more objective recording methods than in any variants of the Gustafson technique. There is therefore considerable room for development. There is also no reason why dental methods should not be combined with methods based on the skeleton, including macroscopically assessed features such as the pubic symphysis, and estimates based upon bone histology that seem to have a similar accuracy to dental histological methods.

Age estimation from the extent of root dentine sclerosis

Root dentine sclerosis (page 193) is the basis of several ageing methods, but its main difficulty lies in recording the extent of the area of sclerosis, and standardizing it for overall root size. Johanson (1971) assessed it qualitatively from ground sections (Figure 9.2) and derived a regression formula for age estimation, with results that compared favourably with alternatives. Bang and Ramm (1970) measured the average extension of the sclerotic zone, in ground sections, and in intact roots held over a bright light. They calculated regression formulae for estimation of age from the average extension, in different classes of teeth and different roots in multi-rooted teeth and, although the standard errors of estimate were large and variable, an independent test on 24 individuals gave results that compared favourably with other methods (Table 9.1). Lucy and colleagues (1994) obtained similar results in an independent test using ground sections. The intact root method has been applied to Norwegian archaeological material (Stermer Beyer-Olsen & Risnes, 1994; Kvaal *et al.*, 1994) and a translucent zone is clearly visible in many specimens when transilluminated with a bright light.

Azaz and colleagues (1977) measured instead the cross-sectional area of the transparent dentine zone, relative to the overall cross-section of the root. Drusini *et al.* (1991) tested a densitometric technique (based on an image analyzer) against straightforward measurements of transparent area in section, but found that there was little to choose between them. The real difficulty with using any single section to measure the extent of the transparent area lies in its complex three-dimensional structure (page 193). Volume measurements based on many serial sections, as a proportion of overall root volume (Table 9.1), may be a better basis for estimating age (Vasiliadis *et al.*, 1983a). This procedure is very time consuming and an alternative may be to use fewer, carefully selected, section planes and estimate the three-dimensional extent of the translucent zone by the mathematical methods known as stereology.

Secondary dentine as an age indicator

Solheim (1992) tested Gustafson's (Figure 9.1) and Johanson's (Figure 9.2) methods for scoring secondary dentine (page 193) deposition qualitatively, against a system for measuring the decreasing size of the pulp chamber and root canal in a single section. The measurements showed a low correlation with age but, when combined in multiple regression formulae, estimated ages could be calculated (Table 9.1). Similar measurements have been made on radiographs of archaeological material (Kvaal *et al.*, 1994). The pulp chamber, the root canal and their secondary dentine lining all have a complex three-dimensional form, which might better be measured volumetrically with serial sections.

Conclusions

Age estimates based on dental histology compare very favourably with alternative methods based on dental attrition and skeletal changes. When tested on independent samples, the best of them are at least as good as the modified summary age of Lovejoy and colleagues (1985). Where it is important to age isolated teeth, there is every reason to employ histological methods, but there is much scope for further development. The biological basis of some methods is still unclear, and there must be differences in rate of development between populations. One approach may be to classify individuals into a set of broad age categories, rather than expect a direct relationship with age in years.

Recently, there has been discussion of the statistical methods used to arrive at age estimates (Lucy & Pollard, 1995; Lucy *et al.*, 1996). Most studies fit a least squares linear regression model to data in which the measure or score of a particular histological change is the independent variable (or controlling variable) and age is the dependent (or response) variable. Such models make assumptions: that both variables are continuous or at least ordinal, that they are linearly related, and that the distribution of age values is normal, with unchanging variance, for each value of the independent variable. Not all the histological changes recorded as indicators of age fit these assumptions very well. For example, whereas the size of the transparent dentine zone can be measured as a continuous variable, the Gustafson methods record changes by a system of arbitrary scores that are better described as ordinal, and are not necessarily linearly related to age. Similarly, the age at which measures or stages of most age related phenomena are attained becomes more variable as the phenomenon progresses further along its scale of development (e.g. attrition, page 236). Linear regression models may therefore not be ideal, and it is clear that alternative statistical approaches should be considered (Lucy *et al.*, 1996). Furthermore, one of the reasons that wide confidence intervals need to be attached to age estimates is the attempt to estimate a single figure for age. If estimates were instead expressed as a probability that age was greater than (or lesser than) a given figure, this would better express the true nature of age estimation techniques and would result in rather tighter confidence limits.

10

Biochemistry of dental tissues

Enamel, dentine and cement are all mineral/organic composites (Table 10.1). Mature enamel is almost entirely mineral, but dentine and cement are about one-quarter organic. Enamel close to the crown surface is more heavily mineralized than the inner enamel, which also has a higher carbonate concentration – factors which must make the inner layer of enamel less resistant to damage through percolation of ground water in buried material. The initial matrix of predentine and precement is entirely organic, whereas enamel matrix has crystallites seeded into it from the start (Table 10.1) and, protected in a bony crypt, may survive in some archaeological contexts.

The inorganic component

Calcium phosphate minerals

The inorganic component of living bone, cement, dentine and enamel is composed almost entirely of calcium phosphates from the apatite family of minerals, the bulk of it hydroxyapatite, a form that is not found outside mammalian tissues. Composition of apatites varies widely through substitutions and vacancies within the crystal lattice, and through adsorption of ions onto the surface (Table 10.2). The commonest substitutions are carbonate and fluorine, and the first crystallites of enamel matrix are particularly rich in magnesium and carbonate, forming the centre of the crystals which grow during the process of maturation (page 148). Solubility of apatites varies with composition, and increases when dental tissues are immersed in solutions of low pH, or in the presence of a calcium chelating agent such as EDTA (page 301). Other calcium phosphate minerals may be found in association with the alternation of demineralization and remineralization that characterize dental caries, dental

Table 10.1. *Composition of mineralized tissues*

	Developing enamel		Mature enamel		Dentine		Bone[b]
	Weight	Volume	Weight	Volume	Weight	Volume	Weight
Inorganic %	37	16	≥96	88 (80–100)	72	50	70
Organic %	19	20	<0.2–>0.6	0.3	20 (18 collagen)	30	22 (21 collagen)
Density g/cm³	1.45		2.9–3		2–2.3		2–2.05
Calcium %			34–40		26–28		24
Phosphorus %			16–18		12.2–13.2		11.2
Ca/P ratio (weight)			1.92–2.17		2.1–2.2		2.15
Ca/P ratio (molar)[a]			1.5–1.68		1.6–1.7		1.66
CO₂ present as carbonate %			1.95–3.66		3–3.5		3.9
Sodium %			0.25–0.9		0.7		0.5
Magnesium %			0.25–0.56		0.8–1		0.3
Fluorine ppm			<25–>5000 (surface)		50–10000		5000
Iron ppm			8–218		60–150		
Zinc ppm			152–227		200–700		
Strontium ppm			50–400		100–600		

Amount of water is variable. Figures for mature enamel, dentine and bone are dry weight or volume.

[a] Compare with Ca/P ratio (molar) 1.667 for pure hydroxyapatite.
[b] Cement analysis not available, but similar to bone.

Sources: Brudevold & Söremark (1967), Rowles (1967), Williams & Elliot (1989).

Table 10.2. *Minerals in hard tissues*

APATITES

Hydroxyapatite – $Ca_{10}(PO_4)_6(OH)_2$

Substitutions of ions into hydroxyapatite lattice, and vacant sites

Ca^{2+} sites: Sr^{2+}, Ba^{2+}, Pb^{2+}, Ra^{2+}
                    Na^+, water, vacancy (less commonly)
                    K^+, Mg^{2+} (uncommon)
PO_4^{3-} sites: AsO_4^{3-}
                    HPO_4^{2-}, CO_3^{2-}, HCO_3^- (less commonly)
OH^- sites: Cl^-, F^-, Br^-, I^-, CO_3^{2-}, O^{2-}
                    water, vacancy (less commonly)

Fluorapatite – $Ca_{10}(PO_4)_6F_2$
Fluorhydroxyapatite – $Ca_{10}(PO_4)_6(OH,F)_2$
Carbonate-containing apatite – $Ca_{10}(PO_4)_6CO_3$

In poisoning by Pb, Ra, Sr or ASO_4 ions are first deposited in the mineral phase of hard tissue and then slowly released.

Surface bound ions

Ca^{2+}, PO_4^{3-}, HPO_4^{2-}, CO_3^{2-}, HCO_3^-, Mg^{2+}, K^+, citrate, water

OTHER CALCIUM PHOSPHATE MINERALS

Whitlockite – related to β-tricalcium phosphate, β-$Ca_3(PO_4)_2$, with impurities Mg^{2+}, Mn^{2+} or Fe^{2+}
Monetite – $CaHPO_4$
Brushite – $CaHPO_4.2H_2O$
Octacalcium phosphate – $Ca_8(HPO_4)_2(PO_4)_4.5H_2O$

POSSIBLE SECONDARY MINERALS

Calcite – $CaCO_3$
Goethite, lepidocrocite, limonite – $FeO.OH$
Pyrite – FeS
Vivianite – $Fe_3P_2O_8.8H_2O$

Sources: Trautz (1967), Limbrey (1975), Williams & Elliot (1989).

calculus (page 257) or burial in the ground. These include whitlockite, brushite and octacalcium phosphate, all of which have a higher solubility than apatite.

Secondary minerals

Minerals, deriving from the groundwater of the burial matrix, may be deposited on surfaces and within spaces in preserved dental tissues. Little is known about these processes, but calcite (Table 10.2) is deposited in this way and teeth are often stained with brown or ochre colours, suggesting iron mineral deposition. One particularly startling form of mineralization is vivianite, which turns a vivid blue on exposure to air.

Fluorine chemistry of dental tissues

Fluorine enters the body in drinking water, food, and nowadays through tablets or drops. Its concentration in drinking water is controlled by geology, hydrology and artificial supplementation, and varies from less than 0.8 ppm up to 45–53 ppm in parts of South Africa and Tanzania (Pindborg, 1982). It causes enamel defects (page 171) in concentrations above 1 ppm (artificially fluoridated water is 0.8–1 ppm) and pronounced defects above 6 ppm. Once ingested, fluorine is absorbed rapidly into the blood and a proportion is taken up by mineralized tissues at their forming surfaces, or surfaces nourished by the bloodstream, to accumulate steadily with age in bone, cement and dentine. Enamel cannot accumulate further fluorine in this way after it has been formed, but it absorbs some additional fluorine from saliva and foodstuffs passing through the mouth.

Fluorine composition varies through a bone or tooth (Figure 10.1). In dentine (Yoon *et al.*, 1958; Jackson & Weidmann, 1959; Yoon *et al.*, 1960; Rowles, 1967) highest concentrations, higher than enamel or bone, are found next to the pulp chamber. Levels decrease into the central part of the dentine and, in the root, rise again markedly into the cement. Secondary dentine has an even higher fluorine concentration than primary dentine. The bulk of enamel shows much lower concentrations, but there is a thin zone of high fluorine concentration at the crown surface, established from the start of maturation. When the tooth is first erupted, highest fluorine levels are found at the occlusal cusp tips but, with wear, these lose their surface zone rapidly whereas approximal and cervical zones are more protected. With higher environmental fluorine, the overall enamel concentrations rise, and the contrast between surface zone and deeper enamel grows.

Where fluorine has been analysed in archaeological specimens (Steadman *et al.*, 1959; Robinson *et al.*, 1986), both the pattern and levels of enamel fluoride are preserved in the expected form, although higher than expected levels have been found at the root surface, perhaps due to selective diagenetic effects. Much higher levels of fluorine are found in older fossil bones and teeth (> 1%), due to continued substitution and adsorption of fluorine from groundwater percolating around and through the specimen. This is so widely recognized that it forms the basis of a relative dating technique (Oakley, 1969; Aitken, 1990), most famously applied to the 'Piltdown man' fraud. Fluorine

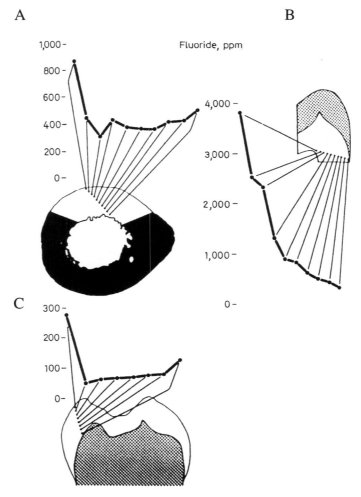

Figure 10.1 Fluorine distribution in bone, dentine and enamel sections. Reproduced from Weatherell, J. A., Deutsch, D., Robinson, C. & Hallsworth, A. S. (1977) Assimilation of fluoride by enamel throughout the life of the tooth, *Caries Research,* **11**, 85–115, with kind permission from S. Karger AG, Basel. A, bone from periosteum to endosteum; B, dentine from pulp chamber to root surface; C, enamel from EDJ to crown surface.

dating was popular at a time when radiocarbon dating required much larger samples of bone, but the method makes the unsafe assumptions of a constant fluorine accumulation rate and an even distribution through the specimen. More recently, this difficulty has been tackled by recording the profile of fluorine through the thickness of a bone or tooth, using a nuclear microprobe technique (Coote & Sparks, 1981; Coote & Molleson, 1988). The typical profile evens out with increased burial time as more fluorine is accumulated, so that changes in the profile rather than the absolute level of fluorine can be used to derive relative age. The method is relatively non-destructive, carried out on sections, and thus could form part of a histological study of ancient teeth.

Uranium and dating

The uranium content of fresh dental tissues and bone is less than 0.1 ppm (Aitken, 1990), but fossil bones and teeth contain up to 1000 ppm. Uranium accumulates steadily from groundwater, even in specimens exposed for a few months on the surface (Williams & Marlow, 1987), and continues building up especially in a thin surface layer (Henderson *et al.*, 1983), although the mechanism is complex (Millard & Hedges, 1995). This process has been exploited as a relative dating indicator in parallel with fluorine (Oakley, 1969; Demetsopoullos *et al.*, 1983), with the advantage that uranium content can be determined radiometrically without damaging the specimen, as for example in *Gigantopithecus* teeth from China.

The radioactive properties of uranium are further exploited in uranium series dating. Two radioactive isotopes, ^{238}U and ^{235}U, give rise to two separate chains of radioactive daughter isotopes, which decay one into another until finally a stable isotope is reached. Dating is based on two radioactive isotopes from the ^{238}U series (^{234}U and ^{230}Th or thorium), and two from the ^{235}U series (^{235}U itself and ^{231}Pa or protactinium). Both thorium and protactinium are insoluble in water and are unlikely to be deposited during life or burial, so they can only accumulate through radioactive decay in the uranium series. If it is assumed that most uranium arrives in the specimen during the early stages of fossilization, with little addition or loss after that (a closed system), then the ratios ^{230}Th/^{234}U and ^{231}Pa/^{235}U can be used to determine the time elapsed since burial. The original method of measurement used 1–2 g samples and produced dates in the range 5000–35 000 BP for the ^{230}Th/^{234}U method, or 5000–250 000 BP for the ^{231}Pa/^{235}U method, but more recently mass spectrometry has permitted the use of smaller samples and increased the date range. It is possible to test for the presence of a closed system by comparing ^{230}Th/^{234}U and ^{231}Pa/^{235}U dates, and to check for consistency with alternative dating methods. Chen and Yuan (1988) carried out uranium series determinations on a large group of non-human teeth from Chinese Palaeolithic sites and, where the ^{230}Th/^{234}U and ^{231}Pa/^{235}U ages agreed, the derived dates showed a good concordance with independent radiocarbon dates. Aitken (1990) suggested that, with mass spectrometry, there might be an advantage in using the more resistant enamel instead of dentine.

Lead in dental tissues

In the modern world, lead enters the body through inhaled air or cigarette smoke, from drinking water that passes through lead piping, or food contaminated by solder in cans, glaze on pots, and pewter vessels. Lead is lost slowly from the body, mostly by excretion into the urine, and almost all that remains gradually accumulates in the skeleton and dental tissues. The pattern of lead distribution in enamel and dentine (Brudevold & Söremark, 1967; Rowles, 1967) is similar to that of fluorine (page 220), and lead accumulates slowly at the enamel surface with age, presumably due to interaction with saliva, food and drink. Dentine, especially secondary dentine (Shapiro *et al.*, 1975), should better represent the body load of lead because it remains cellular throughout life, whereas enamel (under the thin surface layer) can represent only the lead accumulated during crown formation.

Lead has been studied in both living and ancient populations, particularly through the analysis of bone (Aufderhiede, 1989). Difficulties include variation in reporting of analyses, variation between teeth from one individual, gradual accumulation with age, and diagenetic effects that sometimes produce anomalously high levels (Waldron, 1983). In both enamel (Steadman *et al.*, 1959) and dentine (Kuhnlein & Calloway, 1977; Grandjean *et al.*, 1979), ancient specimens usually contain much less lead than specimens derived from modern urban populations. These contrasts may well represent differences in air and water pollution, but diagenetic effects must also be borne in mind.

Diet and trace elements

Reviews of general trace element literature include Aufderheide (1989) and Sandford (1992, 1993). The best known element in relation to diet is strontium (Sillen & Kavanagh, 1992), which occurs at similar concentrations in both bone and dental tissues: commonly 150–250 ppm in bone ash (Aufderhiede, 1989), 100–200 ppm in enamel (Brudevold & Steadman, 1956) and 100–600 ppm in dentine (Rowles, 1967). Strontium content varies little through a tooth, in contrast with lead, uranium and fluorine. It enters the body in the diet but the mammalian gut absorbs calcium more readily than strontium, an effect that is multiplied when carnivores eat other mammals, so that strontium levels should represent the proportions of animal and plant food. There are, however, inconsistencies: the gut varies in its selective absorption, background strontium levels vary from place to place and some foods, for example shellfish, concentrate strontium by many times, whereas milk is notably deficient. In an attempt to demonstrate the dietary relationship of strontium, Elias (1980) collected extracted human teeth from a locality near Calcutta, including seven lifelong vegetarians and seven omnivores. Elias found no significant difference in enamel strontium between the two dietary groups, concluding that the low strontium levels in meat were masked by large vegetable consumption even in the non-vegetarians. This result may have been complicated by the use of enamel, which incorporates strontium only during amelogenesis and this, for most teeth, takes place in early childhood when a lot of low strontium milk is consumed. One reason for the initial excitement about strontium was its apparent independence from diagenetic change (Steadman *et al.*, 1959; Parker & Toots, 1980; Stack, 1986), but these assumptions have also been questioned (Sillen & Kavanagh, 1992).

In recent years most bone studies have analysed groups of elements together, rather than single elements (Buikstra *et al.*, 1989; Sandford, 1992). A strong relationship with diet has been suggested particularly for zinc, which is found in most foods, but reaches highest concentrations in meat and seafood. Zinc is stably bound in mineralized tissues, with a somewhat similar distribution to lead, and there is little evidence for major diagenetic effects (Steadman *et al.*, 1959).

Stable isotope analysis of apatites

Stable isotope analysis has concentrated mostly on the organic component of bone (page 229), and the possibility of applying it to dental apatites has only recently been tested (Schwarcz & Schoeninger, 1991; Katzenberg, 1992; van der Merwe, 1992; Ambrose, 1993). Two stable isotopes of carbon exist – ^{13}C (1.1% of carbon) and ^{12}C (98.9%) – and the difference in ratio between tissue samples and a standard, measured in parts per thousand (per mil or ‰) as $\delta^{13}C$, varies between different groups of plants and animals. One of the largest contrasts is created by the two main pathways of photosynthesis in plants. In C3 plants (all trees and shrubs, root crops, temperate grasses, including domesticated grasses such as wheat, barley and rice) $\delta^{13}C$ averages −26.5 ‰, whereas in C4 plants (tropical and sub-tropical grasses, incorporating the domestic crops sorghum, millets, maize and sugar cane) $\delta^{13}C$ averages −12.5 ‰. When animals eat plants and incorporate the carbon that they contain into the apatite of bones and teeth, further fractionation of the isotopes occurs, so an animal that predominantly eats C3 plants should have an apatite $\delta^{13}C$ of −14.5 ‰, whilst an animal eating predominantly C4 plants should average −0.5 ‰. Predators eating these herbivores fractionate the carbon isotopes still further. Lee-Thorp and colleagues (1989) analysed carbon isotopes in the enamel of 1.8-million-year old baboon teeth from the South African site of Swartkrans, and were able to distinguish between C3-eating browsers on trees and shrubs and C4-eating grazers on savannah grasses. Van der Merwe (1992) used a similar technique with fossil hominid enamel, suggesting a large C4 derived component in their diet through eating savannah grazing animals. Carbon isotope studies using apatite have been controversial because of the possibility of diagenetic change through carbonate deposition from groundwater, but contamination can be minimized by the use of enamel, careful preparation techniques (Krueger, 1991; Lee-Thorpe & van der Merwe, 1991), and the measurement of oxygen isotope ratios to check for carbonate exchange (Schwarcz & Schoeninger, 1991).

Electron spin resonance dating of enamel apatites

For reviews see Aitken (1990), Ikeya (1993), Grün and Stringer (1991). The basis of the ESR technique is the trapping of free electrons by defects in the apatite crystal lattice. Free electrons are generated by nuclear radiation that derives from radioactive elements in the burial matrix, skeletal and dental tissues. The defects are characterized by vacancy of a positively charged ion, attracting and trapping the free electrons that diffuse through the lattice. Whilst the specimen lies buried, progressively more free electrons are generated and trapped so that, if the radiation dose rate is constant, the number of trapped electrons represents the time elapsed since burial. Trapped electrons generate a characteristic signal in an ESR spectrometer, from which the accumulated dose of radioactivity is determined and, assuming a figure for dose rate, the date calculated. Estimation of dose rate is the largest problem in ESR dating and usually involves chemical analysis of the radioactive elements present. It is preferable to work with thick enamel layers from large teeth, and various models have been proposed for uranium uptake. ESR uses small samples (10–100 mg) and yields dates from a few hundred years to over 300 000 BP.

The organic component

Organic component of dentine, cement and bone

Most of the organic component of dentine, cement and bone is protein (Table 10.1). Survival of proteins in buried bones and teeth is highly variable, but a bone may be considered well preserved when it has 5% or more dry weight of protein, compared with > 20% in the fresh tissue (Schwarcz & Schoeninger, 1991). Protein content in archaeological dentine shows little relationship with date, but is closely related to the structural integrity of the dentine (Beeley & Lunt, 1980). Softer specimens often have only 0.1% – 8% by dry weight of protein, whereas harder specimens have 11% – 19%.

Collagen

The great bulk of protein in fresh bone, cement and dentine is collagen. Each collagen molecule is 280–300 nm long, and is built from three α-chains, each consisting of > 1000 linked amino acid residues with a combined molecular weight of 100 kD. There are four types of α-chain, distinguished by minor differences in amino acid sequence but, in dentine, cement and bone, all are Type I chains. These chains are built from 20 different amino acids, combined in proportions that make collagen an unusual protein: glycine comprises about one-third of the amino acid residues, and there are 10% each of proline and its derivative hydroxyproline. Hydroxyproline is unique to collagen and makes a useful marker. Collagen molecules are twisted together into five-stranded fibrils, about 10–200 nm in diameter and, when appropriately stained and examined in the transmission electron microscope, these fibrils exhibit a characteristic 67 nm banding.

Collagen is very stable, but can be broken down by boiling in dilute acid to produce gelatins, which are composed of short lengths (peptides) of α-chain. Gelatins dissolved in water make the old-fashioned hide glue that gives collagen its name (Greek; *kolla*). Collagen is also broken down into peptides in buried specimens and the peptides further broken down into their component amino acids, which may then be leached from the specimen by percolating groundwater. The extent to which all this actually happens varies, even within one specimen, and does not necessarily reflect the surface appearance of preservation. In dentine with 11% protein surviving, the proportions of different amino acid residues may differ little from those of fresh dentine, whereas they diverge substantially in specimens with less than 1% protein (Beeley & Lunt, 1980). Dentine collagen, however, seems to be more resistant than bone collagen (Masters, 1987). The chemical basis of collagen breakdown in the soil is in any case not well understood, but is dependent upon temperature, groundwater and its pH. Microorganisms must be involved and are inhibited by cold, dry, or waterlogged conditions. The Lindow Man bog body, found in Cheshire, England, had only the collagen of its dentine and bone preserved, after the mineral had been leached away (Stead *et al.*, 1986). The time elapsed since burial has little effect on collagen preservation, and Pleistocene fossils have shown the 67 nm banding apparently intact (Wyckoff, 1972), although it is possible that the banding may have been mimicked by a mineralization pattern, because fibrils are very intimately mineralized in life, with apatite crystallites initiated in spaces within their structure.

Other organic components

Proteoglycans are the largest non-collagen fraction of dentine, cement and bone (Table 10.1). They are large molecules, with a protein core linked to carbohydrate chains (glycosaminoclycans or acid mucopolysaccharides). Other protein components of dentine include phosphoproteins, and proteins containing γ-carboxyglutamic acid. In bone, the latter are called osteocalcin, and have been extracted successfully from archaeological material (Ajie *et al.*, 1990). Other proteins, normally found in serum, are also present in dentine and include albumin, which has been demonstrated immunologically in a mastodon bone dated to 10 000 BP (Lowenstein & Scheuenstuhl, 1991), and from archaeological human bone (Cattaneo *et al.*, 1992). Non-collagenous proteins (NCP) appear to survive better than collagen in diagenetically altered bone (Masters, 1987). Lipids are also present as a small component of fresh dentine, and smaller organic molecules include citrate and lactate. The survival of minor organic components in bone is currently the subject of much research (Eglinton & Curry, 1991; Hedges & Sykes, 1992). Little work has been done on dentine but, where it is protected by the enamel cap of the crown, there must be considerable potential for preservation of organic material.

The organic component of the enamel

The two main stages of enamel development, secretion of enamel matrix and maturation (page 148), have different organic components.

The organic component of the enamel matrix

Ninety per cent of the organic component is amelogenin, a protein unique to enamel matrix. Various sizes of amelogenin molecules are present, but the largest, from which the others are derived, has a molecular weight of 20 kD. Amelogenin is a highly unusual protein, with proline making up about one-quarter of its amino acid residues, followed by glutamic acid, leucine and histidine. Hydroxyproline, the characteristic component of collagen, is conspicuous by its absence. During maturation, rapid mineralization is matched by similarly rapid protein removal. This is preceded by a cascade of amelogenin degradation in which enzymes cleave the protein at particular sites to form peptides. Human amelogenin is produced by a single gene, with two copies; one on the X and one on the Y chromosome (AMGX and AMGY). Both copies of the gene are expressed, although 90% of amelogenin derives from AMGX (Fincham *et al.*, 1991). Differences between AMGX and AMGY lead to differences in the amino acid sequence of the resulting amelogenins (Salido *et al.*, 1992), so that amelogenins and their breakdown products may indicate sex. The partly mineralized matrix and transitional zone sometimes survives in archaeological material (page 149) and may contain preserved amelogenins.

About 5% of enamel matrix organic component consists of non-amelogenin proteins (Glimcher *et al.*, 1990). The bulk of these are normally found in serum, particularly albumin, together with some usually found in saliva. Some are rich in proline, and are described as the proline-rich non-amelogenins. Others are related immunologically to the tuft proteins of mature enamel and are known as tuftelin (Robinson *et al.*, 1975; Deutsch *et al.*, 1991).

Organic matrix of mature enamel

Fully mature enamel has a tiny organic component, consisting mostly of proteins and peptides up to 5 kD in molecular weight, but mostly 1–1.5 kD or 8–12 amino acid residues long (Glimcher *et al.*, 1990). Their distribution varies through the enamel, and is concentrated in the tufts – small 'fault planes' of organic material that radiate out from the EDJ. The amino acid sequences are unknown, but the proportions of residues differ sharply from those of amelogenins, with about half the proline, less leucine and less histidine, resulting in a composition more akin to that of the non-amelogenins. The organic component of mature enamel is tightly bound into the mineral structure and is exceptionally resistant to degradation. Proteins and peptides survive in fossil enamel many thousands or millions of years old, and Doberenz *et al.* (1969) demonstrated them in a late Pleistocene mastodon molar from Rancho La Brea, California. Glimcher and colleagues (1990) analysed enamel from a fossil Cretaceous crocodile (about 105 million years BP), which yielded 0.02% by weight of protein/peptides, with molecular weights and amino acid compositions very similar to those of modern enamel.

Racemization and epimerization of amino acids

Molecules of amino acids such as aspartic acid may exist in two alternative three-dimensional structures; the L-enantiomer and the D-enantiomer. All amino acids in animal proteins are formed as L-enantiomer, but gradually a proportion spontaneously changes to the D-enantiomer in the process known as racemization. Strictly speaking, racemization refers to molecules that have only one centre of asymmetry. Some amino acids, such as isoleucine, have two centres with alternative forms that are known as the L-diastereomer and D-diastereomer, and conversion from L to D is called epimerization. The rate of racemization or epimerization is constant under conditions of stable temperature, and it follows that the D : L ratio in an ancient specimen increases with its date. Such ratios have been measured in bone, dentine and enamel (Masters, 1986b; Aitken, 1990), giving a maximum date range of about 100 000 BP for aspartic acid analysis in bone, to

900 000 BP for isoleucine in teeth. Enamel is a better material than dentine or bone, but the main difficulty is the assumption of constant burial conditions. Both racemization and epimerization are highly temperature dependent, and degradation of proteins may release free amino acids, with a different racemization rate to protein-bound amino acids. Amino acid dates must therefore be calibrated with radiocarbon. Reviews of the method include those by Hare, (1980), Masters (1986b) and Aitken (1990).

Dating assumes that negligible changes in D : L ratios have taken place before death and, in specimens several thousand years old, this is reasonable because the post-mortem change is so much larger. Racemization does, however, proceed fast enough at body temperature for detectable amounts of D-enantiomer to accumulate during life, so that it is possible in some circumstances to use D : L ratios to determine age at death (Masters, 1986a; Gillard *et al.*, 1990). Frozen Inuit remains from Alaska (Masters & Zimmerman, 1978; Masters, 1984) were so little affected by post-mortem racemization that the D : L ratios for aspartic acid in dentine yielded ages at death that matched well with alternative estimates. A small group of individuals from a 1000-year-old cemetery in Czechoslovakia also yielded racemization age determinations close to independent estimates (Masters & Bada, 1978), but here a correction for post-mortem racemization was determined by measuring the D : L ratio for aspartic acid in a young individual, whose age could be determined reliably from other dental and skeletal characteristics. Gillard *et al.* (1990) found that similar methodology worked well with dentine from modern teeth although, when applied to known age archaeological specimens from Christ Church, Spitalfields, London, the results were less successful – presumably due to diagenetic effects.

Isotopic studies of organic components

Protein and peptides extracted from bone are the material of choice for radiocarbon dating by the accelerator mass spectrometry (AMS) method (Aitken, 1990). Very small bone samples (< 0.5 g) can be used, but dating of the small volumes of dentine in human teeth remains a challenge, as 0.5 g still represents around 1 cm^3 of tissue. Studies of the stable isotopes (page 224) of carbon, nitrogen and other elements are mostly based on collagen extracted from archaeological bone (De Niro, 1987; Keegan, 1989; Schwarcz & Schoeninger, 1991; Katzenberg, 1992; van der Merwe, 1992; Ambrose, 1993). Collagen extracted from dentine has been little studied but, once again, its better preservation may prove an advantage. Using carbon and nitrogen isotope values together, it is possible to elucidate a number of factors in ancient diet. The clearest archaeological horizon in such studies is the establishment of maize agriculture in the Americas, and similarly millet cultivation in China (both C4 plants appearing in otherwise C3 diets, see page 224). It is also possible to identify a diet that is based upon marine resources, as opposed to terrestrial, and to determine the level of carnivory relative to herbivory. Another possibility is the detection of weaning through a comparison of isotopic analyses in the dentine of different teeth from one individual (Dr Gert Jaap van Klinken, personal communication).

Nucleic acids

Considerable evidence has now accumulated to show that human DNA survives in forensic and archaeological bone (Hagelberg *et al.*, 1989, 1991a; Hagelberg & Clegg, 1991; Hagelberg *et al.*, 1991b; Hedges & Sykes, 1992; Hagelberg, 1994). About 1–10 µg of DNA can normally be extracted from 1–2 g of bone, but the bulk of this is not human DNA and probably arises from micro-organisms. The DNA is in small fragments, strongly bound to the bone apatite. The tiny human DNA component in archaeological bone is demonstrated by using the polyermase chain reaction (PCR) to amplify a selected part of the base sequence of the DNA molecule. Most PCR work has been carried out on human mitochondrial DNA. The mitochondrion is an organelle found in all cells, up to 2000 or more per cell, each of which carries its own DNA, complete with a well-known sequence 16 569 base pairs (bp) long, of which there are 10 000–100 000 copies in each cell. This gives a much better chance for amplifying a particular part of the base sequence than nuclear DNA, in which many genes have very few copies per cell (Strachan & Read, 1996). Typically, 100–400 bp mitochondrial DNA sequences have been amplified from well preserved archaeological bone. From forensic material only a few years or tens of years old, > 1000 bp sequences have been amplified, and it has also proved possible to amplify human nuclear DNA. One of the most spectacular achievements has been to identify the remains of the Romanov family; the last Tzar, Tzarina, their children, physician and servants (Gill *et al.*, 1994), which were discovered in 1991, in a shallow grave in Ekaterinburg, Russia. In order to establish the sex of these individuals, parts of the amelogenin (page 227) gene (AMGX and AMGY) were amplified, including sequences of just over 100 bp. Similar methods have now been used to sex 700-year-old archaeological material from Illinois (Stone *et al.*, 1996).

Bone is currently the material of choice for extraction and amplification of ancient DNA, but there has been some investigation of alternatives. Bone preserves DNA best when it shows few microscopic signs of disruption (Hagelberg *et al.*, 1991a). Preservation of bone is, however, more variable than dentine protected under its cap of enamel (Richards *et al.*, 1995), and dentine has yielded DNA (Professor Bryan Sykes, personal communication).

11

Tooth wear and modification

Dental attrition is wear produced by tooth-on-tooth contact, between neighbouring teeth or opposing teeth, and it produces wear facets on the occlusal surface or at the contact points between teeth. Dental abrasion is wear that is not produced by tooth-on-tooth contact and does not produce clear facets, but is seen as a general loss of surface detail.

Abrasion

Types and distribution of abrasion

Surface detail is lost through contact with abrasive particles carried on the cheeks, tongue and food. Approximal areas are protected from abrasion (Scott & Wyckoff, 1949), and cervical crown surfaces are often protected by calculus deposits (page 255) which, when stripped away, reveal a pristine surface underneath. In modern humans, abrasion is much more pronounced than in archaeological material, due to toothbrushing with abrasive toothpaste. Over the years this produces a very smooth surface with a high gloss, more pronounced in some people depending on the vigour of brushing and, in older individuals when periodontal disease exposes the roots, toothbrush abrasion wears away the softer cement and dentine to create a step below the crown. Various other foreign objects may come into contact with teeth, including toothpicks, smoking equipment, blades and jewellery, and can also cause abrasion.

Microscopic appearance of abrasion

Deliberate cutting and grinding of teeth are distinguished by strong striated markings, all running parallel with the direction of cut, or the twist of the

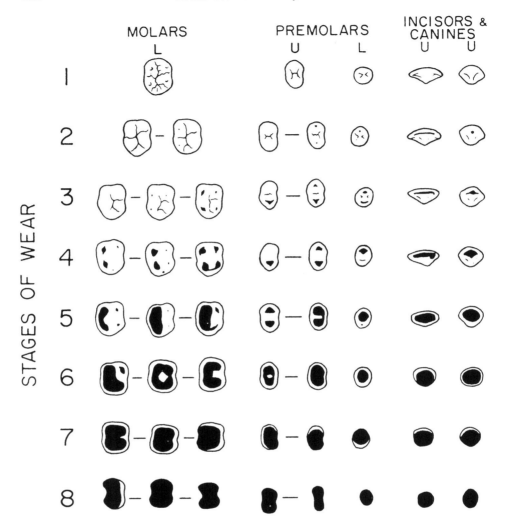

Figure 11.1 Occlusal attrition stages based on the larger diagrams of Murphy (1959a). Reprinted from Smith, B. H. (1984) Patterns of molar wear in hunter-gatherers and agriculturalists, *American Journal of Physical Anthropology*, **63**, 39–56, Copyright © 1984, John Wiley & Sons Ltd. Reprinted by permission of John Wiley & Sons Ltd. Wear stages for molars can be used for both upper and lower permanent dentitions. Premolars, canines and incisors have separate stages.

drill (Milner & Larsen, 1991), but most abrasion does not produce characteristic features and instead progressively reduces the sharpness of normal microscopic crown surface detail (Scott *et al.*, 1949; Scott & Wyckoff, 1949). A newly erupted tooth crown is densely covered with sharply defined Tomes' process pits (page 163) and perikymata. The shallow Tomes' process pits along the perikyma ridge crests are abraded away first, then the ridges decrease in height until deeper Tomes' process pits in the perikyma grooves are also lost. Cheek teeth retain pronounced surface features for longer than anterior teeth, and mesial and distal surfaces retain them much longer than lingual, labial or buccal. Pedersen and Scott (1951) compared rate of surface detail loss between American Whites and Inuit, but found no marked differences.

Occlusal attrition

Scoring and measuring occlusal attrition

The general patterns of wear on different tooth crowns are described in Chapter 2, and several detailed recording schemes have been developed. The best known is that of Murphy (1959a), who devised a series of 'modal forms', graded from a to h, illustrated by standard diagrams showing the pattern of dentine exposure. Smith (1984) produced a summary diagram of Murphy's system, which is now widely used for age estimation purposes (Buikstra & Ubelaker, 1994, and see Figure 11.1). The system of Molnar (1971) was based upon similar distinctions to Murphy, with the addition of secondary dentine exposure, and separate scores for attrition facet orientation and the form of worn surface (Figure 11.2). Scott (1979a) devised a more complex system for permanent molars, deliberately excluding secondary dentine, because the size of pulp chamber varies between individuals. Scott's system divided the occlusal surface into quadrants, each scored separately using a 10-point scale (Figure 11.3). Despite its complexity, this method shows little intra- and inter-observer variability (Cross *et al.*, 1986). Dreier (1994) developed yet another molar tooth crown wear score (MTCW) using detailed descriptions of a variety of features, based on experimental grinding studies.

These scoring systems can be used rapidly on large collections, and clinical indices based on similar scores have also been established (Dahl *et al.*, 1989; Johansson *et al.*, 1993). There are, none the less, problems of subjectivity, and it is also difficult to take into account the possibility that larger teeth wear more slowly than smaller teeth (Walker *et al.*, 1991). Several types of measurement have therefore been used to record attrition in molars. These include crown height, from cervical margin to buccal cusp tips (van Reenen, 1982) or the edge of the worn occlusal surface (Walker *et al.*, 1991), or cusp height relative to the central fossa (Tomenchuk & Mayhall, 1979; Molnar *et al.*, 1983a). The angle of the attrition facet may be measured using a modified protractor (Smith, 1984; Walker *et al.*, 1991). Area of wear facets can be

A

Category of Wear	Incisor and Canine	Premolar	Molars
1	Unworn.	Unworn.	Unworn.
2	Wear facets minimal in size.	Wear facets, no observable dentine.	Wear facets, no observable dentine.
3	Cusp pattern obliterated, small dentine patches may be present.	Cusp pattern partially or completely obliterated. Small dentine patches.	Cusp pattern partially or completely obliterated. Small dentine patches.
4	Dentine patch (Minimal).	Two or more dentine patches, one of large size.	Three or more small dentine patches.
5	Dentine patch (Extensive).	Two or more dentine patches, secondary dentine may be slight.	Three or more large dentine patches, secondary dentine, none to slight.
6	Secondary dentine (Moderate to Extensive).	Entire tooth still surrounded by enamel, secondary dentine moderate to heavy.	Secondary dentine moderate to extensive, entire tooth completely surrounded by enamel.

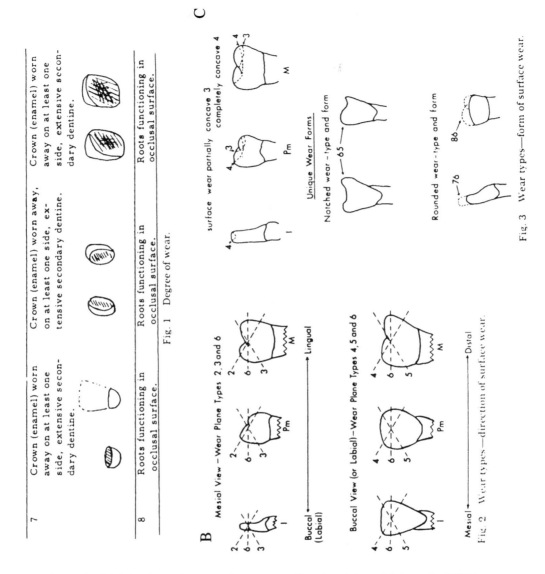

Figure 11.2 Molnar's attrition scoring system. Reprinted from Molnar, S. (1971) Human tooth wear, tooth function and cultural variability, *American Journal of Physical Anthropology*, **34**, 175–190, Copyright © 1984, John Wiley & Sons Ltd. Reprinted by permission of John Wiley & Sons Ltd. A, degree of wear for upper and lower dentitions. B, direction of wear surface – 1, natural form; 2, oblique (buccal–lingual direction); 3, oblique (lingual–buccal direction); 4, oblique (mesial–distal direction); 5, oblique (distal–mesial direction); 6, horizontal (perpendicular to the long axis of the tooth); 7, rounded (buccal–lingual direction); 8, rounded (mesial distal direction). C, form of wear – 1, natural form; 2, flat surface; 3, one-half of surface cupped; 4, entire surface cupped; 5, notched; 6, rounded.

Attrition scoring technique

Score	Description
0	No information available (tooth not occluding, unerupted, antemortem or postmortem loss, etc.)
1	Wear facets invisible or very small
2	Wear facets large, but large cusps still present and surface features (crenulations, noncarious pits) very evident. It is possible to have pinprick size dentine exposures or "dots" which should be ignored. This is a quadrant with *much* enamel.
3	Any cusp in the quadrant area is rounded rather than being clearly defined as in 2. The cusp is becoming obliterated but is not yet worn flat.
4	Quadrant area is worn flat (horizontal) but there is no dentine exposure other than a possible pinprick sized "dot."
5	Quadrant is flat, with dentine exposure one-fourth of quadrant or less. (Be careful not to confuse noncarious pits with dentine exposure.)
6	Dentine exposure greater: more than one-fourth of quadrant area is involved, but there is still much enamel present. If the quadrant is visualized as having three "sides" (as in the diagram) the dentine patch is still surrounded on all three "sides" by a ring of enamel.
7	Enamel is found on only two "sides" of the quadrant.
8	Enamel on only one "side" (usually outer rim) but the enamel is thick to medium on this edge.
9	Enamel on only one "side" as in 8, but the enamel is very thin — just a strip. Part of the "edge" may be worn through at one or more places.
10	No enamel on any part of quadrant — dentine exposure complete. Wear is extended below the cervicoenamel junction into the root.

Figure 11.3 Scott's attrition scoring system for molars. Reprinted from Scott, E. C. (1979) Dental wear scoring technique, *American Journal of Physical Anthropology*, **51**, 213–218, Copyright © 1984, John Wiley & Sons Ltd. Reprinted by permission of John Wiley & Sons Ltd. Procedure for scoring: the occlusal surface is divided visually into four equal quadrants and each is scored 1–10. Scores 1–4 represent the removal of enamel features, whereas 5–10 covers the progressive exposure of dentine.

measured on photographs of the occlusal surface using a planimeter, a digitizing tablet/computer system (Walker, 1978; Richards & Brown, 1981a; Molnar *et al.*, 1983a; Richards, 1984, 1990), or a computer-based image analysis system (Kambe *et al.*, 1991). In casts, where it is not possible to distinguish between dentine and enamel exposed in the facet, the overall outline of each facet is measured. It is often difficult to define facet edges from photographs but, in colour photographs of original teeth, it is possible to measure the area of exposed dentine instead. The current ultimate in three-dimensional precision is the computer-based system of Krejci *et al.* (1994), directly measuring attrition facets with 1 μm resolution.

Patterns and gradients of occlusal attrition

There is usually little difference in wear between sides of the dentition, and most studies score attrition for the left side alone, scoring the right only when teeth are missing (Lavelle, 1970; Lunt, 1978; Scott, 1979a; Walker *et al.*, 1991). Lower molars are slightly in advance of upper, and the progressively later eruption times of first, second, and then third molars produce an attrition gradient between them (Murphy, 1959b; Pal, 1971). Lavelle (1970) found a range of mean differences between molars in a variety of populations, but Lunt (1978) argued that the arithmetic mean was not an appropriate statistic and used a non-parametric methodology instead, proposing that inter-population comparisons should be by gradients related to the attrition grade of first molars. Lunt (1978) and Pal (1971) found no significant differences between males and females, in absolute attrition grades or gradients between teeth, whilst Molnar and colleagues (Molnar, 1971; Molnar *et al.*, 1983b; McKee & Molnar, 1988) found significantly heavier wear in females. Attrition gradients vary between populations and hunter–gatherers show rapid wear, with heavy anterior tooth attrition, when compared with agriculturalists (Hinton, 1981, 1982).

Wear gradients have also been studied by the principal axis method (Scott, 1979b; Richards, 1984; Benfer & Edwards, 1991), which expresses the relationship between pairs of wear scores as two parameters – slope and intercept. If scores for first molars (on the y-axis) are plotted against scores for second molars (on the x-axis), the intercept represents the stage of wear usually reached by first molars at the point when second molars first came into wear (the difference in eruption time). The slope expresses the difference in average wear rate, so a high slope figure suggests that the first molar is wearing rapidly relative to the second.

The plane of occlusion and wear

Occlusal surfaces, both worn and unworn, show varying degrees of inclination and define a curved occlusal plane (Figure 11.4). The Curve of Spee describes the occlusal plane in lateral view, curving upwards to distal along the tooth row, whereas the Sphere of Monson describes the plane of the molars with the lower occlusal surfaces inclined to lingual, and the upper inclined to buccal. In worn dentitions a helicoidal plane is often seen, in which the worn facets of the first molars follow a Reversed Curve of Monson, the less worn occlusal surfaces of the third molars retain the classic Curve of Monson, and the second molars are worn flat. This helicoidal plane has been the subject of some discussion:

1. Attrition pattern. In molars, the supporting cusps (lingual cusps of upper molars and buccal cusps of lower molars) wear more rapidly than the rest of the tooth, so that progressive wear converts the classic Curve of Monson first into a flat plane, and then a Reversed Curve. At any stage of attrition, the first molars are further advanced in this sequence than the second molars

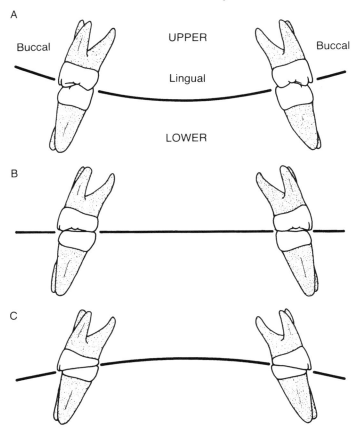

Figure 11.4 Development of the helicoidal plane of attrition. Adapted from Osborn (1982). A, normal Curve of Monson; B, flat occlusal plane with moderate occlusal wear; C, reversed Curve of Monson with advanced wear.

which, in turn, have progressed further than the third. At a moderate degree of attrition, this results in the helicoidal plane (Osborn, 1982) but, if attrition proceeds, the Curve of Monson is reversed for all three molars.

2. Axial inclination. Unworn occlusal surfaces of third molars are more inclined than second, followed by first (Smith, 1986b). The greater degree of tilt ensures that third molars retain a classic Curve of Monson longer.

3. Third molar crossbite. Tobias (1980) suggested that lingual crossbite (page 111) was normal for third molars in *Homo*, but not australopithecines, and would automatically reverse the occlusal curve. Smith (1986b) however showed that the cusps remained in normal occlusal relationship even when third molars where heavily tilted.

4. Enamel thickness. Macho and Berner (1994) found that first molar enamel

was thinner than that of second and third molars, which would accelerate the tendency of the first molar to reverse the Curve of Monson.

Thus, a variety of factors may together generate the helicoidal plane. Smith (1984) studied the changing angle of occlusal wear in lower first molars with increasing attrition in agriculturalists and hunter–gatherers. The curve of Monson was reversed at a faster rate in the agriculturalists, whereas the hunter–gathers had a markedly flat pattern of wear in their molars.

Occlusal attrition and age estimation

It is clear that attrition must in general be an age-affected phenomenon, although there have been few studies of known age material. Hojo (1954) described a general age relationship in a series of modern Japanese dentitions, but showed that a variety of wear states could be found in any age group, and Santini *et al.* (1990) found only a poor relationship between wear and age in Chinese skulls. Tomenchuk & Mayhall (1979) discovered moderate to high correlations between cusp height and age in living Inuit, and Richards & Miller (1991) found high correlations between the size of wear facets and age amongst young aboriginal Australians. Molnar and colleagues (1983a) were able to classify most individuals correctly into both their age group and sex solely on the basis of their attrition pattern. Such studies suggest that attrition does make a good age-at-death indicator, but clinical studies (Johansson *et al.*, 1993) stress the multifactorial causes of abnormally heavy wear, with significant contributions from age, sex and other factors, so that the pattern and degree of wear become more variable with increasing age.

The most widely used attrition ageing system is that of Brothwell (1963b), even though the research on which it was based has not been published (Brothwell, 1989). It is a table (Figure 11.5) showing the range of dentine exposure expected in permanent molars for four different age groups, devised for use in prehistoric to early medieval British material but since applied throughout the world. The Miles (1962, 1963, 1978) method is instead based on the idea that the rate of wear can be calibrated against dental eruption. Miles undertook a study of the teeth in a group of Anglo-Saxon burials from the site of Breedon-on-the-Hill, England, which can still be seen at the Odontological Museum of the Royal College of Surgeons, in London. Miles constructed a diagram (Figure 11.6) comparing wear in first, second and third molars, and fitted an age scale from a baseline group in which the molars were still erupting, assuming that first molars erupted at 6 years, second at 12 and third at 18, with wear rate ratios between them of 6 : 6.5 : 7. The resulting scale of years since eruption, fitted to the chart, allowed an

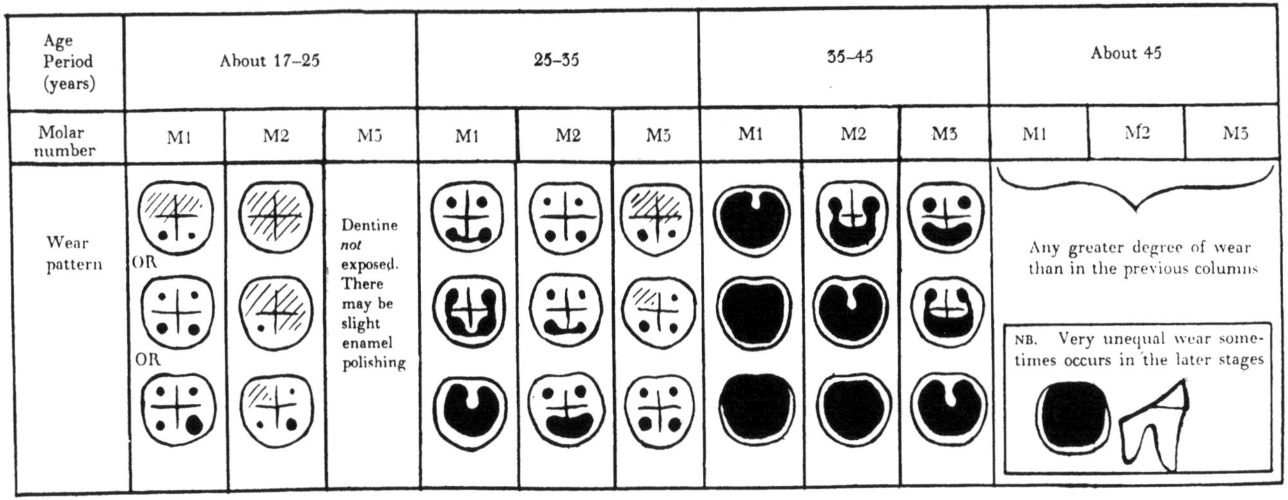

Figure 11.5 Brothwell's system for age estimation from attrition. Reprinted from Brothwell, D. R. (1981) *Digging up bones*, 3rd edn., London & Oxford: British Museum & Oxford University Press, with kind permission from Professor Don Brothwell, Archaeology Department, University of York, 88 Micklegate, York, Y01 1JZ.

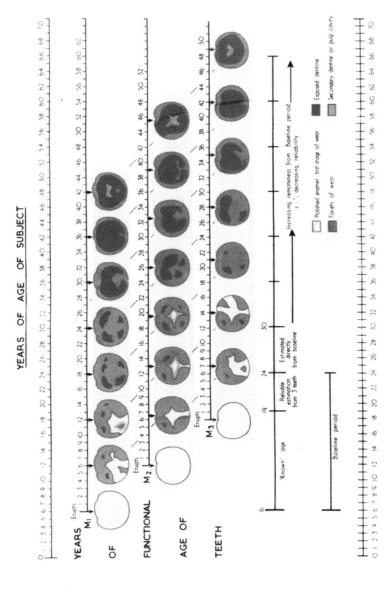

Figure 11.6 Miles' system for age assessment from attrition. Reprinted from Miles, A. E. W. (1962) Assessment of the ages of a population of Anglo-Saxons from their dentitions, *Proceedings of the Royal Society of Medicine*, **55**, 881–886, with kind permission from Professor Loma Miles and the Royal Society of Medicine 1 Wimpole Street, London W1M 8AE. The top and bottom scales are age in years from birth. The separate scales for the three molars represent the functional ages of the teeth, with the stages for first molars marked at 6-year intervals, for second molars at 6.5-year intervals and for third molars at 7-year intervals (see text).

extrapolated age scale to be estimated along the extended wear series. Tests against independent skeletal age estimation methods (Nowell, 1978; Lovejoy, 1985), and known age material (Kieser *et al.*, 1983; Lovejoy *et al.*, 1985) suggest that Miles' method performs at least as well as the alternatives for adult material (Table 9.1).

Approximal (interproximal or interstitial) attrition

Small movements between neighbouring teeth, during clenching of the jaws, produces wear facets around the contact points. Recording is difficult when teeth are still held in the jaw, and one method is to estimate the rate of tooth row reduction from mesiodistal diameters (Begg, 1954; Murphy, 1964; Wolpoff, 1970). Another possibility is to measure the breadth of the approximal wear facet at the occlusal surface (Hinton, 1982) and plot it against occlusal attrition stage, whilst Van Reenen (1992), Whittaker and colleagues (1987) also devised schemes to describe the shape of approximal facets. Extensive approximal wear is seen in all hominoids, living and extinct (Wolpoff, 1970), although living human populations in North America and Western Europe are much less affected. Wolpoff suggested that the main mechanism for approximal wear was mesial tilting of teeth under occlusal loads, and the pattern of worn grooves on approximal facets confirms that the movement is up-and-down (Kaidonis *et al.*, 1992). In a study of Native Americans, Hinton (1982) found heavier approximal wear in Archaic than in Woodland and Mississippian groups and such wear seems to be a particular feature of hunter–gatherer subsistence.

Factors affecting attrition

It is usually assumed that attrition takes place during chewing, when large forces may indeed be exerted (Wolpoff, 1970), but chewing occupies only a small fraction of each 24 hours, and the teeth are more forcibly brought together at other times. Bruxism, the grinding or tapping of the teeth, either when asleep or unconsciously whilst awake, generates much greater forces than those applied during chewing. Clinical studies have focused particularly on the aetiology of abnormally heavy wear (Johansson *et al.*, 1991; Johansson, 1992), and one question is the controlling effect of temporomandibular joint (TMJ) form. Some studies (Owen *et al.*, 1991) show a relationship between the size and shape of the mandibular condyles and extent of wear, whilst others (Whittaker *et al.*, 1985a) do not. Similarly, some studies show a relationship between degeneration of the TMJ and tooth wear (Richards & Brown, 1981b; Richards, 1990), whilst others show no such relationship (Seligman *et al.*, 1988; Sheridan *et al.*, 1991; Pullinger & Seligman, 1993).

Dental microwear

For summaries see Gordon (1988) and Teaford (1988a; 1991). Microwear studies include the reconstruction of jaw movements and chewing cycle in relation to occlusion, with their origin in the work of Gregory (1922) on cusp morphology amongst primates, but most recent microwear studies relate to the reconstruction of diet, a suggestion that may be traced to Dahlberg and Kinzey (1962) and Walker (1978).

Methods of study

Most work involves the use of epoxy replicas (page 299), examined in the SEM using the standard ET detector configuration (page 314). Some studies use magnifications of $\times 100$–$\times 200$, which allow a large area of wear facet to be sampled, whilst others use $\times 300$–$\times 500$, allowing measurements of smaller features to be taken. One practical approach is to take the original photographs at low magnification, use a contact print for counting, and a photographic enlargement for detailed measurement. Another factor is the orientation of the wear facet relative to the SEM's primary electron beam and detector. The standard configuration is for the facet to be tilted and the mesiodistal axis of the tooth to be aligned with the detector, so that distribution of apparent shadows in the image (page 314) emphasizes features that diverge from a mesial/distal orientation. Topographic BSE imaging would allow different directions of apparent illumination to be achieved without moving the specimen (page 315).

Such difficulties may be left behind if profilometers are used instead. Optical interferometry involves no contact with the surface to be measured, and some instruments have a vertical resolution of better than 1 nm, with a horizontal resolution of 0.1 µm (Walker & Hagen, 1994). Several profiles can be taken and then used to reconstruct orientations and sizes of features by the application of stereology, but profilometry may be limited in the maximum area and height differences that can be reported. Another possibility is to use a CLSM and image analysis system (page 311). One further approach is stereophotogrammetry, in which a three-dimensional model is produced from stereo-pairs of SEM photographs, allowing detailed measurements to be taken (Boyde, 1974, 1979).

Assessment and quantification of features

Initial examination involves a qualitative assessment of the distribution of wear in different facets and distribution of features, so that the smaller field of view for more detailed measurement can be selected. Categories for qualitative description of wear patterns have been defined (Rose & Harmon, 1986).

Most quantitative work is carried out on photographic prints of SEM images. Features may be traced onto an acetate overlay (Grine, 1987; Pastor, 1992) to keep track of the features counted and measurements made. Another approach, designed for rapid recording, is to overlay the photograph with a grid (Rose & Tucker, 1994) and to count features that intersect with grid lines, to obtain a measure of feature density. More commonly, the outlines of features are entered from the photograph as coordinates into a computer system, using a digitizing tablet. Another possibility is to use a scanner to digitize the image and then to enter coordinates using a cursor on the computer screen (Ungar *et al.*, 1990; Ungar, 1995). Image analysis systems would seem to offer a ready solution, but the complex pattern of wear features makes it difficult to isolate individual features (Teaford, 1991), although Fourier analysis (Kay, 1987; Grine & Kay, 1988) can be used to express mathematically the orientation and spacing of repeated wear features in the image.

Patterns of microwear

Wear facets

If comparisons are to be made, then it is necessary to standardize the wear facets being studied, including their position, role in the chewing cycle and degree of wear. The chewing cycle of primates occupies one side of the mouth at a time and is divided into three strokes (Hiiemae, 1978). The Closing Stroke closes the mouth from maximum gape to bring the tips of the cusps into contact with one another. This is followed by the two phases of the Power Stroke, which is when the crushing and grinding of food takes place. In Phase I, the cusps of the molars slide past one another in a shearing action, to end up in centric occlusion (Figure 11.7) and bring the lingual surfaces of the upper molar cusps into contact with the buccal surfaces of the lower molar cusps. Phase II is a movement to lingual of the centric position, grinding the lingual surfaces of the buccal lower molar cusps against the buccal surfaces of the lingual upper molar cusps. The final part of the chewing cycle is the Opening Stroke, when the jaw is opened again to maximum gape. Most studies contrast the wear facets in contact during Phase II of the Power Stroke, with those produced by Phase I of the chewing cycle. Each facet is numbered, according to a system (Figure 11.8) which originates in the work of Kay and colleagues (Kay & Hiiemae, 1974; Kay, 1977, 1978; Gordon, 1984a,b); Phase I facets on buccal cusps are labelled 1–4, and on lingual cusps 5–8, whereas Phase II facets on buccal cusps are labelled x, 9 and 10n. Gordon added wear

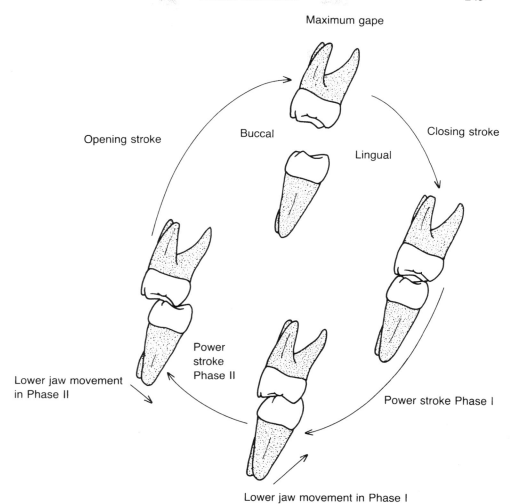

Maximum gape

Opening stroke   Buccal   Lingual   Closing stroke

Lower jaw movement
in Phase II

Power
stroke
Phase II

Power stroke Phase I

Lower jaw movement in Phase I

Figure 11.7 The chewing cycle in molars (see text for details).

facets at the cusp tips of lower molars (Figure 11.9) to the list. During the
chewing cycle, the cusp tip facets are involved in a crushing action, the Phase
I facets in a shearing action and the Phase II facets in a grinding action, so that
this classification should allow homologous facets to be compared between
individuals and species, although the details must be highly variable.

Pits and scratches

Wear features are divided into pits and scratches (Figure 11.10 and Table
11.1). Scratches are variously defined as linear features whose length:breadth
ratio exceeds 2:1 (Gordon, 1988), or 4:1 (Grine, 1986; Grine & Kay, 1988),

Dryopithecus (Sivapithecus)

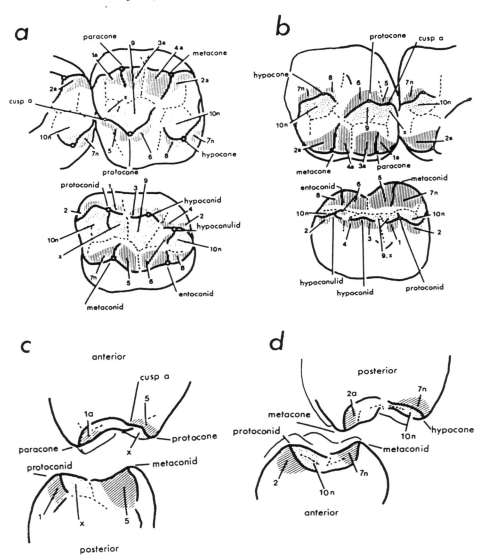

Figure 11.8 Labelling for attrition facets, using *Drypopithecus* as a model. Reprinted from Kay, R. F. (1977) The evolution of molar occlusion in the Cercopithecidae and early catarrhines, *American Journal of Physical Anthropology*, **46**, 327–352, Copyright © 1984, John Wiley & Sons Ltd. Reprinted by permission of John Wiley & Sons Ltd. *a* Occlusal views of upper left first molar and lower right second molar; *b* lingual view of upper left first molar and buccal view of lower right second molar. Hatching shows Phase I facets and stippling shows Phase II.

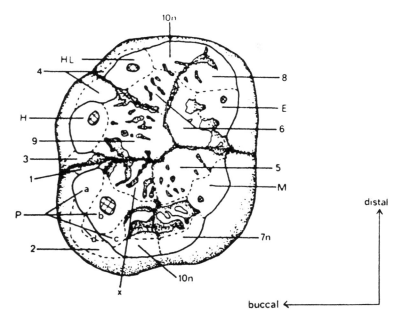

Figure 11.9 Labelling for attrition facets in lower right first molar (of chimpanzee). Reprinted from Gordon, K. D. (1984) Orientation of occlusal contacts in the chimpanzee, *Pan troglodytes verus*, deduced from scanning electron microscopic analysis of dental microwear patterns, *Archives of Oral Biology*, **29**, 783–787, copyright 1984, with kind permission from Elsevier Science Ltd, The Boulevard, Langford Lane, Kidlington OX5 1GB, UK. Facets labelled as in Figure 11.8, with the addition of cusp tip facets: P, protoconid; H, hypoconid; HL, hypoconulid; E, entoconid; M, metaconid.

or 10:1 (Teaford & Walker, 1984). There is clearly a continuum and any division is arbitrary. Microwear feature density is lowest on Phase I facets, higher on cusp tips and highest on Phase II facets (Table 11.2). Scratches are most abundant and longest on Phase I facets, and mostly orientated parallel, whilst they are less common, shorter and more randomly orientated in cusp tip facets, and least common and shortest on Phase II facets. Teeth in wear for only a short time show a greater feature density than teeth that have been in wear for longer.

Distinguishing wear features from artefacts

Some wear is generated in the ground, during excavation, cleaning, storage and handling of specimens, and it is clearly important to distinguish between the features produced by these processes and the wear produced during life. The main sign of post-mortem wear is a departure from the expected pattern

A   B

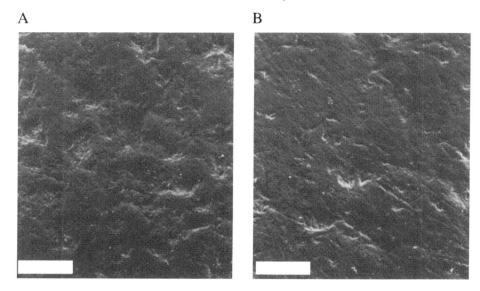

Figure 11.10 Dental microwear on facet 9 (Phase II) of permanent upper right second molars. A, an area with a predominance of pits. B, an area with a predominance of scratches. Buccal–lingual axis is vertical. Epoxy resin replica, examined in ET mode in the SEM. Scale bar 100 μm. Reproduced with the kind permission of Stan Bloor, Institute of Archaeology, University College London.

Table 11.1. *Definitions of microwear pits and scratches*

	Proportions Length:Breadth	Range for length (μm)	Range for breadth (μm)	Orientation
Pits	1:1 to 2:1	1–30	1–30	No
Scratches	> 2:1, > 4:1, or > 10:1	10–800	1–30	Yes

Source: Gordon (1984a, 1984b).

on different wear facets. Thus, Phase I facets without long, parallel scratches should be viewed with suspicion (Teaford, 1988*b*). Cleaning by dental scalers leaves highly characteristic patterns of scratches that can readily be distinguished, and it is usually possible to see the remnants of conservation treatments. Care needs to be taken in cleaning specimens for microwear analysis, using gentle methods and appropriate solvents (page 296).

Table 11.2. *Contrasts between different facets in chimpanzee lower molars*

	Scratches:pits (%)	Mean scratch length (mm)	Feature density (count per mm²)	Scratches aligned parallel (%)
Different facets – all molars pooled				
Facets 1 & 2 (Phase I)	80:19	192	400	96
Cusp tip facets	63:37	141	649	77
Facets 10*n* & *x* (Phase II)	55:44	108	825	55
Different molars – all facets pooled				
1st molar	70:29	164	602	71
2nd molar	64:35	140	621	68
3rd molar	54:45	114	783	81

Source: Gordon (1984a, 1984b).

Interpretations of dental microwear

The pattern of microwear changes rapidly during life (Teaford, 1991, see Figure 7). Microwear may only record the food eaten just before death, and this 'Last Supper' (Grine, 1986) phenomenon causes difficulties in reconstruction of diet when the food supply is highly seasonal. Features produced by hard foods may also predominate over less marked features produced by soft foods. Most work concentrates on comparison of non-hominid primates with widely differing diets, and Teaford and Walker (1984) suggested that orang-utan, and capuchin and mangabey monkeys, which include hard items in their diet, could be distinguished from feeders on softer foods (gorilla, and colobus and howler monkeys) by increased densities of pits and broader scratches. Chimpanzees had an intermediate pattern of wear and ate an intermediate diet whereas *Sivapithecus,* an extinct form from Pakistan, also had intermediate wear features, suggesting that it was not exclusively a hard object or soft object feeder. Further work suggested a more complex picture because microwear patterns varied between different species of colobus and capuchin, and between capuchins from different types of forest or sampled at different seasons.

Grine (1986; 1987) found that *Paranthropus* molars showed a significantly greater density of wear features than *Australopithecus*, a significantly higher proportion of pits relative to scratches on Phase I facets and a weaker parallel alignment of scratches. These results suggested that *Paranthropus* diets generated more crushing and grinding than *Australopithecus* diets. For recent *Homo sapiens,* Teaford (1991) quoted a comparison of pre-contact native populations along the Georgia and Florida coast of the USA with the people of post-contact Spanish mission communities. The pre-contact people had more pitting, slightly larger pits, and wider scratches than the post-contact material, perhaps reflecting the change from hunter–gatherer subsistence to greater dependence on maize agriculture. Rose *et al.* (1991) similarly found that a reduction in microwear evidence for abrasive diet, at around A.D. 1000 in the Central Mississippi valley, was paralleled by increased caries rate and stable isotope (page 229) evidence for maize consumption. At the site of Tell abu Hureyra in Northern Syria (Molleson *et al.*, 1993), Mesolithic and earlier Neolithic teeth had significantly larger pits than those of the later, pottery-using Neolithic, and Pastor (1992, 1994) interpreted qualitative differences between prehistoric groups from the Indus valley in relation to intensification of agriculture.

Mechanisms of wear

Little is known about the detailed mechanisms of tooth wear. It is normally assumed that particles harder than enamel are involved, but food itself rarely contains them, even though stone grinding of plant foods and airborne dust may introduce mineral grains (Leek, 1972). If parallels are drawn with engineering, then three types of wear mechanism can be distinguished:

1. Sliding wear. All surfaces have a pattern of microscopic high points known as asperities and, as one surface is pressed against another and moved sideways, the asperities catch against one another. In a material like enamel, the resulting forces are released by elastic deformation, which often ends in cracking along boundaries between crystals. If the sliding contact is repeated many times, increasing numbers of cracks are formed, and ultimately intersect to cause a slow loss of material.
2. Abrasive wear occurs when there are hard particles between surfaces. If the particles are harder than the surface and have sharp corners, enormous forces are concentrated that result in plastic deformation, even in a brittle material like enamel, to form sharp-edged scratches. Cracks may additionally be produced under the scratches, so that material fractures away when the load is released.
3. Erosion is wear resulting from the impact of hard particles, carried along in a fluid. The impacts create indentations, through plastic deformation or fracturing.

Under intense abrasive or erosive wear, the structure of enamel produces a pattern of surface relief, with crystallites more parallel to the worn surface being removed more rapidly than those at a greater angle. This may partly be because crystals deform and crack differently, depending upon the direction of the applied force, but the accumulation of cracks may also isolate a crystallite lying parallel to the surface more readily than one with its long axis buried in the enamel. Prism structure must therefore have an effect on microwear, but it is difficult to assess its impact on individual features (Maas, 1991, 1994). The appearance of most microwear pits suggests that they are due to the accumulation of cracks from repeated loading, so they may well not involve either abrasive or erosive wear. By contrast, the scratches are evidence of plastic deformation, cutting across enamel structure, and are most likely due to abrasion by hard particles. Lubrication and chemical effects may also be important in microwear.

Other modifications of teeth

A thorough review is given by Milner and Larsen (1991).

Deliberate mutilation of teeth

Today, the most common form of modification relates to dental work involving drilling, filling with amalgam or resin, fitting of crowns and inlays. This is really beyond the scope of this book, but details may be found in clinical and forensic dentistry texts (Cottone & Standish, 1981). Dental work is the best indicator of identity in forensic studies of fragmentary remains. Dental mutilation has, however, been practised for thousands of years, in the Americas, parts of Eastern Asia and Oceania, and Africa. The classic study of the subject is that of Romero (1958; 1970), on pre-hispanic material from the Americas, which documents an array of cuts, file marks and drill holes, largely to upper anterior teeth, and some including stone or gold inlays. Romero's chart is frequently used as the basis for classifying mutilations (Figure 11.11).

Approximal (interproximal) grooving

Grooves on mesial or distal surfaces of teeth are a common find in archaeological material, especially in molars (Ubelaker *et al.*, 1969; Schulz, 1977; Berryman *et al.*, 1979; Frayer & Russell, 1987; Eckhardt & Piermarini, 1988; Brown & Molnar, 1990; Formicola, 1991; Frayer, 1991). They normally form a broad depression running from buccal to lingual, along the cervix, with fine, parallel scratches running along their axis. Such grooves may be caused by the use of tooth picks or similar objects to clean between teeth. It is also possible that they result from the stripping of animal sinews between the clenched cheek teeth, as approximal grooving has been recorded for

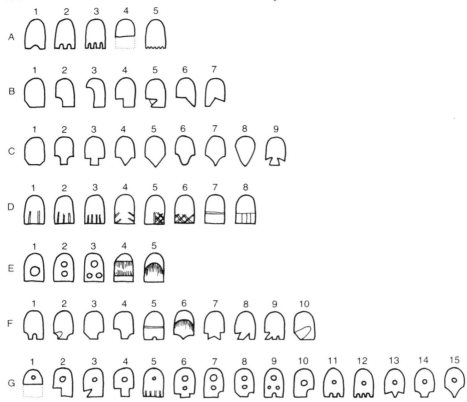

Figure 11.11 Classification of mutilations to upper incisor teeth. After Romero (1958, 1970).

Aboriginal Australians (Brown & Molnar, 1990) where sinew stripping is well documented.

Anomalous wear of the anterior teeth

The anterior teeth may be used in various ways that cause anomalous patterns of wear. One classic form is the circular notch that results from gripping clay pipe stems (Figure 11.12), and some Native North Americans wore labrets – plugs of ivory, bone or stone fitted into slots in the lips (Dumond, 1977), which abraded the labial or buccal crown surfaces. Early examples of labret facets come from the North West coast, dating back to 2000 BC or even earlier, although the labrets themselves are rarely found with these burials, suggesting that they were passed on to others after death (Cybulski, 1994). At some sites, both males and females bear labret facets but, at others, only one of the sexes is involved – male or female. Lingual wear in upper anterior teeth has also been noted in archaeological material (Irish & Turner, 1987;

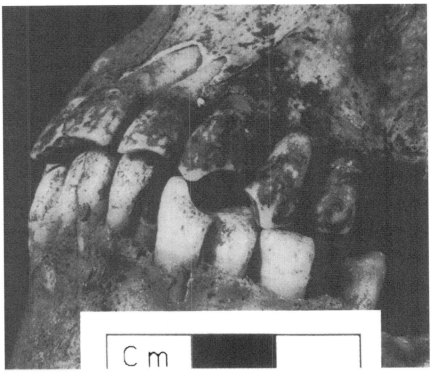

Figure 11.12 Clay pipe wear facet in a post-medieval skull from London. A circular aperture has been worn in the upper and lower left canines and first premolars.

Lukacs & Pastor, 1988), perhaps related to diet, or the use of teeth in food processing or crafts. Larsen (1985) described a pattern of coarse grooves crossing the occlusal wear facets of heavily worn anterior teeth, perhaps from holding sinews or plant fibres clamped in the teeth during processing.

12

Dental disease

Dental plaque

The most common diseases to affect the teeth once they have erupted are related to dental plaque, a dense accumulation of micro-organisms on the tooth surface. These diseases have been responsible for the majority of tooth loss in human populations for many thousands of years. Reviews of plaque biology include those of Lehner (1992), Marsh and Martin (1992).

The tooth surface as a habitat for micro-organisms

The lips, palate, cheek, tongue and gums are colonized by bacteria, fungi, yeasts, viruses, protozoa and other microbes, but their ability to adhere is limited by constant shedding of the mucosa (page 260) surface. Teeth are unique in their hard, non-shedding surface, which allows luxuriant microbial communities to build up especially in fissures, approximal areas and gingival crevices, protected from the rinsing and sweeping of saliva, lips, cheek and tongue. Saliva and gingival crevice fluid (page 260) coat the crown surfaces with an organic layer known as pellicle, and the predominant plaque organisms are bacteria with specialist mechanisms for adhering to it. Plaque bacteria also obtain their nutrients mainly from saliva and gingival crevice fluid (GCF), including proteins, peptides, amino acids and glycoproteins (page 227), which can all be broken down by the combined forces of different bacteria. The human diet passing through the mouth is a less important source of nutrients, but plaque bacteria do metabolize the fermentable carbohydrates (starches and sugars) and casein, a protein that is found in milk and dairy products.

The microbial communities of the plaque

Oral bacteria are divided by the way in which they take a stain – Gram-positive and Gram-negative – and by their capacity to exist without oxygen – aerobic and anaerobic. Further divisions are made in terms of shape, including cocci (little spheres), rods and filaments. Gram-positive cocci dominate normal plaque, especially the streptococci, which are split into the *Streptococcus mutans*, *S. salivarius*, *S. milleri* and S. *oralis* groups. Gram-positive filaments and rods include *Actinomyces*, another large component of the plaque, and *Lactobacillus*. Gram-negative bacteria similarly include cocci such as *Neisseria* and *Veillonella*, and rods and filaments such as *Haemophilus*, *Prevotella*, *Porphyromonas* and *Fusobacterium*.

Adhesion between bacteria and to the acquired pellicle is brought about by factors on the bacterial cell walls, together with adhesives such as the extracellular polysaccharides that are produced from dietary sugar by various species of *Streptococcus*, *Actinomyces*, *Lactobacillus* and *Neisseria*. A plaque deposit is therefore strongly structured, with a definite surface membrane, into which nutrients diffuse selectively. It has its own intercellular fluid whose chemistry changes through the deposit, with less oxygen, lower pH, lower nutrient levels and higher concentrations of the waste products of metabolism in its deeper layers. Sugars diffuse into the plaque fluid, either directly from the diet or as they are released from the breakdown of starches and glycoproteins by bacterial and salivary enzymes. All these sugars cross the bacterial cell walls and are metabolized to produce organic acids, which are excreted into the plaque fluid. A variety of proteins, peptides and amino acids also becomes incorporated into the plaque, and bacteria together have a battery of enzymes that metabolize them to produce alkaline waste products that have a balancing role in the plaque fluid pH.

The mature plaque flora varies between different sites on the teeth. Plaque of cheek teeth fissures has a restricted microbial community, which is dominated by the streptococci and suggests a limited supply of nutrients, deriving mainly from saliva. Plaque at approximal sites is dominated by Gram-positive rods and filaments, although streptococci are present together with Gram-negative forms, and many are anaerobes, which suggests lower oxygen levels than at fissure sites. Plaque of the gingival crevice (page 260) has a notably diverse flora, dominated by obligate anaerobes, including anaerobic streptococci, *Actinomyces,* Gram negative rods, and spirochaetes.

Dental calculus

Calculus is mineralized plaque, which accumulates at the base of a living plaque deposit, and is attached to the surface of the tooth. The mineral is deposited from plaque fluid, but ultimately derives from the saliva (Driessens & Verbeeck, 1989) and the sites closest to the ducts of the salivary glands – lingual surfaces of anterior teeth and buccal surfaces of molars –

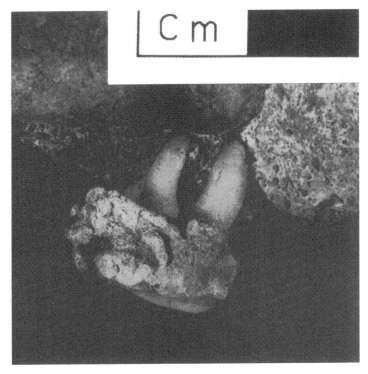

Figure 12.1 Large deposit of supra-gingival calculus in a post-medieval upper
first molar from London.

show the most abundant calculus formation. It is still unclear how plaque
mineralization is initiated, although bacteria probably have an important role
(Scheie, 1989), even though mineralization actually starts first in the extracellular matrix (Friskopp, 1983).

Types of calculus

Two forms of calculus are recognized; supra-gingival and sub-gingival. Supra-
gingival calculus is principally attached to the enamel of the cervical crown
as a band marking the position of the gingival margin (page 260), whereas
more established deposits extend higher up the crown or may even develop
into an elaborate, overhanging outgrowth (Figure 12.1), sometimes on just a
single crown but occasionally throughout the dentition (Hanihara *et al.*, 1994).
The surface is rough, and cream to light brown in colour, although it is sometimes darkly stained. In life, supra-gingival calculus is firmly attached to the
crown but, in archaeological material, the attachment is loosened and deposits
are easily lost. Sub-gingival calculus is deposited on the root surface as the

level of gingival attachment recedes (page 262). It is thinner and harder than supra-gingival calculus, with a similar colouration that is not always easy to distinguish from the normal cement surface in archaeological specimens. Sometimes a more darkly stained line marks the apical boundary of the deposit.

Composition of dental calculus

Calculus varies widely in mineralization (Driessens & Verbeeck, 1989), but sub-gingival calculus is more heavily mineralized (46–83% by volume) than supra-gingival (16–80% by volume). The minerals include apatite, whitlockite, octacalcium phosphate, and brushite (page 219), all of which have been found in archaeological specimens (Swärdstedt, 1966). Brushite is prominent during the earlier part of calculus deposition, whilst mature supra-gingival calculus has more apatite and brushite, and sub-gingival calculus has abundant whit-lockite. The pattern of fluorine composition in both forms is similar to enamel (page 220), with highest concentrations towards the outermost surface of the deposit. The organic component of calculus varies, and includes amino acids, peptides, proteins, glycoproteins, and carbohydrates (Embery, 1989).

The microscopic structure of calculus

Calculus is studied by SEM examination of fractured surfaces, compositional BSE or confocal microscopy of sections (pages 311 and 315). It is more heavily mineralized than dentine or cement, but less so than enamel, and presents an irregular appearance with layerings, voids and clefts. Outlines of bacteria (Figure 12.2) are represented as voids with mineralized shells – the filamentous forms as 2 μm diameter tubules, and shorter rods or cocci as globular out-lines – and similar outlines have been demonstrated in calculus from English medieval human remains (Dobney & Brothwell, 1986). The surface of some deposits shows micro-organisms being engulfed by mineralization, whereas other areas are smooth and burnished. Two patterns of mineralization within calculus have been described (Jones & Boyde, 1972; Lustmann *et al.*, 1976; Jones, 1987):

1. Growth of irregular, spherical, very finely mineralised masses, with a sponge-like form in which the spaces represent bacterial outlines.
2. Precipitation of coarser crystals around dense patches of well-ordered rods and filaments.

There are consistent structural differences between sub-gingival and supra-gingival calculus (Bercy & Frank, 1980; Friskopp & Hammarstrom, 1980; Friskopp, 1983). Supra-gingival deposits have less mineralized matrix or

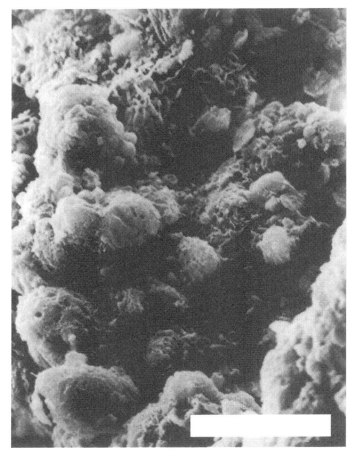

Figure 12.2 The mineralized outlines of micro-organisms in calculus from the same specimen as Figure 12.1, representing coccal forms. Fractured preparation, examined in ET mode in the SEM, operated by Sandra Bond. Scale bar 10 μm.

coarsely crystalline material, more unmineralized patches, and more mineralized bacterial outlines, particularly tunnel-like filamentous forms that run perpendicular to the deposit surface. Sub-gingival calculus has fewer (and more variable) bacterial outlines, and more even mineralization. Food debris is not an abundant component of either type of calculus, even though British prehistoric specimens have yielded pollen grains, phytoliths, fragments of chaff and animal hairs (Dobney & Brothwell, 1986).

Initial calculus formation outlines pkg and Tomes' process pits (page 163) on the enamel surface, and fills resorption hollows or the ends of extrinsic fibres (page 205) at the cement surface. There is some continuity of mineraliz-

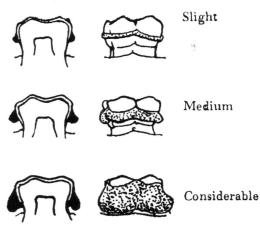

Slight

Medium

Considerable

Figure 12.3 Scoring system for supra-gingival calculus. Reprinted from Brothwell, D. R. (1981) *Digging up bones,* 3rd edn., London & Oxford: British Museum & Oxford University Press, with kind permission from Professor Don Brothwell, Archaeology Department, University of York, 88 Micklegate, York, Y01 1JZ.

ation with both enamel and cement, so that calculus deposits are very intimately attached (Canis *et al.*, 1979; Jones, 1987; Hayashi, 1993).

Methods for recording calculus

Most clinical scoring methods are inappropriate for anthropology and the normal procedure is to use a simple three-point scoring system (Figure 12.3) for supra-gingival calculus in each tooth (Brothwell, 1963b), although it can be scored separately for different surfaces (Manji *et al.*, 1989), or an even finer division of zones (Dobney & Brothwell, 1986). Sub-gingival calculus deposits have also been measured (Powell & Garnaick, 1978) and supra-gingival deposits weighed (Dobney & Brothwell, 1986). Archaeological supra-gingival calculus is easily detached but sufficient vestiges usually remain to allow reconstruction of the approximate original extent.

Factors affecting the distribution of calculus

Plaque deposits are required for calculus to be formed, but thick plaque deposits do not necessarily lead to large calculus deposits. Initiation of mineralization is, however, linked to the extent of plaque, and thus also to those factors that lead to increased plaque accumulation, like poor oral hygiene or carbohydrate consumption. There is little evidence that malocclusions, which create protected niches for plaque deposition, predispose to calculus formation (Buckley, 1981). In living populations, men show more and heavier supra-gingival calculus deposits than women, and the frequency and extent of deposits increase with age (Beiswanger *et al.*, 1989). Extent of sub-gingival calculus has a clear connection with periodontal disease (page 262) and follows the development of sub-gingival plaque, spreading down the exposed root surface with the deepening periodontal pocket (Powell & Garnick, 1978).

Dental caries involves progressive local demineralization of tooth surfaces, whereas calculus involves mineralization and the two conditions should therefore be mutually exclusive (Manji *et al.*, 1989). But both conditions are frequently seen on the same tooth, where arrested carious lesions (page 269) in crown fissures are often covered with calculus (Thylstrup *et al.*, 1989), and active caries may be seen in cement or dentine underneath calculus deposits (Jones, 1987). At a population level there may however be a slight inverse relationship between caries and calculus frequencies.

Periodontal disease

The periodontal tissues that surround and support the tooth include the bone of the jaws, the periodontal ligament, cement, gingivae and mucosa (Figure 12.4). In both jaws, the arch of bone which holds the dentition is called the alveolar process, because it incorporates the *alveolae* (Latin; little holes) or tooth sockets, and the periodontal ligament surrounds each root, holding it into its socket with fibres embedded into both bone and cement (page 198). The alveolar process is covered by a layer of soft tissue, the mucosa, which is gathered up into a cuff around the base of each tooth crown. The mucosa that forms the cuff itself is called the gingivae, and is divided by the gingival groove into the attached gingivae and the free gingivae. It folds over at the gingival margin and is attached to the base of the crown to create a crevice known as the gingival sulcus. The teeth and mouth are bathed with saliva, which is secreted by glands with major ducts emerging under the tongue and inside the cheeks, whilst the gingival sulcus produces its own Gingival Crevice Fluid (GCF). Texts on periodontal tissues and their diseases include those by MacPhee and Cowley (1975) and Schluger *et al.* (1990).

Plaque and the immune system

The body's defences against plaque micro-organisms can be divided into innate and acquired immunity. Innate immunity involves factors that are present at all times, such as the barrier of the oral mucosa, bactericides in saliva and GCF, and phagocytes (cells able to engulf particles and bacteria) including neutrophils, which circulate in the blood, and macrophages distributed through the connective tissues. Acquired immunity, by contrast, is a response by cells

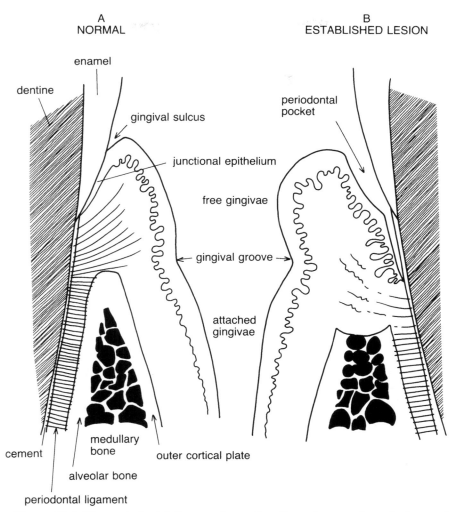

A
NORMAL

B
ESTABLISHED LESION

dentine

enamel

gingival sulcus

periodontal
pocket

junctional epithelium

free gingivae

gingival groove

attached
gingivae

medullary
bone

cement

outer cortical plate

alveolar bone

periodontal ligament

Figure 12.4 The periodontal tissues. A, section through normal tooth, jaw and gingivae. B, section through the same tissues with an established lesion of periodontal disease.

called lymphocytes to a specific antigen. Bacteria produce many different antigens, which sensitize lymphocytes on their first exposure to them so that, when exposed again to the same antigen, they respond in two ways. B-lymphocytes transform into plasma cells and produce free antibodies that bind to specific antigens to neutralize them. T-lymphocytes react either by transforming into killer cells, which bind to foreign cells and destroy them, or by releasing factors that change the behaviour of other cells such as phagocytes. Summaries of oral immunology may be found in the textbooks of Lehner (1992), Marsh and Martin (1992).

Aetiology of periodontal disease

Some bacteria are particularly associated with periodontal disease – including Gram-negative forms and a few Gram-positive streptococci and filaments, although none is solely responsible for the condition. Bacteria in plaque produce numerous factors that activate innate immunity, act as antigens for acquired immunity, or that interfere with the immune response once it is activated. These include enzymes that affect structural proteins, and factors that deactivate antibodies or attract macrophages. Much of the damage in periodontal disease is actually caused by inappropriate triggering of the immune response by such factors.

'Inflammation' describes the succession of changes that occur in tissues in response to injury or irritation. The site at which the changes take place is known as the lesion and, in periodontal inflammation, its development is divided into four stages: initial lesion, early lesion, established lesion and advanced lesion. The first three stages are classed as gingivitis, implying an inflammation that involves the gingivae alone. The last stage is classed as periodontitis, and is a deeper lesion that involves all the periodontal tissues. T-lymphocytes, a few B-lymphocytes and/or plasma cells are normally present in the gingivae, as a response to the presence of plaque antigens, and the initial lesion represents only a slightly elevated level of response, when tiny blood vessels in the gingivae become dilated, allowing fluid to exude out and swell the gingivae. Macrophages accompany the exudate and, following a chemotactic gradient of plaque factors, insinuate themselves through the junctional epithelium (Figure 12.4) to tackle the bacteria in the sulcus. The initial lesion evolves gradually into the early lesion over a week or so. Exudation and macrophage migration rise further, until the junctional epithelium is disrupted and proliferates into the connective tissue underneath, where T-lymphocytes accumulate and collagen fibres become disrupted. After a few weeks, the early lesion develops into the established lesion, where B-lymphocytes or plasma cells predominate and high antibody levels are found in the connective tissue and junctional epithelium. There is a continued disruption of collagen fibres as the area occupied by the lesion expands, often accompanied by a loss of attachment at the periodontal ligament. This causes a periodontal pocket to develop at the side of the tooth, exposing cement on the root surface and allowing sub-gingival plaque to accumulate. Such established lesions often stabilize, and remain stable for many months or even reverse themselves, but episodically they escalate into an advanced lesion, with further disruption of the collagen fibres in the gingivae and periodontal ligament, extension of the periodontal pocket, and progressive resorption of alveolar bone. The lesion does not progress continuously, and resting periods alternate with more active phases over many years.

The pattern just described is known as adult type periodontitis and is very common in adults over 30 years of age. Other forms are uncommon, such as prepubertal periodontitis, which occurs on eruption of the deciduous dentition, and juvenile periodontitis, which occurs around puberty in permanent first molars and incisors. Rapidly progressive periodontitis occurs in late childhood or early adulthood, destroying bone around most teeth. These distinctive and rare forms are unlikely finds in archaeological material.

Bone loss associated with periodontal disease

For reviews see publications by Hildebolt and Molnar (1991), Clarke and Hirsch (1991). Bone loss involves four elements of the jaws (Figure 12.4); the alveolar bone lining the tooth sockets, the outer cortical plates on buccal and lingual sides, and the underlying medullary bone. Neighbouring tooth sockets are separated by an approximal wall composed of the two sheets of alveolar bone, sandwiching a thin layer of underlying medullary bone, and the buccal and lingual walls have a similar sandwich construction provided by the alveolar bone and the outer cortical plates. Two patterns of bone loss are recognized: horizontal and vertical. They are defined largely from radiographic practice (Whaites, 1992; Goaz & White, 1994) and, in dry bone specimens, some difficulties of interpretation arise.

'Horizontal bone loss' describes simultaneous loss in height of all walls surrounding the tooth roots – approximal, buccal and lingual – and also implies that several neighbouring teeth are involved, or even the whole dental arcade (Figure 12.5). It is difficult to record but, at its simplest, the extent of root exposure can be scored (Alexandersen, 1967; Costa, 1982) or measured. In radiographs, the measurement is made between the CEJ and the crest of the approximal walls (CEJ–AC height), whereas Davies *et al.* (1969) devised a Tooth Cervical Height index based on direct caliper measurements of archaeological jaws, and Watson (1986) made similar observations using a graduated probe. Lavelle and Moore (1969) instead used the apex of the root, seen in radiographs, as their reference point. The difficulty with all of these approaches is that the teeth continue to erupt, slowly but steadily, throughout adult life so that neither the CEJ nor the root apex is a stable point from which to take measurements (Whittaker *et al.*, 1982, 1985b; Danenberg *et al.*, 1991). The inferior alveolar canal and lower border of the mandible (Figure 12.7) have been used as alternative reference points in radiographs (Levers & Darling, 1983; Whittaker *et al.*, 1985b; Whittaker *et al.*, 1990), and the AC-canal distance remains constant through a group of variously aged specimens

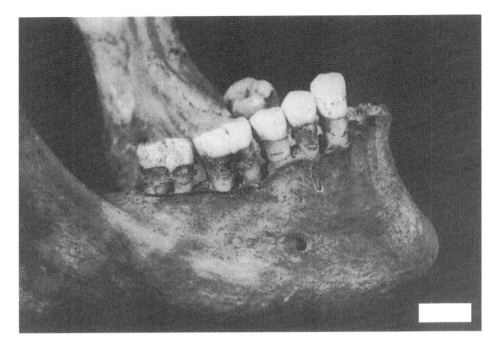

Figure 12.5 Horizontal bone loss in a post-medieval lower jaw from London.
There has been preferential removal of alveolar bone, to leave relatively sharp
edges, but bone loss has been even around the dental arcade. The lower right
second molar has a trench type vertical bone loss defect as well (Table 12.1).
The exposed roots of all teeth have remnants of sub-gingival calculus still
remaining, and the first and second molars have root surface caries lesions. The
second molar has an enamel extension. Scale bar 1 cm.

whilst the CEJ-canal distance increases, suggesting that continuous eruption
took place without alveolar bone loss (Whittaker *et al.*, 1990). But even the
inferior alveolar canal and inferior border vary and do not really provide homologous points for measurements (page 68). Furthermore, the radiological literature is in no doubt that horizontal bone loss does occur in association with
periodontitis in living patients, so that the main point established by continuous
eruption studies is that any attempt to record horizontal bone loss must take
into account the ages of the individuals studied. Any measurements should
therefore be calibrated with reference to such age indicators as attrition itself,
or pubic symphysis changes (page 207).

‘Vertical bone loss’ is localized around individual teeth, or pairs of neighbouring teeth, to form an ‘intrabony defect’ surrounded by high walls of unaffected bone (Figure 12.6). It occurs with or without horizontal bone loss, and
is clearly distinct from periapical changes (page 284). The approximal wall

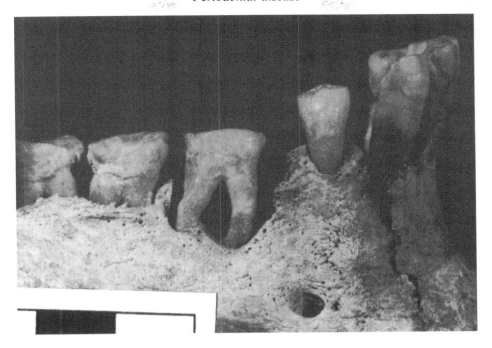

Figure 12.6 Vertical bone loss in a post-medieval lower jaw from London. The lower right first molar has a pronounced ramp type defect (Table 12.1), although there is irregular bone loss around all teeth. None of the teeth in this jaw bears any sign of dental caries or of periapical inflammation related to heavy attrition. Marks on scale bar at 1 cm.

may be lost without affecting the buccal and lingual walls, or the alveolar bone layer alone may be lost, creating a narrow deep pocket. Karn and colleagues (1984) produced a comprehensive classification of such defects (Table 12.1) and other authors (Saari *et al.*, 1968; Larato, 1970; Costa, 1982) have used similar measures. Muller and Perizonius (1980) devised a system, complete with recording form, to record 'interalveolar defects', CEJ–AC distance, root furcation involvement, fenestration (an opening in the buccal or lingual plate), or dehiscence (a deep dip in the alveolar crest). Another change in early periodontitis is the thinning of cortical bone at the crest of the approximal wall. It develops an irregular border with more rounded angles, and a pitted and porotic surface (Costa, 1982; Clarke *et al.*, 1986), for which Kerr (1991; Fyfe *et al.*, 1993) developed a scoring scheme (Table 12.2).

Most specimens show a clear pattern of bone loss that tends to be symmetrical, and affects particularly first and second molars, and to some extent incisors, but less often involves premolars and canines. Teeth are eventually lost as well when all their surrounding alveolar bone has gone. The overall pattern

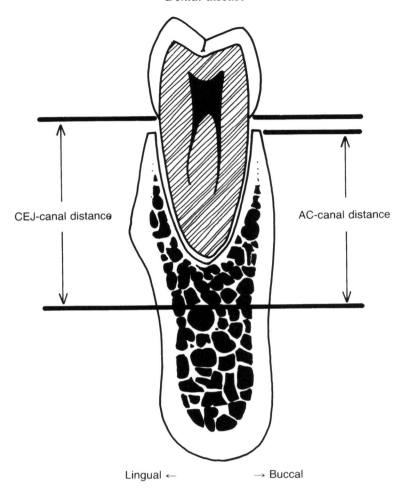

CEJ-canal distance

AC-canal distance

Lingual ←   → Buccal

Figure 12.7 Measurements for detecting horizontal alveolar bone resorption and continuous eruption (explanation in text). Based on a section of the mandibular body at the lower first molar. The superior margin of the mandibular canal is the lower reference point.

of bone loss is the sum total of active periodontitis episodes. Very irregular vertical defects presumably represent recent episodes, because the jaw remodels once the periodontal ligament attachment is lost, reducing the irregularity of defects and erasing all sign of sockets once the teeth have been lost.

The epidemiology of periodontal disease

In large clinical studies, there is a clear relationship between periodontitis and age. Children are rarely affected before puberty. After this point, there is a

Table 12.1. *Classification of non-uniform defects of the alveolar process*

Loss of alveolar bone lining the tooth socket
1. Crater. Loss from only one side of a tooth (classified m, d, b or l).
   Subdivided into one-surfaced crater (just involving one tooth) and
   two-surfaced (involving the interproximal wall between neighbouring
   teeth).
2. Trench. Loss from two or three sides (classified mb, ml, db, bl, mbd, mld,
   bml, bdl).
3. Moat. Loss from all sides of a tooth.

Loss of alveolar bone and outer cortical plates
1. Ramp. A deformity with its margins at different levels (labelled by the
   aspect which is most apical; m, d, b, l, md, db, ml, dl).
2. Plane. A deformity with margins all at roughly the same level.

Combinations
1. Cratered ramp. A crater with part of its buccal or lingual wall missing.
2. Ramp into crater. Other combinations of ramp and crater.

All defects labelled by the teeth involved.
m, mesial; d, distal; b, buccal; l, lingual.
Source: Karn *et al.* (1984).

gradual increase in the proportion of the population affected up to 40 or 50 years of age, paralleled by an increase in severity. The same pattern is seen in the extent of bone loss from the jaws, in dry bone specimens representing living populations (Larato, 1970; Tal, 1985) and in archaeological material (Swärdstedt, 1966; Goldberg *et al.*, 1976; Costa, 1982; Watson, 1986; Kerr, 1990). Most clinical studies find that periodontitis is more common amongst men than women but, in terms of bone loss, there is little consistent difference between males and females (Swärdstedt, 1966; Costa, 1982).

There is evidence from twin studies (Corey *et al.*, 1993) and family studies (van der Velden *et al.*, 1993) for an inherited element in the prevalence of periodontal disease. There are also differences between populations and groups within populations (Fox, 1992; Loe *et al.*, 1992; Harris *et al.*, 1993) but these are usually interpreted in terms of oral hygiene practices and availability of dental treatment, and must also involve the plaque-encouraging nature of the diet. Leigh (1925) showed differences between groups of native Americans with varying patterns of subsistence, and found that bone loss was most marked in those with greatest reliance on maize agriculture. Fyfe *et al.* (1994) found high levels of alveolar bone loss amongst sixteenth–seventeenth centry AD Solomon Islanders, which they interpreted in relation to carbohydrate

Table 12.2. *Classification of approximal wall defects*

Category	Definition of defect at coronal margin of wall	Implication
0	Unrecordable. Neighbouring teeth lost ante-mortem or wall damaged post-mortem	
1	Wall contour convex in incisor region, grading to flat in molar region. Cortical surface smooth and virtually uninterrupted by foramina or grooves	Healthy
2	Wall contour characteristic of region (above). Cortical surface shows many foramina and/or prominent grooves or ridges. Occasionally gross disruption of cortical layer, still with normal contour	Inflammation in overlying soft tissue and a clinical diagnosis of gingivitis
3	Breakdown of contour, with a broad hollow, or smaller discrete areas of destruction. The bone defect characteristically has a sharp and ragged texture	Acute burst of periodontitis
4	Similar breakdown of contour to 3, but defect surfaces are rounded, with a porous or smooth honeycomb effect	Previously acute periodontitis that has reverted to a quiescent phase
5	A deep intra-bony defect, with sides sloping at > 45°, and depth > 3 mm. Surface sharp and ragged or smooth and honeycombed	More aggressive periodontitis in either acute or quiescent phase

Adapted from Kerr (1991).

consumption. Lavelle and Moore (1969) found a significant difference in bone loss between seventeenth century and Anglo-Saxon English mandibles, but Kerr (1990) by contrast found little difference between a Medieval group and living people. Several cases have been reported amongst fossil hominid material, although it is not always possible to distinguish between periodontal bone loss, periapical changes, and diagenetic changes (Alexandersen, 1967). One of the most striking examples of bone loss, which is probably due to periodontal disease, is the Neanderthal skull from La Chapelle-aux-Saints, where the cheek teeth were lost ante-mortem. Overall, the epidemiology of periodontal disease is problematic, because it is multi-factorial in origin and involves inheritance, environment, diet and hygiene together.

Dental caries

Dental caries is a destruction of enamel, dentine and cement resulting from acid production by bacteria in dental plaque (page 255), ultimately leading to the formation of a cavity in the crown or root surface. Summaries can be found in publications by Silverstone *et al.* (1981), Menaker (1980), Pindborg (1970) and Soames and Southam (1993). Usually, caries progresses slowly (chronic caries) and arrested or remineralizing phases alternate with more active phases, so that a cavity may remain stable for months or years (arrested caries). Rapidly progressive destruction (rampant caries) is rare and characteristically results in the loss of most erupted tooth crowns in a child's mouth.

Development of carious lesions

Enamel caries

The first sign of caries in an enamel surface is the appearance of a microscopic white or brown opaque spot, whilst the surface itself remains smooth and glossy. Spot lesions are clearest in the enamel of dry extracted or archaeological teeth, because their increased pore space is filled with air to provide a large contrast in RI (refractive index) with the sound enamel (page 311), and for similar reasons they are accentuated under ultraviolet illumination (Shrestha, 1980). The dark pigmentation of a brown spot comes from bacterial or food stains and is taken to be a sign that the lesion is arrested. As the lesion develops, a white spot grows until it can be seen clearly with the naked eye and, finally, the smooth surface starts to break down, first becoming rough, and then developing into a cavity. Caries is detectable as a radiolucency in dental x-rays, but this is not a sensitive technique, and spot lesions are seen at the surface before they are recognizable radiographically.

The crown surface changes are preceded by changes seen only by microscopy (Darling, 1959, 1963; Silverstone *et al.*, 1981). In transmitted light microsopy of ground sections, mounted in Canada balsam or quinoline (page 312), the early spot lesion occupies a triangular area, with its base at the crown surface and apex facing the EDJ, and includes four characteristic zones (Figure 12.8):

1. 'Intact' surface layer, 10–30 μm thick.
2. Body of the lesion – the central part.
3. Dark zone – surrounding the body, and well developed in arrested caries.

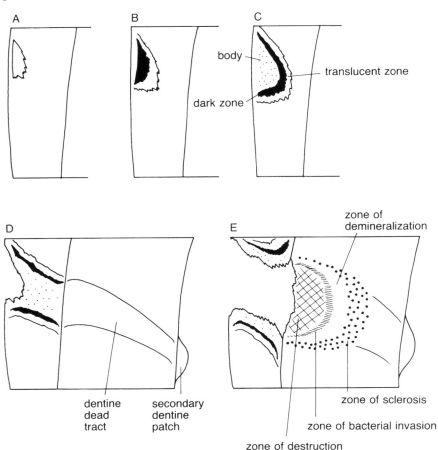

Figure 12.8 The progression of a carious lesion as seen in a crown section. A, formation of a translucent zone in the enamel; B, appearance of a dark zone; C, appearance of the body of the lesion; D, cavitation and development of a dead tract in the dentine; E, dentine caries with a zone of destruction (cross hatched), zone of bacterial invasion (fine lines), zone of demineralization (clear), zone of sclerosis (dotted), dead tract in normal dentine, and patch of reparative secondary dentine.

4. Translucent zone – on the deep side of the dark zone (not always visible).

These zones result from demineralization, which enlarges pore spaces at prism boundaries, cross striations and brown striae (page 157). Pore spaces occupy about 0.1% of intact enamel volume, whereas the body of the lesion has 5–25% pore spaces, which imbibe mounting medium to reduce RI contrasts and render the body more translucent (pages 311–312). The dark zone similarly has 2–4% pore spaces, but some of these are too small to admit the large molecules of mounting medium, and trap air to produce large RI contrasts that scatter light as it passes through. In microradiographs, a marked radiolucency corresponds to the body of the lesion, with prism boundaries and incremental structures strongly outlined, and compositional BSE similarly shows a low-density zone (Jones & Boyde, 1987). SEM studies of the apparently intact surface zone show that it is penetrated by tiny channels, following prism boundaries (Haikel *et al.*, 1983).

In summary, an enamel lesion (Figure 12.8) goes through six stages of development:

1. Appearance of a small translucent zone just below the crown surface.
2. Enlarged translucent zone, with a dark zone at its centre.
3. Further enlargement of lesion, and development of the body within the dark zone.
4. The lesion becomes large enough to be detected as a white or brown spot from the surface, but the surface layer is still intact.
5. The surface starts to break down, becomes chalky and can be scratched by a probe.
6. A cavity forms, making a clear indentation in the crown surface.

Cement caries

Carious lesions are initiated in the thin cement coating of roots exposed by periodontal disease, and rapidly penetrate to the dentine (Nyvad & Fejerkov, 1982; Schüpbach *et al.*, 1989). The initial lesion has a hypermineralized surface, but the demineralized zone beneath exploits the layered cement structure (page 204), so that the cement breaks up along its layers and the lesion spreads sideways, into broad, shallow craters around the circumference of the root (sometimes encircling the exposed root).

Dentine caries

Once enamel or cement lesions reach the EDJ or CDJ, they spread into the dentine. Under the microscope, active dentine lesions have six zones, from the edge of the pulp chamber towards the EDJ (Figure 12.8):

1. Reparative secondary dentine. At an early stage, odontoblasts lay down a reparative patch to cover the pulpal ends of dentinal tubules, and protect the pulp from migration of chemicals and/or bacteria down the tubules.
2. Normal dentine. When the lesion has not progressed right through the dentine, a zone of normal appearance remains between the lesion itself and the patch of secondary dentine.
3. Zone of sclerosis. The lumen of each tubule is narrowed, or obliterated by

deposition of large crystals (page 193) reducing migration of chemicals and micro-organisms.

4. Zone of demineralization. The dentine is soft, but the organic component remains relatively intact, with normal structure.
5. Zone of bacterial invasion. Micro-organisms are present (sometimes densely packed) in the tubules. There are scattered small foci of proteolytic activity, where clusters of micro-organisms distend the dentine structure into bulging masses. Less altered tubules are compressed in between.
6. Zone of destruction. Larger and more closely packed foci of destruction contain large masses of bacteria.

Arrested dentine lesions lack zones of destruction and bacterial invasion, but show a prominent sclerotic zone with signs of remineralisation, and accumulations of large crystals in the tubules of the zone of demineralization (Schüpbach *et al.*, 1992). Some dentine lesions show both active and arrested zones simultaneously. Dentine is affected even before a carious lesion penetrates the overlying enamel, as the chemical products create a dead tract (page 194) that is sealed off by sclerotic dentine and secondary dentine.

Sites of carious lesions

Lesions develop very differently in different parts of the crown or root, and it is important to record their sites of initiation separately.

Pit and fissure sites

Fissure systems of molars and premolars are common caries sites in living populations with a westernized diet (Figure 12.9). Carious lesions are initiated deep in the fissure or pit, in the walls just above the floor, starting as separate foci that usually spread through the system. In section, each developing lesion spreads out towards the EDJ as a triangular area which, even at the stage of cavitation, may not be apparent at the crown surface. Some teeth show a slight widening of the fissure pattern, occasionally accompanied by a dark stain, but often the first sign is a deep cavity that is created when the enamel is undermined, and collapses. Diagnosis of fissure caries is therefore difficult. A sharp probe sticking in the fissure may be diagnostic, but is unreliable (Penning *et al.*, 1992) and, in radiography, the complex folds of enamel in the occlusal surface mask most enamel lesions, so that a radiolucency is apparent only when the dentine is affected (van Amerongen *et al.*, 1992; Espelid *et al.*, 1994; Goaz & White, 1994).

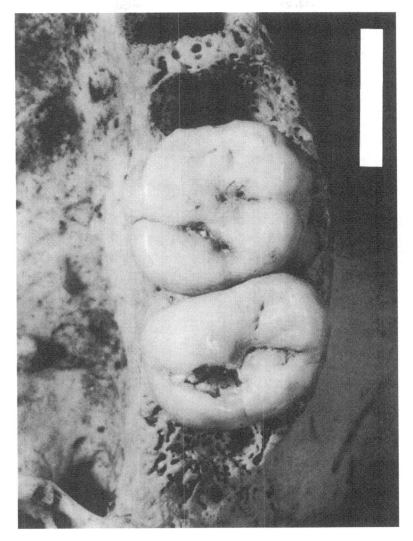

Figure 12.9 Fissure caries, with marked cavities, in the crowns of an upper left first and second molar from Post-Medieval London. Scale bar 1 cm.

Smooth surface sites in the crown

Smooth surfaces are defined as the sides of the crown, outside the fissures or pits and not bordering the cervical margin. Plaque accumulates at protected sites along the buccal or lingual gingival margins and between the approximal surfaces. Smooth surface buccal and lingual sites rarely develop caries, although white or brown spot lesions are sometimes found just above the

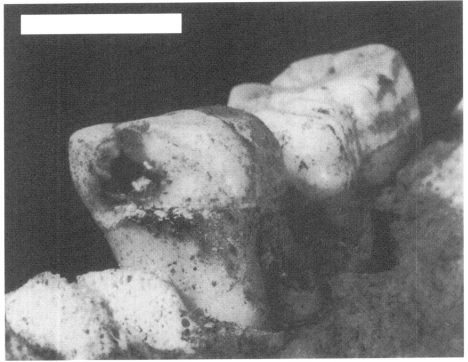

Figure 12.10 Approximal smooth surface caries, with a deep cavity at the lower margin of the mesial approximal wear facet of a lower left first molar from Post-Medieval London. Scale bar 1 cm.

gingival margin in newly erupted teeth. These lesions often fail to develop beyond the spot stage, because the level of gingival attachment recedes with age (page 263) to leave the lesion in an area of more effective natural cleaning, and any cavities that do form tend to be shallow. At approximal smooth surface sites, white or brown spot lesions are commonly initiated just below the contact points of the teeth, and frequently develop into a deep cavity (Figure 12.10) although these do not necessarily involve the dentine underneath. Approximal caries is difficult to see in living patients. A sharp probe may be used to feel for the roughened, softened, or cavitated surface of the lesion or, in radiographs, the enamel lesion is seen as a tiny notch-like radiolucency and, if the dentine does become involved, a broader triangular translucency is apparent. In archaeological and museum specimens, there is more opportunity for manipulating teeth so that approximal surfaces can be seen.

Root surface sites

Root surface caries is associated with root exposure through periodontitis (page 262), although the extent of root exposure is not necessarily related to the

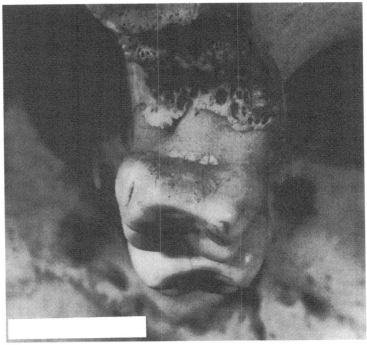

Figure 12.11 Root surface caries along the CEJ on the mesial side of an upper left first molar from Post-Medieval London. This is a very characteristic lesion in much archaeological material. Scale bar 1 cm.

initiation of caries (Vehkalahti & Paunio, 1994). Root surface lesions develop as broad, shallow craters that extend around the circumference of the root (Figure 12.11). The lesion may be initiated in the exposed cement surface alone, or at the CEJ (Figure 12.12), where it undermines the enamel and may ultimately be difficult to distinguish from a lesion initiated in the cervical crown. Root surface lesions may involve any surface of any tooth root (but especially cheek teeth) and they develop slowly through adult life, with many arrested phases. It is normally assumed that lesions with a hard or leathery surface and dark staining are in an arrested state (Nyvad & Fejerskov, 1982; Fejerskov *et al.*, 1993), but colour is not always a reliable guide (Lynch & Beighton, 1994) even though these observations fit well with histological evidence (Schüpbach *et al.*, 1989, 1990, 1992). In dental radiographs, root caries lesions that have penetrated the CDJ are seen in the dentine as scooped radiolucencies (Goaz & White, 1994) that can be difficult to distinguish from cervical burnout (page 13).

The plaque flora and dental caries

For a review, see Marsh and Martin (1992). Experiments with laboratory animals have shown that several species are capable of inducing caries on

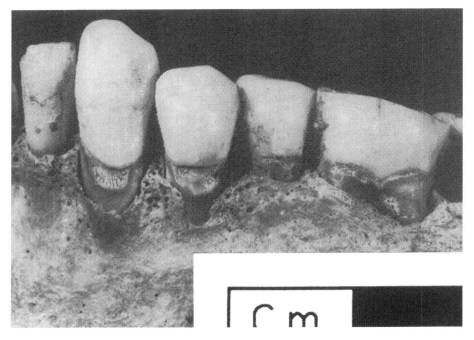

Figure 12.12 Root surface caries just below the CEJ on the buccal side of lower left canine, premolars and first molar from Post-Medieval London.

their own. The *Streptococcus mutans* group are the most cariogenic, followed by members of the *S. oralis, S. milleri* and *S. salivarius* groups, *Actinomyces naeslundii, A. viscosus,* and lactobacilli. Studies on living people do not implicate any one species alone, although *Streptococcus mutans* and lactobacilli are prominent in lesions passing towards cavitation. These species transport and process sugars rapidly, produce acid rapidly and can continue their metabolic functions under conditions of acidity that few other bacteria could tolerate. Both are uncommon in normal plaque and are at a competitive disadvantage with a low sugar diet but, when sugars figure prominently, they have the advantage and increase it by generating low pH conditions.

Diet and dental caries

The cariogenic potential of diet is monitored experimentally in two ways:

1. By testing the effect of different diets on caries frequency. This has involved groups of children living in institutions, including the Vipeholm study in Sweden (Gustafsson *et al.*, 1954) and the Hopewood House study in Australia (Harris, 1963). Similar experiments have been carried out

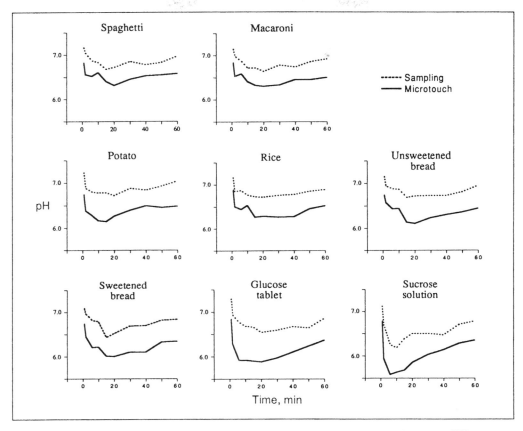

Figure 12.13 Stephan curves for different carbohydrate foods, using two different methods of measurement. Reprinted from Lingström, P., Birkhed, D., Granfeldt, Y. & Björck, I. (1993) pH measurements of human dental plaque after consumption of starchy foods using the microtouch and the sampling method, *Caries Research*, **27**, 394–401, with permission from S. Karger AG, Basel.

using laboratory animals (Mundorff *et al.*, 1990; Mundorff-Shrestha *et al.*, 1994).

2. By monitoring the pH of plaque fluid in response to different foods. This can be done either by taking small samples of plaque at timed intervals, or using fine electrodes attached to a pH meter. Depending on the nutrients available to the plaque bacteria, there are continuous pH fluctuations during the day, and the effect of a single food item can be expressed as a plot of time versus pH, known as a Stephan (1944) curve (Figure 12.13). The critical point is pH 5.5, and plaque pH phases lower than this are accompanied by demineralization, whilst neutral phases are accompanied by remineralization from calcium and phosphate in the plaque fluid. The

nett loss of mineral in caries depends upon the balance of low versus neutral pH phases over a long period.

Carbohydrates

The role of carbohydrates (starches and sugars) in dental caries has been reviewed by Sheiham (1983) and Navia (1994). Sugar's importance was conclusively indicated in the Vipeholm experiment, which was conducted in Sweden during the late 1940s and divided children into groups with different diets. The Control Group diet contained little sugar, and was eaten only at specific mealtimes, whereas the Sucrose Group ate the same diet but included sugary foods at mealtimes, and the 24 Toffee Group additionally ate toffees between mealtimes. The Control Group had a low caries rate and the 24 Toffee Group a high rate, with the Sucrose Group in between. The Vipeholm experiment thus emphasized not only the importance of sugars, but also the frequency of sugar consumption. Following the administration of a single sucrose dose, plaque pH falls within minutes, then recovers slowly to neutral over the next hour (Figure 12.13), but repeated administration of sucrose depresses plaque pH again before it has fully recovered, keeping it at critically low levels for longer periods. All common dietary sugars are capable of producing a rapid pH depression, including glucose, maltose, fructose and lactose, but sucrose has a particularly large effect. Sucrose is also the raw material for extracellular polysaccharide production (page 255) and is the commonest sweetening agent in foods.

The role of starches in the diet is complicated. The Vipeholm experiment and Hopewood House study demonstrated that starches do not on their own give rise to high levels of caries in children, but experiments with rats (Mundorff *et al.*, 1990; Mundorff-Shrestha *et al.*, 1994) have shown a slight relationship between starch and caries. Starches also produce a pH depression in Stephan curves (Lingström *et al.*, 1993), but this is less marked and less rapid than for sugars (Figure 12.13) although the pH is held low for longer. Starch molecules are too large to diffuse into plaque, and are broken down by salivary and bacterial enzymes to release maltose, with less dense foods such as potatoes being broken down more rapidly than more solid foods like spaghetti. Starchy foods thus have a low cariogenicity, especially when they are dense and do not cling to the teeth but, if the food contains both starch and sugar, it combines marked pH depression with a longer duration (Mörmann & Mühlemann, 1981; Lingström *et al.*, 1993), and the Vipeholm study confirmed that such foods were highly cariogenic.

Proteins and fats

The role of dietary protein and fat in dental caries is not well known. Proteins from saliva and GCF (page 260) are broken down by a range of different bacteria, but there is little evidence that dietary protein or fats are metabolized, with the exception of casein, a protein found in milk and dairy products. Milk and cheese do seem to have a protective effect (Bowen & Pearson, 1993), partly related to casein metabolism (page 255), partly to their coating of crown surfaces, which inhibits the adherence of food, and partly to the presence of calcium and phosphorous. Mundorff and colleagues (1994) found that protein, fat, phosphorous and calcium in the diet were associated with a low dental caries rate in laboratory rats. There is some supporting evidence from human populations. In their traditional way of life, the Inuit ate a diet dominated by proteins and fats, and enjoyed an exceptionally low caries rate, but caries has become much more common since their adoption of a westernised, carbohydrate-rich diet (Pedersen, 1966; Møller *et al.*, 1972; Costa, 1980).

Fluorine and caries

Fluorine is incorporated into enamel matrix (page 220) and confers protection from the start, because it decreases the solubility of dental tissues. When applied at the surface, as food and drink, in mouthwashes or in toothpastes, it has a number of actions; inhibiting bacteria, enhancing remineralization, and by becoming incorporated into exposed surfaces. It is important to consider environmental fluorine in any study of caries epidemiology, but this is difficult to establish for archaeological teeth (page 171).

Scoring and recording dental caries

Epidemiological surveys are based upon direct examination, using the naked eye and a sharp dental probe. Good lighting is essential and, with museum specimens, it is possible to use a low power stereomicroscope with ultraviolet illumination, but there are still difficulties with white and brown spot lesions and most studies count only clear cavities (Moore & Corbett, 1971; Lunt, 1974; Whittaker *et al.*, 1981; Kerr *et al.*, 1990). Particular care must be taken with archaeological material, because diagenetic effects may mimic the presence of caries (Poole & Tratman, 1978), and it is important to check a proportion of diagnoses by sectioning (Swärdstedt, 1966; Whittaker *et al.*, 1981; O'Sullivan *et al.*, 1993).

Statistical evaluation of caries rates is problematic, particularly in fragmentary archaeological material. The most common statistic is the DMF index (Klein *et al.*, 1938), determined by counting the number of decayed, missing or filled teeth and summing them together for each individual, so that the caries experience of a population can be expressed as its mean DMF score. In its original form, the DMF index is based upon whole teeth (DMFT index), and one difficulty is that it fails to take more than one lesion per tooth into account. This is addressed through the DMFS index, which records lesions per surface of the tooth, but any DMF index has the further limitation of assuming that all teeth are lost as a result of dental caries. This is clearly not the case, because teeth are as commonly lost through periodontal disease or, in archaeological material, lost post-mortem. Costa (1980) attempted to apply a correction factor for post-mortem loss, but formidable difficulties remain in comparing archaeological material with studies of living populations. Another deficiency of the DMF approach is that it does not distinguish between different caries initiation sites. An alternative is the Moulage system (Rönnholm *et al.*, 1951), used in Scandinavian countries for clinical and archaeological studies (Swärdstedt, 1966), which consists of a series of plaster models describing most initiation sites and forms of dental caries. Work with archaeological material has itself resulted in other methods that provide a more detailed level of recording. Moore and Corbett (1971) developed a scheme for scoring caries at different initiation sites (Table 12.3), and this has been followed by several other studies (Whittaker *et al.*, 1981; Kerr *et al.*, 1990; O'Sullivan *et al.*, 1993). To allow for post-mortem tooth loss, caries frequency can be expressed as a percentage of the total number of those sites surviving, carious or non-carious (Kerr *et al.*, 1990), and a rapid recording system of this type is given in Table 12.3. Despite these developments, most archaeological studies still express the count of carious teeth only as a percentage of the teeth present. This is unsatisfactory because, if tooth classes are not separated, the overall percentage is influenced by the pattern of post-mortem tooth loss. Dental caries is more common in molars than anterior teeth, and molars are less likely to be lost post-mortem, so that archaeological collections with poor anterior tooth preservation will have artificially inflated caries rates.

Epidemiology of dental caries

In living human populations (Menaker, 1980; Silverstone *et al.*, 1981), caries has a characteristic pattern. For all types of carious lesions, molars are most commonly affected, followed by premolars and then anterior teeth. In coronal caries, the most frequently affected sites are pits and fissures of cheek teeth, followed by approximal crown surfaces of cheek teeth and anterior teeth, whereas buccal/labial and lingual lesions are uncommon. Coronal caries is a disease of children, rising steadily to 15 or so years of age, and then falling away in early adulthood. It is more common in girls than boys, but earlier

Table 12.3. *Initiation sites and recording for dental caries*

Initiation sites
1. Fissure or pit. Different parts of a pit and fissure system are not distinguished as the defects spread along them. Count as one site per molar or premolar (and upper incisor when lingual pit present).
2. Approximal surface, just below contact point. Count two sites per tooth, and one for the most distal tooth of each quadrant. Take account of maloccluions.
3. Crown surface just above location of healthy gingival margin. Most common site is labial surface, but may affect all sides.
4. Cement–enamel junction. Small lesions lying on the CEJ, commonly on mesial and distal sides, but may affect all sides.
5. Root surface.
6. Gross. Divided into gross cervical (cavity too large to see if 3, 4 or 5 above), gross coronal (too large to see if 1, 2 or 3 above), or just gross when the crown is so destroyed that no site of initiation can be deduced.

Sources: Moore & Corbett (1971), Whittaker *et al.* (1981).

Rapid scoring sheet for counting caries in relation to initiation sites

Divide dentition into permanent anterior, permanent cheek, deciduous anterior, deciduous cheek. For each division, record counts of sites present and sites carious as follows:

Sites	**present**	**carious**	
Fissure and pit sites	☐	☐	
Approximal sites	☐	☐	
Cervical sites	☐	☐	enamel caries
		☐	CEJ caries
		☐	root surface caries
		☐	gross cervical caries
Tooth count	☐	☐	gross crown caries
		☐	gross gross caries

Lesions are expressed as percentages of sites present.

Count 1 fissure & pit site per cheek tooth, if any vestige of fissure system present
Count 2 approximal sites (mesial & distal) per tooth and 1 for most distal tooth in quadrant
Count 1 cervical site per tooth, if any vestige of CEJ is present

dental eruption in girls (page 140) exposes their teeth to risk for longer. Root surface caries also particularly affects the approximal surfaces of cheek teeth, but is instead a disease of adults. The pattern of dental caries is similar in members of the same family over several generations, perhaps due to inherited factors, but environmental factors such as dental treatment and diet also have a large role. The clearest single factor in caries epidemiology is however sugar, as shown by the decrease in caries rate during sugar rationing in Japan, Norway and the Island of Jersey during the 1939–45 war, which was followed by a rise once normal supplies were resumed. Caries is moderately common amongst wild great apes, particularly chimpanzees, whose diet includes a lot of fruit and therefore sugar. Gorillas, which eat considerably less fruit, have much lower caries rates (Miles & Grigson, 1990; Kilgore, 1995), whereas orang-utans seem to occupy an intermediate position between chimpanzees and gorillas (Stoner, 1995).

Archaeological and museum collections of jaws from people without a carbohydrate-rich diet show a very different pattern, in which coronal caries was much less common and its place was taken by cervical lesions, most of which probably originated in root surface sites. Caries was very uncommon amongst fossil hominids, into Palaeolithic and Mesolithic contexts. Nevertheless, a celebrated example is the rampant caries in the Middle Pleistocene skull from Broken Hill, Zambia (Brothwell, 1963c), and coronal caries has also been noted in *Australopithecus* (Robinson, 1952; Clement, 1958) and *Paranthropus* (Grine *et al.*, 1990). In European material (Brothwell, 1959; Hardwick, 1960; Brothwell, 1963c; Tattersall, 1968; Tóth, 1970) there was a gradual increase from very low caries rates in Palaeolithic, Mesolithic, Neolithic, Bronze and Iron Age contexts, to a rapid rise through Medieval and modern times. In parallel, the number of carious teeth per mouth increased, with more pit and fissure caries, less cervical caries, and more children affected. Similar general trends have been demonstrated for Egypt and Nubia (Armelagos, 1969; Greene, 1972; Hillson, 1979)

Some of the most detailed studies have been carried out on British material. Moore and Corbett (1971, 1973, 1975; Corbett & Moore, 1976) found a consistent pattern of dental caries in Neolithic, Bronze Age, Iron Age, Romano-British and Medieval material. Molars were the most commonly affected teeth, whereas canines and incisors were least and, in adults, the initiation sites were most often along the CEJ on the distal or mesial side. Caries increased in frequency with age and with resorption of the alveolar bone, with roughly equal frequencies in males and females. Caries was uncommon in younger individuals, where crown fissures and pits were the main lesion sites. A similar pattern has been noted in other Romano-British material (Whittaker *et al.*,

1981), in Medieval Scotland (Kerr *et al.*, 1990), in Medieval Sweden (Swärdstedt, 1966) and Finland (Varrela, 1991), and a large study of deciduous teeth from similar periods has yielded comparable results (O'Sullivan *et al.*, 1993). Moore and Corbett (1975) further examined material from AD 1665 plague pits, and another study (Corbett & Moore, 1976) examined skulls from a cemetery at Ashton-under-Lyme, in Lancashire, England, which spanned the 19th century. CEJ caries remained constant throughout, but high approximal and fissure sites, with a modern caries pattern, were established during the second half of the century and it is tempting to link this to the historically documented removal of import duties on refined sugars in AD 1845.

In North America, increased reliance upon maize agriculture is a clear cultural horizon (page 229). Several studies have shown an increase in caries rate (Cassidy, 1984; Cook, 1984; Larsen *et al.*, 1991; Pfeiffer & Fairgrieve, 1994) associated with the change from a hunter–gatherer diet with meat and low carbohydrate plant foods, to a diet heavy with starch-rich cereal. Post-European contact archaeological groups also show higher caries rates than pre-contact groups, implying a still greater reliance on cereal agriculture. Similar rises in caries rate have been associated with increased reliance on arable agriculture in studies of material from South America (Kelley *et al.*, 1991; Ubelaker, 1994), South Asia (Lukacs, 1992), Egypt and Nubia (Hillson, 1979; Beckett & Lovell, 1994), and other regions throughout the world (Turner, 1979; Larsen, 1995), although Lubbell and colleagues (1994) found no such rise in association with the boundary between Mesolithic and Neolithic in Portugal. Littleton and Frohlich (1993), studied a range of archaeological material from the Arabian Gulf, and found low caries rates in groups whose subsistence was based on marine resources, or a combination of pastoralism, fishing and agriculture. Groups subsisting on a higher level of agriculture or intensive gardening had high rates of caries. The reverse relationship was demonstrated by Walker and Erlandson (1986), in material from Santa Rosa Island, California, where there was a shift from an economy based upon gathering wild plant foods to one based upon marine resources, with an associated fall in caries rate.

In summary, where starch-rich plant foods form a small part of the diet, caries rate is very low indeed. The introduction of cereal agriculture (and thus dietary starch) has been linked to a rise in dental caries, in which the dominant form was root surface or CEJ lesions, initiated in adulthood. Where sugars have been introduced into the diet, fissure and approximal caries, particularly in children, becomes dominant. Other factors cannot, however, be ignored, and enamel hypoplasia (page 165) provides lines of weakness exploited by developing carious lesions (Mellanby, 1934; Cook, 1990), although there is still little epidemiological evidence for the impact of these defects. The relationship between dental attrition and caries is similarly unclear. Groups with high attrition rates often have low caries rates (Powell, 1985; Larsen, 1995), but this does not necessarily suggest a causal relationship. Occlusal attrition might remove fissure caries if it developed slowly, but the normal pattern at this site is for rapid progression of lesions, and attrition could not much affect caries at cervical sites, which are the predominant form in populations where attrition rates are highest. Finally, the possible effect of fluoride in the water supply should never be forgotten, even if it is difficult to establish for archaeological material (page 220). Lukacs and colleagues (1985) found a low caries rate associated with evidence of fluorosis (page 171) in Neolithic material from Mehrgarh, Pakistan, with the additional evidence of high environmental fluorine.

Inflammatory conditions of the teeth and jaws

Oral pathology and radiology textbooks provide general summaries of inflammatory conditions (Soames & Southam, 1993; Goaz & White, 1994).

Pulpitis

Inflammation is triggered by irritants, emanating from a mixed flora of oral bacteria that enter the pulp through dental caries, attrition or fracture of teeth. It is unlikely that bacteria pass directly down the exposed dentinal tubules, but their toxins do so and, in dental caries, pulpitis often starts before the chamber itself is exposed. Once triggered, the inflammatory response generates exudate which, confined within the tooth, increases pressure inside the pulp chamber to compress blood vessels and cause local pulp death, which in turn leads to suppuration (pus production). This may be contained for some time by a wall of granulation tissue (a special soft tissue that is associated with the healing process) but, ultimately, untreated pulpitis usually leads to death of the whole pulp.

Periapical periodontitis

After pulp death, inflammation passes down the root canal and the bacteria themselves, their toxins or the products of inflammation, emerge from the apical foramen and trigger an inflammatory response in the periodontal tissues. Surrounding bone is resorbed to accommodate a growing mass of granulation tissue called a periapical granuloma. Relatively few bacteria actually invade the bone, and an open root canal is needed to maintain inflammation. Radiographs show that inflammation may remain at a low level for years, with the only sign a

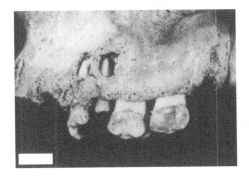

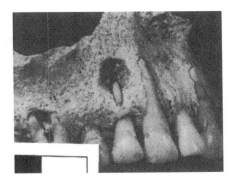

Figure 12.14 Periapical abscesses. A, fistulae emerging from around the apices of upper left first and second premolars, in a post-medieval skull from London. B, occlusal view of the same specimen, showing gross caries in the premolars, which has resulted in exposure of the pulp. C, a lesion in which there has been remodelling around the periapical region of the upper right first premolar (note the way in which the roots of canine and incisor have been exposed by the loss of the thin buccal bone plate – nothing to do with periapical abscessing). Scale bars 1 cm.

widening of periodontal ligament space and the formation of a sclerotic zone of thickened bone trabeculae, with no sign of swelling on the jaw surface. A periapical granuloma is usually diagnosed when a radiolucency (diffuse or sharply defined) involves loss of the lamina dura around the root apex.

Acute periapical abscess

Most acute abscesses develop from a periapical granuloma by the accumulation of pus. Pressure may be relieved by drainage through the root canal, but usually pus passes through the bone of the jaw along a tunnel known as a fistula, which emerges on the buccal side in most cases but may also appear on the lingual side, in the nasal cavity, or in the maxillary sinus. Radiographically, there may be little evidence of a rapidly forming abscess but, in

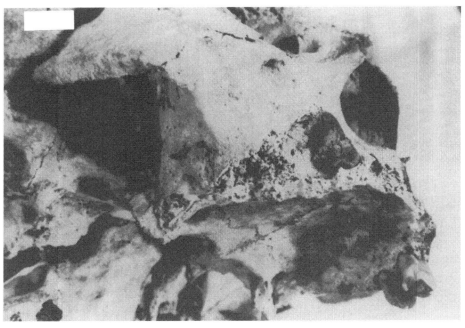

Figure 12.15 A cyst in the region of the upper right canine of a skull from Nubia, shown by a large cavity just below the nasal aperture. Most teeth have also been lost, and the alveolar process has generally remodelled. Scale bar 1 cm.

dry bone specimens, a fistula or a broad pit around the root apex often provide clear signs (Figure 12.14).

Cysts

Cysts are cavities in the jaws, containing fluid or paste in life. By far the most common are radicular cysts, initiated from a previously existing periapical granuloma, and seen radiographically as large radiolucencies over 1.5 cm in diameter (Goaz & White, 1994). Many radicular cysts remain confined within the normal jaw profile, but some grow to form a surface bulge in which the outer layer of bone becomes very thin and eggshell-like (Figure 12.15). Radicular cysts are found in adults, especially in upper anterior teeth, and are usually not painful, so that they often remain undetected. Next most common are dentigerous cysts, associated with unerupted teeth that have been retained in the jaw. They are particularly common in permanent lower third molars, and occur in both children and adults, especially males. For a review see Shear (1992).

Archaeological and anthropological evidence for periapical inflammation and cysts

Cysts are uncommon finds, but periapical abscesses are abundant in archaeological material (Alexandersen, 1967), although care is needed because the thin plate of cortical bone that normally overlies tooth sockets may break away during burial, to mimic the presence of a fistula. This is relatively easy to identify, but where irregular bone loss from periodontal disease (page 260) and periapical inflammation both occur in the same material, they may be hard to distinguish. Radiography may help, although Swärdstedt (1966) found that the majority of periapical radiolucencies were matched by fistulae. One celebrated example is the Middle Pleistocene Broken Hill skull, from Zambia, with most of its teeth both carious and associated with an apical bone resorption. Another well-known case is the Neanderthal from La Ferrassie, in France, in which periapical bone resorption is apparently associated with very heavy attrition. In living people, this is not a common cause of pulpitis, but heavy attrition is a feature of elderly Neanderthals and may have been a major pulp exposing agent in earlier populations, perhaps in association with fractures of teeth (Alexandersen, 1967). Where teeth have been lost and the alveolar process has remodelled it is, however, often difficult to reconstruct the pathological processes involved.

13

Conclusion: current state, challenges and future developments in dental anthropology

Identification and basic sorting

The anatomy of human teeth is very well known, because of the needs of dental surgery, so there is rarely any difficulty in identifying closely even a fragmentary tooth from archaeological or forensic remains. When several individuals form a mixed, fragmentary burial, then the teeth are amongst the best elements for determining how many individuals are present, and which fragments belong to which individual. This can be done simply, by matching morphological variation or state of wear, or it can be done with precision, by matching the sequences of enamel or dentine layers (pages 177 & 189). The challenge is not so much in establishing methodologies although there are still problems, for example, in distinguishing worn cheek teeth, but more in educating anthropologists and archaeologists more widely in identification skills. Recovery of teeth, particularly in juvenile material, is still not always complete and misidentified specimens are still found in museum stores. One major aim of this book is therefore to summarize the main identification criteria that are useful in anthropology.

Metrical variation

The basic diameters of the crown have been measured for many years, and there is general agreement about their definition. They can be recorded reliably, with low rates of observer error, even though they need to be measured to a high level of precision in comparison with bone measurements because teeth are relatively small and comparisons between taxa, populations and sexes often involve mean differences of fractions of a millimetre. Crown diameters are recorded routinely, so that there are large lists of published measurements available and, for example, they have an important place in descriptions of new fossils and in the definitions of hominid taxa. Population differences within living humans have proved difficult to establish, largely because of the extensive overlap in range of diameter measurements between populations, but there has been some success in showing broad differences between groups of populations. The main success, however, has been in demonstrating that studies of permanent and deciduous crown diameters within a collection of material can be used to separate the dentitions of boys and girls with a high level of accuracy (page 82) – an important development because the most difficult remains to sex are those of children. The main challenge to metrical studies of the dentition, however, lies in the question of what to do when the measurement set for a dentition is incomplete due to missing or worn teeth. A larger number of worn teeth could be included with some of the alternative measurements, but the statistical problems of missing data are more intractable. The variable pattern of missing and worn teeth (in both living mouths and archaeological material) is one of the factors that has limited the application of sexing by crown diameters, and population comparisons have been restricted to mean diameters for each tooth class, calculated from the available measurements. The future is likely to hold a range of solutions to the missing data problem, because it also affects other areas of statistics and several possible approaches are being investigated (Scott & Hillson, 1988; Scott *et al.*, 1991). Another major area of challenge is in understanding the mechanisms by which crown diameters vary – not only in terms of their heritabilities, but also in terms of the crown formation processes that generate the overall crown morphology from which the measurement points are defined. New ways to measure and record crown form are being developed (below) that will impinge on the traditional crown diameter approach, and one way of investigating the relationship between crown morphology and processes of formation is to examine the changing form of layers in enamel and dentine.

Non-metrical variation

One of the main achievements of dental anthropology has been the establishment of a standardized methodology for recording morphological variation as a series of discrete grades or categories of variant (page 86). This allows rapid recording and is easily understood, and it has been used widely on large collections. An extensive database, built from studies in many countries on museum material and dental models of living patients, is now available, and has yielded broad morphological groupings that can be interpreted in terms of the migrations and ancestry of human populations. The major challenge is to control the variation between different observers in scoring the variants. No matter how good the series of reference plaques, personal interpretation is needed at the boundaries between grades, and it is important to maintain supporting literature and training materials. Another challenge comes from the way in which non-metrical variants are generated during the process of dental development, and their whole status as discrete, discontinuous features is complicated because they show continuous variation in form, in addition to presence or absence. Non-metrical characters are fast becoming metrical variants through a whole range of new possibilities for measuring crown morphology, which could allow the crown to be broken down into its basic structural elements of cusps and ridges. Such methods are already being applied to fossil hominids (page 103). For these approaches to develop further, they require a more detailed definition of the structural building blocks of the crown and root, very different from the usual definitions of anatomical features, such as the 'fundamental macroscopic units' of Carlsen (1987).

Occlusion

A standard methodology for recording dental occlusion is available, and is applicable both to living patients and museum material, but the difficulty with archaeological specimens is that distortions of skull bones mean that relationships between the jaws cannot be recorded reliably. Nevertheless, occlusal relationships between teeth in each jaw can be measured, and it is possible to make comparisons with casts from living and recent human populations. It is still difficult to demonstrate consistent differences between populations, because of the large range of variation in occlusal anomalies and the overlap between populations, but several studies have shown an increase in malocclusions in relation to the adoption of a more westernised lifestyle and diet (page 115). The challenge is to understand the different factors involved in such an occlusal change – the morphology of the dentition itself, the role of jaw bone remodelling, and the role of attrition.

Dental development and age estimation in children

Dental anthropology has inherited some long established, and well-tried methods for age-at-death estimation based upon the stage of development reached by the dentition, such as the tables of Schour and Massler, and Gustafson and Koch (pages 134 and 142). They are easy to use and, where it has been possible to test them, they seem broadly to work even if they have tended to be used rather uncritically. For more detailed records, there is also a well-established series of grades for scoring dental development in individual teeth (page 127), which allows age estimation by comparison with reference groups of living children in several large, radiographically based studies. Another alternative for ageing young children is the measurement of developing tooth crowns, but the main possibility for an independent check in past populations and extinct hominid forms is the counting of layered structures in enamel and dentine (page 177). This has already led to a vigorous discussion of dental development timing in the australopithecines and early hominines.

The challenge for anthropology in ageing childrens' remains by dental development is justification of the standards used. The chart-based methods, in particular, either do not state their sources or use a large mixture of different studies, but the standards that are apparently more securely based on a single reference population also run into difficulties of compatibility (in methods used and in dental development rate). It is doubtful that age estimates would benefit from yet more large radiographic studies of living children, and it is more important to establish an independent check that can be used on museum and archaeological collections directly. This is where the layered structure of enamel and dentine formation comes in. Much more data are required, to establish the basic geometry of crown growth, followed by studies of whole dentitions from single individuals by both section and crown surface methods.

Hypoplasia and dental development

Developmental defects of enamel are recognized as an important indicator of general health in both living and ancient human populations, and methods for rapid recording of these defects at a macroscopic level are well established. The challenge for studies of hypoplasia is made by a consideration of the defects at the microscopic level, where they are seen as part of a much more complex pattern of growth disruption, and it is no longer clear just what can be inferred from the size and position of defects, as seen with the naked eye. Much future work is likely to be based upon detailed microscopy, both of the crown surface and of sections through the crown. The relationship between surface defects and the internal layering of the brown striae of Retzius (page 161) needs to be investigated, together with the mechanisms by which disruptions to the ameloblasts generate the different forms of defect, and the relationship between the form of the defect and the timing of the growth disrupting event.

Dental age determination methods in adults

Dental attrition is now well established as a consistent performer for age estimation in adult remains, and it has the advantage that teeth often survive better than key bone elements for ageing, such as the pubic symphysis. Several attrition scoring schemes are available, together with methods for measuring the size and orientation of attrition facets, and all show a broad relationship between extent of wear and age, although this relationship is much closer in young adults than it is in older adults. The gradients of wear between different teeth in the dentition are also quite well understood, in relation to the eruption timing of the different tooth classes, and have been used to standardize rate of attrition in the Miles age estimation method (page 239). Attrition-based ageing has probably gone as far as it can, although there are some possibilities of new and more detailed recording schemes, and alternative statistical approaches such as seriation may add to the internal consistency of attrition gradients within collections of material.

Of the various histological age estimation methods for adult dental remains, root dentine sclerosis and cement layering have the greatest potential, with the possible addition of secondary dentine deposition in the pulp chamber. A variety of methodologies has been tested for recording each of these phenomena, many showing a relationship with age that is as good as that with attrition. In other words, dental histology ageing methods seem to work as well as anything else, but there are several challenges. Methods of microscopy are needed, particularly for cement layering, that are appropriate for archaeological specimens as well as fresh material, and there is still room for development in measurement and statistical approaches. Yet another challenge is in understanding the physiological basis of these age-related changes, particularly cement layering, but dental histology ageing methods already seem effective enough to warrant more attention in the future than they currently receive.

Diet, wear and microwear

At a macroscopic level, tooth wear shows some characteristic patterns. Hunter–gatherer groups, in particular, seem to have a distinctive distribution and angle of wear, different to that of early agricultural groups. Most notably in North America, a change in wear pattern is matched by evidence from caries rates and stable isotopes for an increasing reliance on maize agriculture. Even simply recorded with the naked eye, therefore, tooth wear has an important place in the reconstruction of past human subsistence.

Dental microwear studies are still developing, even though much research has already been carried out. Mechanisms of jaw movement and tooth contact are understood, so that homologous facets can be selected for analysis, and the distribution of the main features of microwear can be established. Broad differences in microwear patterns have been demonstrated, but the results are often difficult to interpret. One challenge facing microwear studies is the standardization of techniques for microscopy, recording, measuring and counting features because, whilst the basic methodology is agreed, different studies vary in detail and most recording methods are so slow that each study is based only on a few individuals. Yet another challenge is in the fundamental understanding of the mechanical processes that lead to microwear textures. These are still poorly understood and, for example, the relative effects of enamel microstructure, chemical environment, dietary components and behaviour are unclear. Future developments are likely to include image analysis techniques, to aid standardization and allow faster recording of features, and profilometry or other techniques that produce a three-dimensional model of microwear features, which can then be defined and measured directly.

Plaque-related dental disease

Periodontal disease has been an area of intense research in clinical dentistry over the past ten years. A great deal is known about its aetiology in living populations, although an understanding of its epidemiology has lagged behind, largely due to a lack of simple tests that identify periodontal disease unambiguously, and its chronic and episodic nature. Bone loss as a component of the disease is less well understood, although several methodologies are available for recording it in archaeological specimens and museum collections, as well as in radiographs of living patients. Periodontal disease in adults is an age-related process and the main challenge lies in standardizing the amount of bone loss, because of the lack of stable reference points for measurements (page 263). In addition, it is difficult to interpret the pattern of bone loss, partly because it is not always clear what this represents in terms of the clinical appearance of the disease, and partly because its epidemiology is not well known even in the living.

Dental caries is one of the most consistently recognized dental conditions in anthropological research, and almost all archaeological cemetery reports include at least a basic record. There are now excellent recording systems devised specifically for archaeological material that note not only the presence of carious lesions, but also their location (page 279). With the very large amount of clinical research carried out on caries during recent years, it is possible to use this information to make interpretations of diet, even though the archaeological recording systems are much more detailed than those usually employed for caries epidemiology in the living. There is, however, still a large amount of basic recording carried out on archaeological material and the major challenge is to standardize recording using a more detailed system, so that broader dietary comparisons can be made.

Biochemistry

Biochemistry is regarded by many as the cutting edge of biological science and if, for example, it is possible to extract and sequence nuclear DNA from archaeological or forensic remains, there is little point in attempting to reconstruct biological affinities from phenotypic variants such as dental morphology. Biochemical analysis of ancient remains is, however, a rapidly developing field and it is still not clear what will be possible. One key challenge is an understanding of the processes of diagenetic change in dental and skeletal material. Most work has been carried out on bone, but there is increasing acknowledgement that the highly mineralized nature of enamel offers a protected environment for the survival of biochemical information, both within the enamel itself and in the dentine underneath the crown. Teeth are therefore likely to become an important focus for biochemical research, and this is why a large chapter has been devoted to the subject in this book.

Teeth in anthropology

There is little doubt that teeth are one of the most important parts of any anthropological study of ancient or forensic human remains. In any of the main categories of information required, they perform better, or at least as well as the bony remains of the skeleton. Even a tiny fragment is readily identifiable and teeth are highly resistant to the rigours of most burial environments. Teeth provide by far the best age-at-death estimation methods for young individuals and they yield methods for adults that are as good as any other. Potentially, they also provide an important method for sexing juvenile remains in a collection of material, and they preserve the only record of each individual's growth process which has an independent, built-in clock. They show a large range of variation in morphology, which can be compared directly between ancient remains and living people, and has helped in combination with dental treatment records to identify remains in forensic work. Patterns of microwear and of dental disease offer the possibility of reconstructing diet, and teeth provide one of the most protected environments for preservation of biochemical information on diet and biological affinities.

Anthropology therefore relies heavily upon teeth, but they still suffer from an image problem. Part of this may be to do with childhood memories of the dentist's chair and some may simply find teeth unaesthetic, although on close examination they are really very stylish sculptures that are seen to best advantage in fossilized material, or when coated with gold for electron microscopy. Others may be put off by their small size and sheer complexity, but this is precisely how so much useful information is concentrated into them.

Appendix A

Field and laboratory methods

Excavation of skulls, jaws and dentitions

Further reading

Spriggs & Van Byeren (1984), Watkinson (1978) and Koob (1984).

Excavation

Excavators should be familiar with both adult and young children's remains (see the sequence in Figure 5.9), and the spoil should be sieved through a 5 mm mesh, to catch small fragments. Ancient bone is often soft, and must not be scraped or brushed vigorously, so it is better to pick away the burial matrix gently. The remains should be recorded and removed rapidly, and shielded from hot sun to prevent them from drying out too fast.

1. If the remains are tough enough to lift immediately, they are exposed, lifted out and bagged up (the skull and dentition are best bagged together, checking that all the expected teeth are included).
2. If the remains are too fragile they need to be lifted in their block of burial matrix. A heavy gauge (0.5 mm thick) aluminium foil wrapping may provide sufficient support, but loose burial matrix may need further support. The block may be encapsulated in plaster bandages, after first wrapping with plastic film or aluminium foil, or supported by injecting consolidant emulsion (5–10% weight/volume concentration, page 297) into a series of holes, prodded into the matrix well away from bones and teeth.

Samples for biochemical analysis should be taken at the time of lifting, using clean gloves and tools, together with sterile sample pots (some have a spatula built into the lid, which avoids any hand contact). Material for histology is also best taken at this point, and placed in a sample pot with 70% alcohol.

Initial cleaning on site

Burial matrix should be cleaned away as soon as possible and, if they were tough enough to lift immediately, most bones and teeth may be washed in water without damage (test a few first), avoiding long soaking and vigorous scrubbing, whilst also taking care not to dislodge any calculus deposits (page 256). It is a good idea to sieve the residue with a 1 mm mesh, just in case any small fragments have been missed. All remains are laid out on labelled trays and allowed to dry slowly, under cover from direct sunlight. With juvenile jaws, the developing teeth should be identified, dried and bagged separately as they emerge from their crypts – a job that is usually best done in the laboratory.

Packing and storage

NB: Seek advice from an archaeological conservator.

1. The clean, dry remains are usually stored in stout cardboard boxes, and prevented from rattling about by packing with crumpled acid-free tissue paper. Permanent labelling is vital for long-term storage, on the bones themselves, with the packing, and on the box. Humidity is also important because, if storage conditions are too dry, the bones and teeth gradually crack. Watertight packing may prevent water loss, but may allow mould to develop, and cardboard boxes allow a balance to be maintained. They should ideally be stored in controlled humidity – the recommended relative humidity (RH) for bone and dentine is 50%–65% – and really critical specimens can be protected inside a sealed plastic box containing pre-conditioned silica gel and a humidity indicator card.

2. Waterlogged bones and teeth sometimes dry out without damage, but others must be stored damp. Damp packages can be improvised with polythene bags and wet polyether foam padding. A refrigerator slows the growth of mould.

Laboratory cleaning

Further reading

Rixon (1976) and Hillson (1992b).

Equipment and materials

The essentials are a soft brush, aerosol or bulb-operated puffer, hardwood sticks (avoid bamboo, which contains abrasive particles), 'cotton wool' in the

UK or 'cotton' in the US (wind a whisp of it around the end of a stick to make a bud), dental hygienists' scalers (the push/watch-spring and hoe types are best). A stereomicroscope and bench lamp are useful. Two solvents are normally used; alcohol (100% industrial methylated spirits) and acetone (use both with care). Acetic (ethanoic) acid, 10% in aqueous solution, is used to dissolve carbonate deposits.

Dry cleaning methods

Dry, unconsolidated matrix may be brushed or blown away. More consolidated matrix is dislodged with a pointed hardwood stick, and cemented deposits are removed with steel scalers – flicking the deposit away without touching the tooth surface. To minimize breakages, it is important to apply pressure only along the long axis of the tooth, in an apical direction. Final cleaning may be carried out by spreading impression material (page 299), letting it set and then pulling it gently away with any trapped dirt.

Washing and solvent-based methods

Water is avoided if possible, because delicate specimens may crack as they dry out. Alcohol is applied on home-made cotton buds to soften the matrix, but acetone may be needed to remove organic deposits, or old acrylic consolidants (page 298). If in doubt, a small area should be tested first. If water is needed (for example to loosen a clay burial matrix), it is best used sparingly, by adding a little to the alcohol before applying it with a cottonwool bud. Where thick deposits need to be softened, water is gently sprayed on to them, and they are gently picked away with a hardwood stick. Calcium carbonate concretions are dissolved with dilute acetic acid, but their removal may leave delicate remains weakened.

Consolidation, reconstruction and adhesives

Further reading

Spriggs & Van Byeren (1984), Howie (1984) and Koob (1984).

Materials

The preferred materials are acrylic resins, including solid resin (Paraloid B72), emulsions (Primal AC-61 and AC-634 and Revacryl 452) or colloidal dispersions (Primal WS-24 or WS-50), and cellulose nitrate adhesive.

Procedure

NB: Seek advice from an archaeological conservator before attempting consolidation. Consolidants should be avoided if possible, because it is difficult to ensure that they impregnate the specimen evenly, their long-term effect is unpredictable, and they are difficult to remove. Solid resin is dissolved in acetone to produce a 5–10% weight/volume solution for application to dry bones but, for damper remains, acrylic emulsions or colloidal dispersions are diluted with water to 2%–4% weight/volume and applied with soft brushes, droppers, syringes or sprays. A widely used alternative is polyvinyl acrylate (PVA), but this has poor penetration into bone and high permeability to water. All the resins can be redissolved in acetone, but PVA changes with time and may be difficult to remove from specimens that were treated some years in the past.

Reconstruction of original specimens is rarely justifiable for research purposes, because the reconstruction involves an element of interpretation, and makes it difficult for others to question it afterwards. In fact, there are advantages to fragmentary material, which often exposes the developing dentition inside immature jaws. If reconstruction is deemed essential, then it is best to use adhesives that dissolve readily in acetone (a cellulose nitrate glue or a Paraloid B-72 solution).

Dissection of jaws

Equipment

Dental drills provide the best way to cut bone (the type driven by electric motor is most portable), with straight and contra-angle handpieces, separating discs and mandrels, a selection of burs, and scalers (page 297). At a pinch, a hobby/modelling drill, or an ordinary small hacksaw, will cut bone quite effectively.

Procedure

NB: Cutting bone with a dental drill creates dust, and it should ideally be carried out in a fume cupboard. Jaws are dissected to expose the developing teeth inside their bony crypts. Children's faces consist largely of developing teeth, separated and covered by thin but tough bone walls, so it is necessary to have a good working knowledge of jaw anatomy at different stages of development (Figure 5.9). A clean break is easier to fit together again than a cut surface, and partial openings in archaeological specimens can sometimes be exploited by breaking the jaw gently in the hands. A small cut may provide a similar line of weakness. Where a crypt has to be opened fully, it is best to make a small cut beyond the growing apex of the tooth, just large enough to admit the tip of the hoe scaler, and then carefully pull away the buccal wall. Alternatively, a door can be cut in the wall with a separating disc.

Surface impressions and replicas

Further reading

Larsen (1979), Waters (1983), Hillson (1992b) and Beynon (1987).

Materials

Silicone dental impression materials are widely used in anthropology for finely detailed work, most commonly Coltene President (Coltene AG, Feldwiesenstrasse 20, CH-9450, Altstatten, Switzerland) and 3M Express (3M Dental Products, Division Building 225-4S-11, 3M Center, St Paul, MN 55144-1000, USA). They are available in two forms; a putty and a fine body or wash material, which flows into microscopic surface details. Many less expensive silicone materials are also available, and can be used in larger quantities to take impressions from whole jaws. Replicas are cast in the impressions using a range of epoxy resins, for example Epo-tek 301 (Epoxy Technology Inc., 14 Fortune Drive, Billerica, MA 01821-3972, USA) or various Araldite (CIBA-GEIGY, Plastics Division, Duxford, Cambridge, CB2 4QA, United Kingdom) casting resins.

Procedures

In most common usage, an impression is the negative mould that is taken directly from the specimen surface, whereas a replica is the positive form that is cast from the impression. Most work uses replicas (with the best materials, a resolution of details less than 1 μm in size can be achieved), but direct examination of impressions can be useful, for example, in studies of fissure systems in molars.

Dental impressions may be taken from living people only by clinically qualified dentists, who spread the impression material into a special tray, into which the patient bites. Models of the dentition can be cast in the impression with plaster of Paris, or purpose-made dental 'stone', and large collections of these are available for study by anthropologists. In addition, anthropologists frequently take their own impressions of museum and archaeological specimens, which require different techniques because material is forced between teeth, and into their sockets, making it difficult to release the impression without damage once it has hardened. A great deal of preparation is therefore required, plugging holes and narrow spaces with wax or aluminium foil packing. For large impressions, it may also be necessary to apply a separating fluid (either soapy water or a proprietary make), which may affect the specimen and needs to be discussed with curators. The general rule is that the specimen must not be damaged, and careful consultation is needed at all stages – some famous specimens have had so many impressions taken that they are in danger of collapse. Casting whole jaws is a complex and time-consuming business, best left to specialists.

It is usually easier to take impressions of just a few teeth in the jaw at a time or, even better, single surfaces of single teeth. For well-preserved, fully formed teeth a two-stage impression technique can be used. The tooth or teeth are wrapped in aluminium kitchen foil and pressed into a ball of putty impression material and, when hardened, the tooth and foil are removed. Fine body material is introduced with a syringe and the tooth is pressed into the impression again. More delicate specimens are better treated by spreading fine body material directly onto the surface, draping it over as a sheet, which can be reinforced by adding more material later. Great care is needed in removing the impression, which should then be allowed to dry off for a few minutes, labelled in pencil on a cut surface, and sealed in a dust-proof box or plastic bag. Resin is poured into the open impression (sometimes small dams need to be made with more impression material) to cast the replica, and it is often convenient to fix a stout copper wire to the back with regular quick-set epoxy adhesive, to aid removal and handling. Replicas should then be stored in tightly sealed, labelled containers.

Sectioning

Further reading

Beasley *et al.* (1992), Boyde (1984), Gustafson & Gustafson (1967), Schenk *et al.* (1984), Wallington (1972) and Schmidt & Keil (1971).

Equipment and materials

The most important equipment is an appropriate saw. Simplest (and probably best) is the rotating disc type, with a diamond abrasive-edged blade against which the specimen is held by a counter-weighted arm. Finer cuts are achieved in machines with an annular configuration, where a very thin blade is stretched tight and cutting takes place at its circular inside edge, or with machines based on a spool of wire, which passes continuously to and fro across the specimen surface. Sections are usually finished by hand polishing, but rotary polishers are available, and can save a good deal of work. If thin ground sections are to be prepared, then polishing jigs and a zero bonding jig produce a much more professional finish. The best finishing technique is diamond micromilling, involving the movement of a prepared specimen block underneath a rapidly rotating diamond-armed milling head, but this equipment is expensive and most routine preparations are finished by grinding and polishing techniques.

Polishing is best carried out on sheets of abrasive paper, held down on hard surfaces, followed by diamond abrasives spread onto hard plastic mats that are especially made for the purpose. It is best to avoid water-based lubricants when polishing archaeological material (page 297) and, although proprietary lubricants are available, 70% alcohol works reasonably well in most cases. The most commonly used embedding material is polymethylmethacrylate (PMMA), although several epoxy resin systems are available. The advantage of PMMA is its hardness, and the fact that methylmethacrylate can be introduced into the specimen as a very low viscosity monomer which penetrates even the finest spaces (Table A.1). The most commonly used adhesives for cementing specimens onto glass slides are epoxy resins (e.g. Epotek 301, above) and cyanoacrylates ('Super glue'), and adhesion may be further improved by coating the slide beforehand (Table A.2).

Nitric acid, 5–10%, or 10–15% EDTA (ethylenediamenetraacetic acid) are commonly used to demineralize specimens, followed by wax embedding, microtome sectioning and staining with eosin/hematoxylin. Advice on these techniques is best sought from a hospital histopathology laboratory.

Buehler Ltd, 41 Waukegan Road, Lake Bluff, ILL 60044, USA, manufacture a comprehensive range of cutting, grinding and polishing equipment and supplies, and have offices throughout the world. Struers A\S, Valhojs Alle 176, DK-2610 Rodovre/Kobenhavn, Denmark is also a large, international company with a large range of products. Logitech Ltd, Erskine Ferry Road, Old Kilpatrick, Glasgow, G60 5EU, Scotland, supply specialist polishing equipment and supplies at the top end of the market.

Procedure

The normal technique for the microscopy of archaeological material is to produce cut, ground and polished sections of intact teeth. All these processes can create features (known as artefacts) of their own, and each stage of preparation needs to be kept to a minimum. All mineralized tissues contain harder and softer elements, and the softer elements are worn away faster to produce hollows, whilst harder elements stand up proud, and the edges of holes and cracks are progressively rounded. It is therefore usual to embed the specimen in supporting material beforehand, using material that can be impregnated very intimately into the specimen (Table A.1), and is hard enough not to be polished

Table A.1. *Methylmethacrylate embedding schedules*

Technique for fresh bone and dental tissues

1. Reflux with 50:50 chloroform:methanol in Soxhlet apparatus over 2 weeks continuously to remove fatty material and water.
2. Immerse specimen in monomer mixture[a]. Change monomer mixture every 24 hours, to produce 3 changes in all.
3. Immerse specimen in monomer mixture with added activator (benzoyl peroxide[b]). Place in an oven set to a constant 32 °C until solidified.

[a] Monomer mixture. Methylmethacrylate monomer is flash distilled to remove the added stabilizing agent, 0.01% quinol. Two flasks are connected by a U-tube. Monomer is added to one flask and is frozen by liquid nitrogen. Air is evacuated by pump from flasks and tube. The liquid nitrogen bath is removed from the first flask and the second is immersed instead. Monomer condenses in the latter flask as the former slowly warms up. After distillation 95% (w/v) methylmethacrylate is mixed with 5% (w/v) styrene monomer, for stability under an electron beam.

[b] Benzoyl peroxide is explosive, and must be handled with great care, following the safety guidelines of the institution in which the work is being carried out.

Source: Boyde *et al.* (1990).

Method for archaeological material

1. Immerse in 100% industrial methylated spirit (IMS) for 1 day.
2. Immerse in a mixture of 1:1 IMS:chloroform for up to 7 days.
3. Immerse in a mixture of 1:1 IMS:unwashed methylmethacrylate monomer for 1 day.
4. Immerse in unwashed methylmethacrylate monomer for 1 day.
5. Immerse in a mixture of 19:1 washed methylmethacrylate monomer:styrene, with 0.2% (w/v) 2,2'-azobis 2-methylproprionitrile, for 1 day at 20 °C and then 5 days at 32 °C until polymerized.

Methylmethacrylate monomer is washed with 5% sodium hydroxide solution to remove the 0.01% quinol stabilizer. The sodium hydroxide is then washed out with distilled water and the methacrylate is dried over calcium chloride.

Source: Developed by Sandra Bond at the Institute of Archaeology, UCL.

away much faster. Once embedded, the specimen block is cut in the required section plane, using an appropriate sawing machine, and then ground. The grinding/polishing schedule (Table A.2) gradually passes from coarser to finer abrasives, removing at each stage the scratches produced by the previous stage, but much practice is required to achieve a good surface.

Table A.2. *Grinding and polishing schedule for embedded surfaces and sections*

Glass slide preparation
1. Grind slide to constant thickness, using a polishing jig, on a glass plate with a water dispersed slurry of 600 grit abrasive (aluminium oxide).
2. Wash away abrasive thoroughly with distilled water, and dry.
3. Immerse in xylene for 2 minutes.
4. Immerse in 100% Industrial Methylated Spirit (IMS) for 2 minutes.
5. Immerse in acid alcohol (100:1 70% IMS: concentrated hydrochloric acid).
6. Wash thoroughly with distilled water.
7. Immerse in 100% IMS for 2 minutes.
8. Immerse in APES (3-(triethoxysilyl)propylamine 1% (w/v) solution in alcohol) for 30 seconds.
9. Dip briefly in 100% IMS, to remove excess APES.
10. Wash with distilled water.
11. Dry in oven (50 °C) overnight.

Specimen polishing schedule
1. Hand polishing on abrasive papers, on a hard surface, using a figure-of-eight motion. Start with the coarsest grade, followed by progressively finer grades; 600, 800, 1200 grit. At each grade, polishing continues until scratches are just eliminated (examine in oblique light). Specimens must be carefully washed between grades. Water is used as a lubricant and for washing in fresh material, but alcohol is better for archaeological material (page 301).
2. Polish, with a hard plastic mat fixed to a rotating metal plate, with diamond abrasive using a non-water based spray lubricant, in a series of progressively finer grades; 3 μm, 1 μm, 0.25 μm. Specimens are carefully washed between stages.

Preparation of flat surfaces for compositional BSE imaging
1. Embed specimen in methylmethacrylate (Table A.1).
2. Cut required plane of section, using saw machine (use IMS as a lubricant for archaeological material).
3. Follow normal polishing schedule, but keep each stage as short as possible.

Preparation of ground thin sections
1. Embed with methylmethacrylate (Table A.1).
2. Prepare glass slide (above).
3. Cut required plane of section with saw machine (use IMS as lubricant for archaeological specimens).
4. Polish surface using normal schedule.
5. Wash specimen block very carefully in 100% IMS, and dry carefully, making sure that fingers do not touch the polished surface.
6. Cement the polished surface onto the glass slide using Epotek 301, and a zero bonding jig.
7. Use a slide-holding chuck in the saw machine and cut away the main part of the specimen block, leaving a slice about 250 μm thick attached to the slide.
8. Polish, using the normal schedule, to reduce the slice down to the required thickness (usually 60 μm) during the 600 and 800 grit stages. The finer abrasives are used for finishing, but are too slow to use for reducing thickness. It is possible to use a polishing jig to control the section thickness, but almost as good a result can be obtained by regular checks with a thickness gauge.

Source: Developed by Sandra Bond at the Institute of Archaeology, UCL.

Transmitted light microscopy (page 310) requires a thin, plane-parallel slice (known as a ground section) to be prepared from the tooth, and cemented to a glass microscope slide (Table A.2). With freshly extracted teeth it is possible to produce really thin sections down to 3–4 μm (Fremlin, 1961) but archaeological material may break up as the section becomes thinner and, for most practical purposes, sections are ground to thicknesses between 60 μm and 150 μm. When the thickness of the section is reduced by polishing, the bond with the slide is critical, and it is best to achieve as thin as possible a layer of adhesive (a zero bond).

Demineralized preparations are not often used in dental anthropological research but are sometimes needed, for example, in the study of cement (page 204). Archaeological specimens are difficult, and many do not have sufficient organic material surviving (page 225), so that it may be better first to prepare a ground section and then to demineralize it *in situ* on the slide. Demineralisation using nitric acid takes a few days, but if it takes longer than four days it may damage the staining properties of the tissue. EDTA is regarded as a better method, but may take some weeks to complete demineralization. Standard wax embedding and microtome sectioning techniques are used, together with eosin and hematoxylin staining.

Etching, coating and allied techniques

Equipment

Sputter coaters and carbon coaters are specialist equipment, found in SEM laboratories.

Etching

Etching is used in the study of enamel, to induce artefacts that track along the structures of the tissue. All acids act on the tissue in some depth, exploiting porous spaces under the surface, but the amount of surface relief produced is variable. A strong mineral acid (1 N hydrochloric acid for 15 seconds) (Rose, 1977) produces modest surface relief, because it acts only by dissolving away material. A weak acid (0.5% phosphoric for 30 seconds to 2 minutes) (Boyde *et al.*, 1978) causes the growth of insoluble mineral species, which stand out at the surface and enhance relief, but involves a minimum of tissue disturbance (Boyde, 1984). Maximum relief in enamel is produced with ethylenediamenetetraacetic acid (5% EDTA, buffered to roughly neutral pH) for 5 hours. This acts particularly at the boundaries between enamel prisms (page 149), causing them to stand out at the etched surface in high relief.

Coating

Specimens for SEM work are routinely coated, with carbon (for BSE; page 315), or gold or gold–palladium (for ET; page 314). Particular care needs to be taken when sputter-coating epoxy resin replicas (page 209) with gold, because they may be damaged by heat. In such cases it is best to run the coating unit in short bursts (Beynon, 1987), and Boyde (1984) suggested a schedule of alternately switching on the coater for 1–2 seconds and off for 10–15 seconds. The coatings are very thin and can be removed relatively easily. Gold and palladium can be removed by immersing the specimen in mercury, without wetting it (particularly useful for archaeological material). It is sometimes possible also to use an aerosol organic anti-static agent, such as Duron, instead of coating with gold.

Tooth crowns are too glossy to examine directly by reflected light microscopy. A gold sputter coating may help, or another possibility is to 'smoke' them with ammonium chloride (heated in a test tube over a gas burner to form a vapour through which the tooth is passed). This produces a pale, matte surface that can be wiped away (Goodman & Rose, 1990).

Photography of museum and archaeological specimens

Further reading

Dorrell (1989).

Equipment

Basic requirements include a 35 mm camera, macro lens, cable release, extension tubes or bellows, reversing ring, sturdy tripod or copy stand, photographic lights, black cotton velvet background, reflectors (improvised by gluing crumpled aluminium foil onto card) and diffusers (cut sections of white plastic bottles), scales (drawn on card) and flash system.

Procedure

Teeth are small, so that dental photography is always macro-photography and, with a 35 mm camera, it is important to fill the frame with the main subject. Accepted practice is to photograph bones and teeth in correct anatomical orientation (i.e. the right way up), so that upper teeth are shown crown downwards and lower teeth crown uppermost. A close-up should include enough of the surrounding dentition and jaw to show clearly where it is situated, and there should always be an indication of scale. A black-velvet background is effective for bones and teeth, is easily portable and does not need to be lit. The specimen is raised on small blocks, just high enough to ensure that the pile of the velvet is out of focus. Shape is emphasized by oblique lighting and, by convention, this is arranged so that the illumination appears to come from the top left of the photograph. In the laboratory, it generally is easier to use photographic lamps to achieve this than it is to use a flash system but, in the field, there may be no alternative to flash. When working in the field for any length of time it is, in any case, a good idea to develop films on site, check the results and ensure that the film cannot be accidentally re-exposed. Films can be loaded into light-proof developing tanks inside a sleeved changing bag and processed in an ordinary sink or basin.

It may be helpful to 'smoke' (page 305) teeth before photographing them. Ultraviolet and infrared photography may also be useful (Gibson, 1971), to emphasize early lesions of dental caries (page 269), opacities (page 169) and diagenetic changes to teeth (page 196), but require special lighting arrangements, filters and films.

Radiographic techniques

Further reading

Langlais *et al.* (1995), Goaz & White (1994) and Whaites (1992).

Equipment

Anthropologists normally use a fully enclosed x-ray machine inside a shielded case with all controls outside (Faxitron or similar), with industrial x-ray film (35 mm pre-wrapped roll film can be cut up into 40 mm lengths and sealed with light-proof tape, to approximate standard dental film packets), and polystyrene blocks to support specimens. Some facility for developing the films is needed, but this can be improvised if necessary by using a sleeved changing bag and developing tank (above).

Computed tomography (CT) scanners have been used for important fossils (Conroy & Vannier, 1987) or mummies, but are found only in hospital radiology departments and require their active collaboration. The scanner is integrated with a computer system, and images are normally presented as a series of 'slices' through the specimen. The resolution and spacing of slices depends both upon the operation of the scanner and the computer program used to process the data, but these are limited by committment of time and equipment.

Some programs produce three-dimensional models that can be rotated to give an excellent idea of internal organization.

Principles

An x-ray beam is a stream of x-ray photons that darkens a photo-sensitive film, and the amount of darkening depends upon the number of photons that have impinged, reflecting the balance between the rate at which photons arrive and the exposure time. An object placed between the beam and the film creates a shadow image because it absorbs the photons, and more highly mineralized components absorb more photons and are therefore more radio-opaque than less mineralized components. In a dry jaw specimen, the most radio-opaque material is enamel, followed by dentine, then cement and bone. The more opaque regions cast a heavier shadow, under which the film is darkened less than radiolucent areas, and the contrast is controlled by the characteristics of the film emulsion and the voltage setting of the x-ray tube. Resolution is controlled by other settings of the x-ray tube (usually with minimal adjustment on basic equipment), the closeness of film and specimen (the closer the better), and the film emulsion grain size (industrial films have highest resolution).

Procedure

Radiography is potentially dangerous, so it is important to seek advice, and follow the institution's policy or regulations. Anthropologists themselves only make radiographs of archaeological, museum or forensic specimens, and their main use of radiography is to investigate the developing dentition inside young jaws (it is less useful in dental palaeopathology). The specimen is set up on polystyrene blocks inside the x-ray machine, with the following general guidelines:

1. The beam should pass through the centre of the area of interest in the specimen.
2. The plane of interest within the specimen should be perpendicular to the beam.
3. The film plane should be perpendicular to the beam.
4. The film should be as close as possible to the specimen, and both specimen and film should be as far as possible from the x-ray tube, within the limits of the machine's enclosure.

This is easier to arrange with a mandible than with a skull but, when carried out on a jaw, is equivalent to the 'paralleling technique' used in clinical dental radiography, in which the film is held parallel with the long axis of the teeth. The full dentition can be covered by 21 projections, each centred on a particular tooth, or contact between teeth, with periapical projections including the teeth and periapical region of one jaw at a time, and bitewing projections centred on the occlusal plane between upper and lower teeth. Many important studies of dental development in living children (page 126) have used a different projection, the oblique lateral, which shows the upper and lower cheek teeth together, but more recently this has been replaced by specialized equipment producing panoramic projections of the whole dentition.

Morphology and measurements

The Arizona State University (ASU) system for according dental morphology (page 86) has been described fully by Turner *et al.* (1991). It requires a series of reference plaques that can be obtained at cost from Diane Hawkey, Department of Anthropology, Box 872402, Arizona State University, Tempe, AZ 85287-2402, USA. Photographs of plaques from this system have been included, with permission, in Chapter 3 (Figures 3.4 to 3.8, 3.10 and 3.11).

Measurements of crown diameters are taken with small dial callipers, with their jaws machined to fine points. These can be made from ordinary engineers' callipers, if the appropriate machine tools are available, but pointed versions are made by some manufacturers. Some models include transducers and appropriate circuitry to pass measurements directly into a computer system.

Appendix B

Microscopy

Light microscopy

Further reading

Hartley (1979), Bradbury (1984) and Schmidt & Keil (1971).

The limits of simple observation

Some details of dental histology can be made out with minimal magnification. Unaided, the human eye easily resolves dots 200 μm apart, in a specimen held 250 mm away. At best, the eye resolves 70 μm, and a hand lens or loupe may resolve 10μm – the main limitation is in lighting the object well enough.

Compound microscopes

Serious dental histology, however, relies on compound microscopes, with an objective lens that forms the primary image and provides the resolving power, and an eyepiece lens that provides secondary magnification. A microscope for dental studies might have × 1 or x 4, × 10 and × 40 objectives, with × 10 and × 15 eyepieces, altogether giving a range of magnifications from × 10 to × 600. A moderately good × 10 objective might have a resolving power of 1 μm and a field depth (the thickness of the plane of focus) of 10 μm, and a similar quality × 40 objective might have a 0.4 μm resolving power and 2 μm depth of field.

Most microscopes can be fitted with a purpose-built camera, but it is also possible to obtain fittings for most ordinary 35 mm cameras. An alternative approach is to use a television camera connected to a computer-based image analysis system. Images are stored, filtered and enhanced, and measured. Most microscopes can also be fitted with a drawing attachment (drawing tube) with which tracings may be made directly from the image, and these can be very useful when keeping track of layer counts (page 177), or recording crown morphology or attrition facets.

Transmitted light microscopes

Ground sections of teeth, or demineralized and stained preparations (page 300), are mounted on glass slides and examined with a transmitted light microscope. The section is illuminated by the condenser lens assembly, which is mounted underneath the stage holding the glass slide. Much work can be carried out using the normal microscope configuration, but a range of more specialized types is also used. In dark field microscopy, an opaque circular patch underneath the condenser lens permits only a ring of light to enter, so the specimen is illuminated by oblique rays of light, to produce a dark image in which sharp changes in RI (page 311) are bright. For low magnifications it is possible to improvise this arrangement with a circle of card.

The polarizing microscope is another specialized form, but is a mainstay of dental histology. A ground tooth section is illuminated with polarized light (vibrating only in one plane) from a special filter, the polarizer, which is built into the condenser assembly. The specimen is observed through another polarizing filter, the analyser, fitted above the objective, with a vibration plane perpendicular to that of the polarizer. Brightness, darkness and colour are controlled by the specimen's light refracting properties and orientation, and the section thickness. The microscope has a rotating stage, marked off in degrees, to measure these orientations. Polarizing microscopes may be improvized by cutting filters from sheets of polaroid but, for dental histology, it is best to specify a polarizing microscope in the first place (it can also be used for ordinary transmitted light).

Reflected light microscopes

For low magnifications, a transmitted light microscope can simply be converted by using a lamp to light the surface of the specimen but, for higher powers, a specially designed assembly is needed. These focus light, either through the objective itself or around it, onto the specimen. The former is known as bright field illumination, and the latter dark field.

Another form of reflected light microscope, used as a basic tool in dental anthropology, is the low power stereomicroscope. A good-quality instrument has a wide field of view that can be observed with binocular vision, giving the normal impression of nearness and farness, height and depth. This is strongest at low magnifications (with their larger depth of field), so such microscopes work best at $\times 10$–$\times 50$ or so. Illumination is critical, and it is well worth investing in good-quality fibre-optic lamps that can really direct the light where it is needed.

Confocal microscopy

Confocal microscopes focus a point of light accurately onto a plane within the specimen and observe it through an aperture that excludes light transmitted or reflected by any other plane. The image is built up by scanning the point of light across the plane of focus. In confocal scanning laser microscopes (CSLM), a laser beam is scanned by a system of mirrors whereas, in tandem scanning reflected light microscopes (TSRLM or TSM) the point of light is supplied by apertures in a spinning disk. Both CSLM and TSM are commonly used in a reflected light configuration, with the illuminating beam focused through the same objective as the image. The confocal arrangement allows a very thin plane (1 µm deep) to be imaged, and this can be focused into a translucent specimen, through the intact surface, for several hundred micrometres. A TSM attached to an image analysis system can thus record a series of through-focus images to build a three-dimensional model of sub-surface structure. A CSLM, in a similar way, can produce profiles or three-dimensional models of fine surface detail.

The factors that form images in light microscopy

In an ideal world, sections for transmitted light microscopy would be cut thinner than the structures under examination but, in dental tissues, the finest structures are 4 µm across, and the practical lower limit for most ground sections is 30 µm in thickness (page 304). Most structures are therefore not imaged directly, but detected through their cumulative effect on the light that passes through. Some of these features are due to the scattering of light as it passes through the section, including brown striae of Retzius (page 157), which are dark in transmitted light and bright in reflected light or dark-field microscopy. They represent bands of more porous enamel that either scatter the beam to the side as it passes through, or scatter it back to the observer, depending upon the illumination. It is important to realize that the striae of Retzius are not the result of staining, but that their brown colouration results from the fact that the blue end of the spectrum is scattered more than the red. Other features are more likely to be due to reflection phenomena at the boundaries between materials of different light transmission rate (refractive index or RI). Such boundaries cause a beam of light that is crossing the boundary to bend, and the degree of bending depends upon the contrast in RI. If the materials contrast strongly, or the light beam hits the boundary obliquely enough, then the beam may be reflected back at the boundary instead of passing through. This is seen, for example, in the dead tracts of dentine (page 194) that have air (RI 1) trapped in the dentinal tubules, and thus have a strong contrast with the apatite mineral (RI 1.63) of dentine. Dead tracts are therefore dark in transmitted light, and bright in reflected light. Internal reflection effects at steeply inclined (and therefore obliquely illuminated) boundaries may also explain the visibility of cross-striations and prisms in transmitted light microscopy of enamel (page 155). Much depends on the RI of the medium in which the specimen is mounted, and the extent to which it penetrates the spaces in the tissue. A range of mounting media is available, from quinoline (closest to apatite at RI 1.62), to Canada Balsam (RI 1.54), alcohol (RI *c.*1.36) and water (RI 1.33).

With stained preparations, the images are formed by absorption and fluorescence of light. The dyes in a conventionally stained section absorb some colours of light, but allow others to pass through and create the image seen by the observer. Fluorescent stains (such as tetracyclines) work differently, by absorbing ultraviolet light (which cannot be seen) and fluorescing within the visible part of the spectrum.

The appearance of dental tissue in the polarizing microscope

Polarizing microscopy is based upon the phenomenon of birefringence, shown by most minerals and many large regular organic structures, in which a beam of light is constrained into two groups of waves, each polarized perpendicular to one another, and with different RI. In an apatite (pages 148 and 153) crystal, the lower RI waves (fast waves) are polarized parallel to the long axis of the crystal, whilst the higher RI waves (slow waves) are polarized perpendicular to it – an arrangement known as optically negative. The difference in RI is called birefringence, and it varies with orientation. Seen end on, an apatite crystal has zero birefringence and is dark, whereas maximum birefringence is seen when the crystal is viewed perpendicular to its long axis. When the waves emerge from the crystal, they recombine slightly out of phase, and are resolved by the analyser filter to produce a range of interference colours related to the birefringence and thickness of the section. Apatite has a low birefringence and dull interference colours; greys, whites and yellows in 60–100 μm sections. Rotating the microscope stage also has an effect on the image. When either fast or slow waves coincide in their plane of polarization with the polarizer or analyser, each crystal becomes dark and, as the stage is rotated through 360°, the crystal flashes bright and dark at 90° intervals. This phenomenon is known as extinction. Apatite crystallites are too small to see individually, but they are arranged into groups of similar orientation that extinguish all together, so that the difference in angle between, for example, prism boundaries (page 152) and extinction positions gives the overall crystallite orientation.

Collagen, the fibrous protein of dentine and cement, has a higher birefringence than apatite, with the slow waves polarized parallel to the fibres' long axis (optically positive). It is similarly possible to use extinction to determine fibre orientation. Furthermore, dental tissues are composites of apatite crystals, collagen molecules, other organic molecules, water and air-filled spaces. This mixed body produces its own additional form or textural birefringence. In dentine and cement, the form birefringence is optically positive, aligned with the long axis of collagen fibrils, and much higher than the intrinsic birefringence of apatite. Form birefringence is also seen when the enamel becomes demineralized in dental caries (page 270), which increases pore space and allows the tissue to imbibe more mounting medium.

Scanning electron microscopes

Further reading

Goldstein *et al.* (1992), Watt (1985) and Boyde (1984).

How the SEM works

A scanning electron microscope (SEM) focuses electrons into a fine beam, which it scans across the specimen surface, and an image is produced from the changing effects of the beam on the specimen. These effects are measured by detectors, whose electrical signals are amplified and processed into the image. The operator sees it on a television-like screen (viewing CRT), and a photographic record is made using a camera fitted over a separate screen (recording CRT). Maximum magnification depends upon the design of the microscope, and the way in which it is operated, but is typically about × 200 000, whereas the lowest magnification might be × 10. The main advantage of the SEM, however, is its very large depth of field (the amount of a high relief surface seen in focus at the same time), which again depends upon the operation of the instrument, but may be as much as 4 mm with a low magnification. This, for example, allows the whole crown surface, or a rough fractured surface, to be imaged in focus all at one time.

The two detectors most commonly used in dental histology are the Everhart–Thornley configuration (often called the secondary electron detector), and the solid-state backscattered electron detector. Secondary electrons and backscattered electrons emanate from the surface of the specimen as a result of its interactions with the electron beam, within a tiny pear-shaped area usually extending just 1 μm below the surface:

1. Electrons enter the specimen in this interaction zone and scatter through it. Most come to rest within the specimen, but a few scatter far enough to escape from the surface, and these are the so-called backscattered electrons (BSE). The rate at which BSE escape depends largely upon the composition of the specimen, with higher atomic number elements producing a higher yield than lower number elements, and higher density materials producing a higher yield than less dense materials. In dental specimens, the BSE signal is sensitive to slight variations in the degree of mineralization and in the chemical composition of minerals present. There is also, however, a variation in BSE yield caused by the slope of the specimen surface relative to the axis of the electron beam, or when there is a sharp eminence or edge. The image produced from the BSE signal is therefore the result of a complex mixture of compositional and surface topographical effects.
2. As it scatters through the specimen, each incoming electron may also initiate a number of secondary electrons (SE). Most SE probably do not travel very far, and many do not reach the surface, but enough escape from the surface 10 nm to be detected. The SE yield is partly controlled by the electron beam voltage, and partly by the surface topography (in a similar way to BSE).

If the specimen is an insulator (as dental tissues are) then an electrical charge builds up during operation of the microscope, although this can be minimized by the way in which it is operated. The charge may affect the detectors, or deflect the electron beam, and so is undesirable. Most specimens are therefore coated with a very thin conducting layer of gold or carbon (page 305).

The Everhart–Thornley detector

The 'standard' detector for many SEM applications is the Everhart–Thornley or scintillator type (ET detector). It is particularly efficient at detecting SE and so is often described as a secondary electron detector, but it also detects some BSE. The ET detector produces an image showing mainly the topography of the specimen surface, but there is also a compositional component that should not be forgotten. The observer sees the specimen as though looking vertically down on it, and the image is arranged to give the appearance of illumination from the top of the picture. Surfaces tilted towards the detector produce a stronger signal and thus appear brighter, whereas surfaces tilted away are darker. Superimposed over this, edges, points and ridges are especially bright, and higher atomic number or higher density zones also produce a slightly stronger signal.

Back-scattered electron detectors

Several detectors have been designed to exclude SE and maximize detection of BSE, but the most suitable for dental histology is a solid-state type. Both compositional and topographic attributes of the specimen contribute to the image, and many solid-state detectors are designed to help distinguish them. In topographic mode they give the effect of strongly oblique illumination, and the direction of apparent illumination can be changed by altering the signal processing. Usually, it is arranged so that the illumination appears to come from the top left corner of the image. Alternatively, if specimen composition is the main interest, the first step is to make the specimen as flat as possible, and arrange it perpendicular to the electron beam axis. The detector is then operated in a mode that minimizes the effect of any remaining surface topography, even though it is not possible to filter it out completely. The effect of topography has therefore to be judged by comparing 'compositional' images with 'topographical' images of the same area of the specimen. In general, however, the more heavily mineralized tissues and tissue components show up brighter, so that enamel is much brighter than dentine, which is in turn brighter than cement and bone.

References

Aas, I. H. (1979) The depth of the lingual fossa in permanent incisors of Norwegian Lapps, *American Journal of Physical Anthropology*, **51**, 417–420.

Aas, I. H. (1983) Variabililty of a dental morphological trait, *Acta Odontologica Scandinavica*, **41**, 257–263.

Aas, I. H. & Risnes, S. (1979) The depth of the lingual fossa in permanent incisors of Norwegians. I. Method of measurement, statistical distribution and sex dimorphism, *American Journal of Physical Anthropology*, **50**, 335–340.

Aiello, L. & Dean, C. (1990) *An introduction to human evolutionary anatomy*, London: Academic Press.

Aiello, L. C. & Molleson, T. (1993) Are microscopic ageing techniques more accurate than macroscopic ageing techniques', *Journal of Archaeological Science*, **20**, 689–704.

Aitken, M. J. (1990) *Science-based dating in archaeology*, Longman Archaeology Series, London: Longman.

Ajie, H. O., Kaplan, I. R., Slota, P. J. & Taylor, R. E. (1990) AMS radiocarbon dating bone osteocalcin, *Nuclear Instruments and Methods*, **B52**, 433–438.

Alexandersen, V. (1967) The pathology of the jaws and temporomandibular joint, in Brothwell, D. R. & Sandison, A. T. (eds.) *Diseases in antiquity*, Springfield: Thomas, pp. 551–595.

Alvesalo, L. (1971) The influence of sex chromosome genes on tooth size in man, *Suomen Hammaslääkäriseunan Toimituksia (Proceedings of the Finnish Dental Society)*, **67**, 3–54.

Alvesalo, L. & Tigerstedt, P. M. A. (1974) Heritabilities of human tooth dimensions, *Hereditas*, **77**, 311–318.

Ambrose, S. H. (1993) Isotopic analysis of paleodiets: methodological and interpretive considerations, in Sandford, M. K. (ed.) *Investigations of ancient human tissue. Chemical analyses in anthropology*, Food and Nutrition in History and Anthropology Volume 10, Langhorne, Pennsylvania: Gordon & Breach, pp. 59–130.

Anderson, D. L., Thompson, G. W. & Popovitch, F. (1976) Age of attainment of mineralisation stages of the permanent dentition, *Journal of Forensic Sciences*, **21**, 191–200.

Andersen, V. (1898) Die Querstreifung des Dentins, *Deutsche Monatsschrift für Zahnheilkunde*, **16**, 386–389.

Armelagos, G. J. (1969) Diseases in ancient Nubia, *Science*, **163**, 255–259.

Arya, B. S., Thomas, D. R., Savara, B. S. & Clarkson, Q. D. (1974) Correlations among tooth sizes in a sample of Oregon caucasoid children, *Human Biology*, **46**, 693–698.

Asper, H. (1916) Ueber die ''Braune Retzius'sche Parallelstreifung'' im Schmelz der Menschlichen Zähne, *Schweizerische Vrtljschrift für Zahneilkunde*, **26**, 275.

Aufderheide, A. C. (1989) Chemical analysis of skeletal remains, in Iscan, M. Y. & Kennedy, K. A. R. (eds.) *Reconstruction of life from the skeleton*, New York: Alan R. Liss, pp. 237–260.

Azaz, B., Michaeli, Y. & Nitzan, D. (1977) Ageing tissues of the roots of non-functional human teeth (impacted canines), *Oral Surgery, Oral Medicine, Oral Pathology*, **43**, 572–578.

Bailit, H. L. & Friedlaender, J. S. (1966) Tooth size reduction: a hominid trend, *American Anthropologist*, **68**, 665–672.

Bailit, H. L., Workman, P. L., Niswander, J. D. & Maclean, C. J. (1970) Dental asymmetry as an indicator of genetic and environmental conditions in human populations, *Human Biology*, **42**, 626–637.

Bang, G. & Ramm, E. (1970) Determination of age in humans from root dentine transparency, *Acta Odontologica Scandinavica*, **28**, 3–35.

Baume, L. J., Horowitz, H. S., Summers, C. J., Dirks, O. B., Brown, W. A. B., Carlos, J. P., Cohen, L. K., Freer, T. J., Harvold, E. P., Moorrees, C. F. A., Satzmann, J. A., Schmuth, G., Solow, B. & Taatz, H. (1970) A method for measuring occlusal traits. Developed by the F.D.I. Commission and Classification and Statistics for Oral Conditions (COCSTOC) Working Group 2 on Dentofacial Anomalies, 1969–72, *International Dental Journal*, **23**, 530–537.

Bäckmann, B. (1989) *Amelogenesis imperfecta. An epidemiologic, genetic, morphologic and clinial study*, University of Umeå, Sweden, Departments of Pedodontics and Oral Pathology, doctoral dissertation.

Beasley, M. J., Brown, W. A. B. & Legge, A. J. (1992) Incremental banding in dental cementum: methods of preparation for teeth from archeological sites and for modern comparative specimens, *International Journal of Osteoarchaeology*, **2**, 37–50.

Beckett, S. & Lovell, N. C. (1994) Dental disease evidence for agricultural intensification in the Nubian C-Group, *International Journal of Osteoarchaeology*, **4**, 223–240.

Beeley, J. G. & Lunt, D. A. (1980) The nature of the biochemical changes in softened dentine from archaeological sites, *Journal of Archaeological Science*, **7**, 371–377.

Begg, P. R. (1954) Stone Age man's dentition, *American Journal of Orthodontics*, **40**, 298–312, 373–383, 462–475, 517–531.

Beiswanger, B. B., Segreto, V. A., Mallatt, M. E. & Pfeiffer, H. J. (1989) The prevalance and incidence of dental calculus in adults, *Journal of Clinical Dentristry*, **1**, 55–58.

Bell, L. S. (1990) Palaeopathology and diagenesis: an SEM evaluation of structural changes using backscattered electron imaging, *Journal of Archaeological Science*, **17**, 85–102.

Bell, L. S., Boyde, A. & Jones, S. J. (1991) Diagenetic alteration to teeth *in situ* illustrated by backscattered electron imaging, *Scanning*, **13**, 173–183.

Benfer, R. A. & Edwards, D. S. (1991) The principal axis method for measuring rate and amount of dental attrition: estimating juvenile or adult tooth wear from unaged adult teeth, in Kelley, M. A. & Larsen, C. S. (eds.) *Advances in Dental Anthropology*, New York: Wiley-Liss, pp. 325–340.

Bercy, P. & Frank, R. M. (1980) Microscopie electronique a balayage de la plaque dentaire et du tartre a la surface du cement humain, *Journal de Biologie Buccale*, **8**, 299–313.

Bermann, M., Edwards, L. F. & Kitchin, P. C. (1939) Effect of artificially induced hyperpyrexia on tooth structure in the rabbit, *Proceedings of the Society for Experimental Biology & Medicine*, **41**, 113–115.

Bermudez de Castro, J. M., Durand, A. I. & Ipiña, S. L. (1993) Sexual dimorphism in the human dental sample from the SH site (Sierra de Atapuerca, Spain): a statistical approach, *Journal of Human Evolution*, **24**, 43–56.

Berry, A. C. (1978) Anthropological and family studies on minor variants of the dental crown, in Butler, P. M. & Joysey, K. A. (eds.) *Development, function and evolution of teeth*, London: Academic Press, pp. 81–98.

Berryman, H. E., Owsley, D. W. & Henderson, A. M. (1979) Non-carious interproximal grooves in Arikara Indian dentitions, *American Journal of Physical Anthropology*, **50**, 209–212.

Berten, J. (1895) Hypoplasie des Schmelzes (Congenitale Schmelzdefecte; Erosionen), *Deutsche Monatsschrift für Zahnheilkunde*, **13**, 425–439; 483–498; 533–548; 587–606.

Beynon, A. D. (1987) Replication techniques for studying microstructure in fossil enamel, *Scanning Microscopy*, **1**, 663–669.

Beynon, A. D. (1992) Circaseptan rhythms in enamel development in modern humans and Plio-Pleistocene hominids, in Smith, P. & Tchernov, E. (eds.) *Structure, function and evolution of teeth*, London & Tel Aviv: Freund Publishing House Ltd, pp. 295–310.

Beynon, A. D. & Dean, M. C. (1988) Distinct dental development patterns in early fossil hominids, *Nature*, **335**, 509–514.

Beynon, A. D. & Dean, M. C. (1991) Hominid dental development, *Nature*, **351**, 196.

Beynon, A. D., Dean, M. C. & Reid, D. J. (1991a) Histological study on the chronology of the developing dentition in gorilla and orangutan, *American Journal of Physical Anthropology*, **86**, 189–203.

Beynon, A. D., Dean, M. C. & Reid, D. J. (1991b) On thick and thin enamel in hominoids, *American Journal of Physical Anthropology*, **86**, 295–309.

Beynon, A. D. & Wood, B. A. (1986) Variations in enamel thickness and structure in East African hominids, *American Journal of Physical Anthropology*, **70**, 177–195.

Beynon, A. D. & Wood, B. A. (1987) Patterns and rates of enamel growth in the molar teeth of early hominids, *Nature*, **326**, 493–496.

Biggerstaff, R. H. (1968) On the groove configuration of mandibular molars: the unreliability of the 'Dryopithecus pattern' and a new method for classifying mandibular molars, *American Journal of Physical Anthropology*, **29**, 441–444.

Biggerstaff, R. H. (1969) The basal area of posterior tooth crown components: the assessment of within tooth variations of premolars and molars, *American Journal of Physical Anthropology*, **31**, 163–170.

Biggerstaff, R. H. (1970) Morphological variations for the permanent mandibular first molars in human monozygotic and dizygotic twins, *Archives of Oral Biology*, **15**, 721–730.

Biggerstaff, R. H. (1973) Heritability of Carabelli's cusp in twins, *Journal of Dental Research*, **52**, 40–44.

Biggerstaff, R. H. (1975) Cusp size, sexual dimorphism, and heritability of cusp size in twins, *American Journal of Physical Anthropology*, **42**, 127–140.

Biggerstaff, R. H. (1976) Cusp size, sexual dimorphism, and heritability of maximum molar cusp size in twins, *Journal of Dental Research*, **55**, 189–195.

Black, T. K. (1978) Sexual dimorphism in the tooth-crown diameters of deciduous teeth, *American Journal of Physical Anthropology*, **48**, 77–82.

Black, T. K. (1980) An exception to the apparent relationship between stress and fluctuating dental asymmetry, *Journal of Dental Research*, **59**, 1168–1169.

Blake, G. C. (1958) The peritubular translucent zone in dentine, *British Dental Journal*, **104**, 57.

Blumberg, J. E., Hylander, W. L. & Goepp, R. A. (1971) Taurodontism: a biometric study, *American Journal of Physical Anthropology*, **34**, 243–255.

Bocklage, C. E. (1992) Method and meaning in the analysis of developmental asymmetries, in Lukacs, J. R. (ed.) *Culture, ecology & dental anthropology*, Journal of Human Ecology Special Issue No. 2, Delhi: Kamla-Raj Enterprises, pp. 147–156.

Bolk, L. (1916) Problems of human dentition, *American Journal of Anatomy*, **19**, 163–170.

Bowen, W. H. & Pearson, S. K. (1993) Effect of milk on cariogenesis, *Caries Research*, **27**, 461–466.

Boyde, A. (1963) Estimation of age at death of young human skeletal remains from incremental lines in dental enamel, *Third International Meeting in Forensic Immunology, Medicine, Pathology and Toxicology, Plenary Session 11A*, London.

Boyde, A. (1970) The surface of the enamel in human hypoplastic teeth, *Archives of Oral Biology*, **15**, 897–898.

Boyde, A. (1971) The tooth surface, in Eastoe, J. E., Picton, D. C. A. & Alexander, A. G. (eds.) *The prevention of periodontal disease*, London: Kimpton, pp. 46–63.

Boyde, A. (1974) Photogrammetry of stereo-pair SEM images using separate measurements from the two images, *Scanning Electron Microscopy*, 1974, 101–108.

Boyde, A. (1975) Scanning electron microscopy of enamel surfaces, *British Medical Bulletin*, **31**, 120–124.

Boyde, A. (1976a) Amelogenesis and the development of teeth, in Cohen, B. & Kramer, I. R. H. (eds.) *Scientific foundations of dentistry*, London: Heinemann.

Boyde, A. (1976b) Enamel structure and cavity margins, *Operative Dentistry*, **1**, 13–28.

Boyde, A. (1979) The perception and measurement of depth in the SEM, *Scanning Electron Microscopy*, 1979 II, 67–78.

Boyde, A. (1984) Methodology of calcified tissue specimen preparation for scanning electron microscopy, in Dickson, G. R. (ed.) *Methods of calcified tissue preparation*, Amsterdam: Elsevier, pp. 251–307.

Boyde, A. (1989) Enamel, in Berkovitz, B. K. B., Boyde, A., Frank, R. M., Höhling,

H. J., Moxham, B. J., Nalbandian, J. & Tonge, C. H. (eds.) *Teeth*, Handbook of Microscopic Anatomy, New York, Berlin & Heidelberg: Springer Verlag, Vol. V/6, pp. 309–473.

Boyde, A. (1990) Developmental interpretations of dental microstructure, in DeRousseau, C. J. (ed.) *Primate life history and evolution*, Monographs in Primatology Volume 14, New York: Wiley-Liss, pp. 229–267.

Boyde, A., Fortelius, M., Lester, K. S. & Martin, L. B. (1988) Basis of the structure and development of mammalian enamel as seen by scanning electron microscopy, *Scanning Microscopy*, **2**, 1479–1490.

Boyde, A., Hendel, P., Hendel, R., Maconnachie, E. & Jones, S. J. (1990) Human cranial bone structure and the healing of cranial bone grafts: a study using back-scattered electron imaging and confocal microscopy, *Anatomy & Embryology*, **181**, 235–251.

Boyde, A., Jones, S. J. & Reynolds, P. S. (1978) Quantitative and qualitative studies of enamel etching with acid and EDTA, *Scanning Electron Microscopy*, 1978 II, 991–1002.

Boyde, A. & Martin, L. B. (1984) A non-destructive survey of prism packing patterns in primate enamels, in Fearnhead, R. W. & Suga, S. (eds.) *Tooth Enamel IV*, Amsterdam: Elsevier Science Publications BV, pp. 417–421.

Brace, C. L. (1964) The probable mutation effect, *American Naturalist*, **97**, 39–49.

Brace, C. L. & Mahler, P. E. (1971) Post-Pleistocene changes in the human dentition, *American Journal of Physical Anthropology*, **34**, 191–204.

Brace, C. L., Rosenberg, K. R. & Hunt, K. D. (1987) Gradual change in human tooth size in the late Pleistocene and post-Pleistocene, *Evolution*, **41**, 705–720.

Brace, C. L. & Ryan, A. S. (1980) Sexual dimorphism and human tooth size differences, *Journal of Human Evolution*, **9**, 417–435.

Brace, C. L., Ziang-qing, S. & Zhen-biao, Z. (1984) Biological and cultural change in the European Late Pleistocene and Early Holocene, in Smith, F. H. & Spencer, F. (eds.) *The origin of modern humans*, New York: Alan R. Liss, pp. 485–516.

Bradbury, S. (1984) *An introduction to the optical microscope*, Royal Microscopical Society microscopy handbooks 1, Oxford: Oxford University Press.

Bradford, E. W. (1967) Microanatomy and histochemistry of dentine, in Miles, A. E. W. (ed.) *Structural and chemical organization of teeth*, London: Academic Press, Vol. 2, pp. 3–34.

Brekhus, P. J., Oliver, C. P. & Montelius, G. (1944) A study of the pattern and combinations of congenitally missing teeth in man, *Journal of Dental Research*, **23**, 117–131.

Bromage, T. G. (1991) Enamel incremental periodicity in the pig-tailed macaque: a polychrome fluorescent labeling study of dental hard tissues, *American Journal of Physical Anthropology*, **86**, 205–214.

Bromage, T. G. & Dean, M. C. (1985) Re-evaluation of the age at death of immature fossil hominids, *Nature*, **317**, 525–527.

Brothwell, D. R. (1959) Teeth in earlier human populations, *Proceedings of the Nutrition Society*, **18**, 59–65.

Brothwell, D. R. (ed.), (1963a) *Dental anthropology*, London: Pergamon Press.

Brothwell, D. R. (1963b) *Digging up bones*, 1st edn., London: British Museum.

Brothwell, D. R. (1963c) The macroscopic dental pathology of some earlier human populations, in Brothwell, D. R. (ed.) *Dental anthropology*, London: Pergamon Press, pp. 272–287.

Brothwell, D. R. (1989) The relationship of tooth wear to aging, in Iscar, M. Y. (ed.) *Age markers in the human skeleton*, Springfield: Charles C. Thomas, pp. 303–316.

Brothwell, D. R., Carbonell, V. M. & Goose, D. H. (1963) Congenital absence of teeth in human populations, in Brothwell, D. R. (ed.) *Dental anthropology*, London: Pergamon Press, pp. 179–189.

Brown, T. & Molnar, S. (1990) Interproximal grooving and task activity in Australia, *American Journal of Physical Anthropology*, **81**, 545–553.

Brudevold, F. & Söremark, R. (1967) Chemistry of the mineral phase of enamel, in Miles, A. E. W. (ed.) *Structural and chemical organization of teeth*, London: Academic Press, Vol. 2, pp. 247–278.

Brudevold, F. & Steadman, L. T. (1956) The distribution of lead in human enamel, *Journal of Dental Research*, **35**, 420.

Buckley, L. A. (1981) The relationships between malocclusion, gingival inflammation, plaque and calculus, *Journal of Periodontology*, **52**, 35–40.

Buikstra, J. E., Frankenberg, S., Lambert, J. B. & Xue, L. A. (1989) Multiple elements: multiple expectations, in Price, T. D. (ed.) *The chemistry of prehistoric human bone*, School of American Research Advanced Seminar Series, Cambridge: Cambridge University Press, pp. 155–210.

Buikstra, J. E. & Ubelaker, D. H. (eds.) (1994) *Standards for data collection from human skeletal remains*, Arkansas Archeological Survey Research Series No 44, Fayetteville: Arkansas Archeological Survey.

Bullion, S. K. (1987) *Incremental structures of enamel and their applications to archaeology*, University of Lancaster, PhD dissertation.

Bunon, R. (1746) *Expériences et demonstrations faites à l'Hôpital de la Salpêtriere, et à S. Côme en présence de l'Académie Royale de Chirurgie*, Paris.

Burns, K. R. & Maples, W. R. (1976) Estimation of age from individual adult teeth, *Journal of Forensic Sciences*, **21**, 343–356.

Butler, P. M. (1939) Studies in the mammalian dentition – and of differentiation of the postcanine dentition, *Proceedings of the Zoological Society, London, B*, **107**, 103–132.

Butler, P. M. (1967a) Comparison of the development of the second deciduous molar and first permanent molar in man, *Archives of Oral Biology*, **12**, 1245–1260.

Butler, P. M. (1967b) Relative growth within the human first upper permanent molar during the prenatal period, *Archives of Oral Biology*, **12**, 983–992.

Butler, P. M. (1968) Growth of the human second lower deciduous molar, *Archives of Oral Biology*, **13**, 671–682.

Butler, P. M. (1971) Growth of human tooth germs, in Dahlberg, A. A. (ed.) *Dental morphology and evolution*, Chicago: University of Chicago Press, pp. 3–13.

Butler, P. M. & Joysey, K. A. (eds.) (1978) *Development, function and evolution of teeth*, London: Academic Press.

Calcagno, J. M. (1989) *Mechanisms of human dental reduction. A case study from post-Pleistocene Nubia.*, University of Kansas Publications in Anthropology 18, Lawrence: University of Kansas.

Calcagno, J. M. & Gibson, K. R. (1991) Selective compromise: evolutionary trends and mechanisms in hominid tooth size, in Kelley, M. A. & Larsen, C. S. (eds.) *Advances in dental anthropology*, New York: Wiley–Liss, pp. 59–76.

Calonius, P. E. B., Lunin, M. & Stout, F. (1970) Histological criteria for age estimation of the developing human dentition, *Oral Surgery*, **29**, 869–876.

Campbell, T. D. (1925) *Dentition and palate of the Australian aboriginal*, Publications under the Keith Sheridan Foundation, Adelaide: University of Adelaide.

Canis, M. F., Kramer, G. M. & Pameijer, C. M. (1979) Calculus attachment. Review of the literature and new findings, *Journal of Periodontology*, **50**, 406–415.

Carbonell, V. M. (1963) Variations in the frequency of shovel-shaped incisors in different populations, in Brothwell, D. R. (ed.) *Dental anthropology*, London: Pergamon Press, pp. 211–234.

Carlsen, O. (1987) *Dental morphology*, Copenhagen: Munksgaard.

Carlson, D. S. & van Gerven, D. P. (1977) Masticatory function and post-Pleistocene evolution in Nubia, *American Journal of Physical Anthropology*, **46**, 495–506.

Cassidy, C. M. (1984) Skeletal evidence for prehistoric subsistence adaptation in the central Ohio River Valley, in Cohen, M. N. & Armelagos, G. J. (eds.) *Palaeopathology at the origins of agriculture*, New York: Academic Press, pp. 307–346.

Cattaneo, C., Gelsthorpe, K., Phillips, P. & Sokol, R. J. (1992) Reliable identification of human albumin in ancient bone using ELISA and monoclonal antibodies, *American Journal of Physical Anthropology*, **87**, 365–372.

Chagula, W. K. (1960) The cusps on the mandibular molars of East Africans, *American Journal of Physical Anthropology*, **18**, 83–90.

Charles, D. K., Condon, K., Cheverud, J. M. & Buikstra, J. E. (1986) Cementum annulation and age determination in *Homo sapiens*. I. Tooth variability and observer error, *American Journal of Physical Anthropology*, **71**, 311–320.

Charles, D. K., Condon, K., Cheverud, J. M. & Buikstra, J. E. (1989) Estimating age at death from growth layer groups in cementum, in Iscan, M. Y. (ed.) *Age markers in the human skeleton*, Springfield: Charles C Thomas, pp. 277–316.

Chen, T. & Yuan, S. (1988) Uranium series dating of bones and teeth from Chinese Palaeolithic sites, *Archaeometry*, **30**, 59–76.

Christensen, G. J. & Kraus, B. S. (1965) Initial calcification of the human permanent first molar, *Journal of Dental Research*, **44**, 1338–1342.

Clark, D. H. (ed.) (1992) *Practical forensic odontology*, Oxford: Wright.

Clarke, N. G., Carey, S. E., Srikandi, W., Hirsch, R. S. & Leppard, P. I. (1986) Periodontal disease in ancient populations, *American Journal of Physical Anthropology*, **71**, 173–183.

Clarke, N. G. & Hirsch, R. S. (1991) Physiological, pulpal, and periodontal factors influencing alveolar bone, in Kelley, M. A. & Larsen, C. S. (eds.) *Advances in dental anthropology*, New York: Wiley–Liss, pp. 241–266.

Clarkson, J. (1989) Review of terminology, classification, and indices of developmental defects of enamel, *Advances in Dental Research*, **3**, 104–109.

Clarkson, J. & O'Mullane, D. (1989) A modified DDE index for use in epidemiological studies of enamel defects, *Journal of Dental Research*, **68**, 445–450.

Clement, A. J. (1958) The antiquity of caries, *British Dental Journal*, **104**, 115–123.

Clinch, L. A. (1963) A longitudinal study of the mesiodistal crown diameters of the deciduous teeth and their permanent successors, *European Orthodontic Society*, Report 33, 202–215.

Colby, G. R. (1996) Analysis of dental sexual dimorphism in two Western Gulf of Mexico precontact populations utilizing cervical measurements (abstract), *American Journal of Physical Anthropology.* Supplement 22, 87.

Colyer, J. F. (1936) *Variations and diseases of the teeth of animals,* London: John Bale, Sons & Danielsson.

Commission on Oral Health R. E. (1982) An epidemiological index of developmental defects of dental enamel (DDE Index), *International Dental Journal,* **32**, 159–167.

Condon, K. (1981) *Correspondence of developmental enamel defects between the mandibular canine and first premolar,* University of Arkansas, MA dissertation.

Condon, K., Becker, J., Condon, C. & Hoffman, J. R. (1994) Dental and skeletal indicators of a congenital treponematosis, *American Journal of Physical Anthropology,* Supplement 18, 70.

Condon, K., Charles, D. K., Cheverud, J. M. & Buikstra, J. E. (1986) Cementum annulation and age determination in *Homo sapiens.* II. Estimates and accuracy, *American Journal of Physical Anthropology,* **71**, 321–330.

Conroy, G. C. & Vannier, M. W. (1987) Dental development of the Taung skull from computerized tomography, *Nature,* **329**, 625–627.

Cook, D. C. (1980) Hereditary enamel hypoplasia in a prehistoric Indian child, *Journal of Dental Research,* **59**, 1522.

Cook, D. C. (1984) Subsistence and health in the Lower Illinois Valley: osteological evidence, in Cohen, M. N. & Armelagos, G. J. (eds.) *Palaeopathology at the origins of agriculture,* New York: Academic Press, pp. 235–269.

Cook, D. C. (1990) Epidemiology of circular caries: a perspective from prehistoric skeletons, in Buikstra, J. E. (ed.) *A life in science: papers in honor of J. Lawrence Angel,* Scientific Papers No 6, Center for American Archaeology, SIU, pp. 64–86.

Coote, G. E. & Molleson, T. (1988) Flourine diffusion profiles in archaeological human teeth: a method for relative dating of burials?, in Prescott, J. R. (ed.) *Archaeometry: Australian studies 1988,* Adelaide: Department of Physics, University of Adelaide, pp. 99–104.

Coote, G. E. & Sparks, R. J. (1981) Flourine concentration profiles in archaeological bones, *New Zealand Journal of Archaeology,* **3**, 21–32.

Corbett, M. E. & Moore, W. J. (1976) Distribution of dental caries in ancient British populations: IV The 19th Century, *Caries Research,* **10**, 401–414.

Corey, L. A., Nance, W. E., Hofstede, P. & Schenkein, H. A. (1993) Self-reported periodontal disease in a Virginia twin population, *Journal of Periodontology,* **64**, 1205–1208.

Corruccini, R. S. (1977a) Crown component variation in hominoid lower third molars, *Zeitschrift für Morphologie und Anthropologie,* **68**, 14–25.

Corruccini, R. S. (1977b) Crown component variation in the hominoid lower second premolar, *Journal of Dental Research,* **56**, 1093–1096.

Corruccini, R. S. (1978) Crown component analysis of the hominoid upper first premolar, *Archives of Oral Biology,* **23**, 491–494.

Corruccini, R. S. (1991) Anthropological aspects of orofacial and occlusal variations and anomalies, in Kelley, M. A. & Larsen, C. S. (eds.) *Advances in dental anthropology,* New York: Wiley–Liss, pp. 295–323.

Corruccini, R. S., Handler, J. S. & Jacobi, K. P. (1985) Chronological distribution of enamel hypoplasias and weaning in a Caribbean slave population, *Human Biology*, **57**, 699–711.

Corruccini, R. S., Kaul, S. S., Chopra, S. R. K., Karosas, J., Larsen, M. D. & Morrow, C. (1983a) Epidemiological survey of occlusion in North India, *British Journal of Orthodontics*, **10**, 44–47.

Corruccini, R. S. & Lee, G. T. R. (1984) Occlusal variation in Chinese immigrants to the United Kingdom and their offspring, *Archives of Oral Biology*, **29**, 779–782.

Corruccini, R. S. & McHenry, H. M. (1980) Cladometric analysis of Pliocene hominids, *Journal of Human Evolution*, **9**, 209–221.

Corruccini, R. S. & Potter, R. H. Y. (1980) Genetic analysis of occlusal variation in twins, *American Journal of Orthodontics*, **78**, 140–154.

Corruccini, R. S., Potter, R. H. Y. & Dahlberg, A. A. (1983b) Changing occlusal variation in Pima Amerinds, *American Journal of Physical Anthropology*, **62**, 317–324.

Corruccini, R. S., Sharma, K. & Potter, R. H. Y. (1986) Comparative genetic variance and heritability of dental occlusal variables in US and Northwest Indian twins, *American Journal of Physical Anthropology*, **70**, 293–299.

Corruccini, R. S. & Townsend, G. C. (1990) Occlusal variation in Australian aboriginals, *American Journal of Physical Anthropology*, **82**, 257–265.

Corruccini, R. S. & Whitley, L. D. (1981) Occlusal variation in a rural Kentucky community, *American Journal of Orthodontics*, **79**, 250–262.

Costa, R. L. (1980) Incidence of caries and abscesses in archeological Eskimo skeletal samples from Point Hope and Kodiak Island, Alaska, *American Journal of Physical Anthropology*, **52**, 501–514.

Costa, R. L. (1982) Periodontal disease in the prehistoric Ipiutak and Tigara skeletal remains from Point Hope, Alaska, *American Journal of Physical Anthropology*, **59**, 97–110.

Cottone, J. A. & Standish, S. M. (eds.) (1981) *Outline of forensic dentistry*, Chicago: Year Book Medical Publishers.

Cross, J. F., Kerr, N. W. & Bruce, M. F. (1986) An evaluation of Scott's method for scoring dental wear, in Cruwys, E. & Foley, R. A. (eds.) *Teeth and anthropology*, B. A. R. International Series No 291, Oxford: British Archaeological Reports, pp. 101–108.

Crossner, C. G. & Mansfield, L. (1983) Determination of dental age in adopted non-European children, *Swedish Dental Journal*, **7**, 1–10.

Cybulski, J. S. (1994) Culture change, demographic history and health and disease on the Northwest Coast, in Larsen, C. S. & Milner, G. R. (eds.) *In the wake of contact, Biological responses to conquest*, New York: Wiley–Liss, pp. 75–86.

Dahl, B. L., Oilo, G., Andersen, A. & Bruaset, O. (1989) The suitability of a new index for the evaluation of dental wear, *Acta Odontologica Scandinavica*, **47**, 205–210.

Dahlberg, A. A. (1945) Paramolar tubercle (Bolk), *American Journal of Physical Anthropology*, **3**, 97–103.

Dahlberg, A. A. (1947) The evolutionary significance of the protostylid, *American Journal of Physical Anthropology*, **8**, 15–25.

Dahlberg, A. A. (1949) The dentition of the American Indian, in Laughlin, W. S.

(ed.) *The physical anthropology of the American Indian*, New York: Viking Fund, pp. 138–176.

Dahlberg, A. A. (1961) Relationship of tooth size to cusp number and groove conformation of occlusal surface patterns of lower molar teeth, *Journal of Dental Research*, **44**, 476–479.

Dahlberg, A. A. (1963) Analysis of the American Indian dentition, in Brothwell, D. R. (ed.) *Dental anthropology*, London: Pergamon Press, pp. 149–178.

Dahlberg, A. A. (ed.) (1971) *Dental morphology and evolution*, Chicago: University of Chicago Press.

Dahlberg, A. A. & Kinzey, W. G. (1962) Etude microscopique de l'abrasion et de l'attrition sur la surface des dents, *Bulletin du Groupement International pour les Recherches Scientifique en Stomatologie*, **5**, 242–251.

Dahlberg, A. A. & Menegaz-Bock, R. M. (1958) Emergence of the permanent teeth in Pima Indian children, *Journal of Dental Research*, **37**, 1123–1140.

Danenberg, P. J., Hirsch, R. S., Clarke, N. G., Leppard, P. I. & Richards, L. C. (1991) Continuous tooth eruption in Australian aboriginal skulls, *American Journal of Physical Anthropology*, **85**, 305–312.

Darling, A. I. (1959) The pathology and prevention of caries, *British Dental Journal*, **107**, 287–296.

Darling, A. I. (1963) Microstructural changes in early dental caries, in Soggnaes, R. F. (ed.) *Mechanisms of hard tissue destruction*, Washington: American Association for the Advancement of Science, pp. 171–185.

Dart, R. A. (1925) *Australopithecus africanus*: the man-ape of South Africa, *Nature*, **115**, 195–199.

Davies, D. M., Picton, D. C. A. & Alexander, A. G. (1969) An objective method of assessing the periodontal condition in human skulls, *Journal of Periodontal Research*, **4**, 74–77.

Davies, P. L. (1967) Agenesis of teeth: a sex limited trait, *Journal of Dental Research*, **46**, 1309.

Davis, P. J. & Hägg, U. (1994) The accuracy and precision of the 'Demirjian System' when used for age determination in Chinese children, *Swedish Dental Journal*, **18**, 113–116.

De Niro, M. J. (1987) Stable isotopy and archaeology, *American Scientist*, **75**, 182–191.

De Vito, C. & Saunders, S. R. (1990) A discriminant function analysis of deciduous teeth to determine sex, *Journal of Forensic Sciences*, **35**, 845–858.

Dean, H. T. (1934) Classification of mottled enamel diagnosis, *Journal of the American Dental Association*, **21**, 1421–1426.

Dean, M. C. (1987a) The dental developmental status of six East African juvenile fossil hominids, *Journal of Human Evolution*, **16**, 197–214.

Dean, M. C. (1987b) Growth layers and incremental markings in hard tissues; a review of the literature and some preliminary observations about enamel structure in *Paranthropus boisei*, *Journal of Human Evolution*, **16**, 157–172.

Dean, M. C. (1993) Daily rates of dentine formation in macaque tooth roots, *International Journal of Osteoarchaeology*, **3**, 199–207.

Dean, M. C., Beynon, A., Reid, D. & Whittaker, D. (1993a) A longitudinal study of tooth growth in a single individual based on long and short period incremental markings in dentine and enamel, *International Journal of Osteoarchaeology*, **3**, 249–264.

Dean, M. C. & Beynon, A. D. (1991) Histological reconstruction of crown formation times and initial root formation times in a modern human child, *American Journal of Physical Anthropology*, **86**, 215–228.

Dean, M. C., Beynon, A. D., Thackeray, J. F. & Macho, G. A. (1993b) Histological reconstruction of dental development and age at death of a juvenile *Paranthropus robustus* specimen, SK 63, from Swartkrans, South Africa, *American Journal of Physical Anthropology*, **91**, 401–420.

Dean, M. C., Stringer, C. B. & Bromage, T. G. (1986) A new age at death for the Neanderthal child from Devil's Tower, Gibraltar and the implications for studies of general growth and development in Neanderthals', *American Journal of Physical Anthropology*, **70**, 301–309.

Demetsopoullos, I. C., Burleigh, R. & Oakley, K. P. (1983) Relative and absolute dating of the human skull and skeleton from Galley Hill, Kent, *Journal of Archaeological Science*, **10**, 129–134.

Demirjian, A. & Goldstein, H. (1976) New systems for dental maturity based on seven and four teeth, *Annals of Human Biology*, **3**, 411–421.

Demirjian, A., Goldstein, H. & Tanner, J. M. (1973) A new system of dental age assessment, *Human Biology*, **45**, 211–227.

Demirjian, A. & Levesque, G.-Y. (1980) Sexual differences in dental development and prediction of emergence, *Journal of Dental Research*, **59**, 1110–1122.

Deutsch, D., Goultschin, J. & Anteby, S. (1981) Determination of human fetal age from the length of femur, mandible and maxillary incisor, *Growth*, **45**, 232–238.

Deutsch, D., Palmon, A., Fisher, L. W., Kolodny, N., Termine, J. D. & Young, M. F. (1991) Sequencing of bovine enamelin ('tuftelin') a novel acidic enamel protein, *Journal of Biological Chemistry*, **266**, 16021–16028.

Deutsch, D., Pe'er, E. & Gedalia, I. (1984) Changes in size, morphology and weight of human anterior teeth during the fetal period, *Growth*, **48**, 74–85.

Deutsch, D., Tam, O. & Stack, M. V. (1985) Postnatal changes in size, morphology and weight of developing postnatal deciduous anterior teeth, *Growth*, **49**, 202–217.

DiBennardo, R. & Bailit, H. L. (1978) Stress and dental asymmetry in a population of Japanese children, *American Journal of Physical Anthropology*, **48**, 89–94.

Ditch, L. E. & Rose, J. C. (1972) A multivariate sexing technique, *American Journal of Physical Anthropology*, **37**, 61–64.

Doberenz, A. R., Miller, M. F. & Wyckoff, R. W. G. (1969) An analysis of fossil enamel protein, *Calcified Tissue Research*, **3**, 93–95.

Doberenz, A. R. & Wyckoff, R. W. G. (1967) The microstructure of fossil teeth, *Journal of Ultrastructure Research*, **18**, 166–175.

Dobney, K. & Brothwell, D. (1986) Dental calculus: its relevance to ancient diet and oral ecology, in Cruwys, E. & Foley, R. A. (eds.) *Teeth and anthropology*, B. A. R. International Series No 291, Oxford: British Archaeological Reports, pp. 55–82.

Doran, G. A. & Freeman, L. (1974) Metrical features of the dentition and arches of populations from Goroka and Lufa, Papua, New Guinea, *Human Biology*, **46**, 583–594.

Dorrell, P. (1989) *Photography in archaeology and conservation*, Cambridge Manuals in Archaeology, Cambridge: Cambridge University Press.

Doyle, W. J. & Johnston, O. (1977) On the meaning of increased fluctuating dental asymmetry: a cross populational study, *American Journal of Physical Anthropology*, **46**, 127–134.

Dreier, F. G. (1994) Age at death estimates of the protohistoric Arikara using molar attrition rates: a new quantification method, *International Journal of Osteoarchaeology*, **4**, 137–148.

Driessens, F. C. M. & Verbeeck, R. M. H. (1989) Possible pathways of mineralization of dental plaque, in Ten Cate, J. M. (ed.) *Recent advances in the study of dental calculus*, Oxford: IRL Press at Oxford University Press, pp. 7–18.

Drusini, A., Calliari, I. & Volpe, A. (1991) Root dentine transparency: age determination of human teeth using computerized densitometric analysis, *American Journal of Physical Anthropology*, **85**, 25–30.

Dumond, D. E. (1977) *The Eskimos and Aleuts*, Ancient Peoples and Places, London: Thames & Hudson.

Eckhardt, R. B. & Piermarini, A. L. (1988) Interproximal grooving of teeth: additional evidence and interpretation, *Current Anthropology*, **29**, 668–670.

Eglinton, G. & Curry, G. B. (eds.) (1991) *Molecules through time: fossil molecules and biochemical systematics*, London: The Royal Society.

Elias, M. (1980) The feasibility of dental strontium analysis for diet-assessment of human populations, *American Journal of Physical Anthropology*, **53**, 1–4.

Eliot, M. M., Souther, S. P., Anderson, B. G. & Arnim, S. S. (1934) A study of the teeth of a group of school children previously examined for rickets, *American Journal of Diseases of Children*, **48**, 713.

Embery, G. (1989) The organic matrix of dental calculus and its interaction with mineral, in Ten Cate, J. M. (ed.) *Recent advances in the study of dental calculus*, Oxford: IRL Press at Oxford University Press, pp. 75–86.

Erdbrink, D. P. (1965) A quantification of the *Dryopithecus* and other lower molar patterns in man and some of the apes, *Zeitschrift für Morphologie und Anthropologie*, **57**, 70–108.

Escobar, V., Melnick, M. & Conneally, P. M. (1976) The inheritance of bilateral rotation of maxillary central incisors, *American Journal of Physical Anthropology*, **45**, 109—116.

Espelid, I., Tveit, A. B. & Fjelltveit, A. (1994) Variations among dentists in radiographic detection of occlusal caries, *Caries Research*, **28**, 169–175.

Falin, L. I. (1961) Histological and histochemical studies of human teeth of the Bronze and Stone Ages, *Archives of Oral Biology*, **5**, 5–13.

Falk, D. & Corruccini, R. (1982) Efficacy of cranial versus dental measurements for separating human populations, *American Journal of Physical Anthropology*, **57**, 123–128.

Fanning, E. A. (1961) A longitudinal study of tooth formation and root resorption, *New Zealand Dental Journal*, **57**, 202–217.

Fanning, E. A. & Brown, T. (1971) Primary and permanent tooth development, *Australian Dental Journal*, **16**, 41–43.

Fédération Dentaire Internationale (1971) Two-digit system of designating teeth, *International Dental Journal*, **21**, 104–106.

Fejerskov, O., Baelum, V. & Østergaard, E. S. (1993) Root caries in Scandinavia in the 1980's and future trends to be expected in dental caries experience in adults, *Advances in Dental Research*, **7**, 4–14.

Fejerskov, O., Josephsen, K. & Nyvad, B. (1984) Surface ultrastructure of unerupted mature human enamel, *Caries Research*, **18**, 302–314.

Fejerskov, O., Manji, F, & Baelum, V. (1988) *Dental fluorosis – a handbook for health workers*, Copenhagen: Munksgaard.

Ferembach, D., Schwidetzky, I. & Stloukal, M. (1980) Recommendations for age and sex diagnoses of skeletons, *Journal of Human Evolution*, **9**, 517–549.

Filipsson, R. (1975) A new method for assessment of dental maturity using the individual curve of number of erupted permanent teeth, *Annals of Human Biology*, **2**, 13–24.

Fincham, A. G., Bessem, C. C., Lau, E. C., Pavlova, Z., Shuler, Z., Slavkin, H. C. & Snead, M. L. (1991) Human developing enamel proteins exhibit a sex-linked dimorphism, *Calcified Tissue International*, **48**, 288–290.

Fitzgerald, C. M., Foley, R. A. & Dean M. C. (1996) Variation of circaseptan cross striation repeat intervals in the tooth enamel of three modern human populations (abstract), *American Journal of Physical Anthropology*, Supplement 22, 104–105.

Formicola, V. (1991) Interproximal grooving: different appearances, different etiologies, *American Journal of Physical Anthropology*, **86**, 85–86; discussion 86–87.

Fox, C. H. (1992) New considerations in the prevalence of periodontal disease, *Current Opinion in Dentistry*, **2**, 5–11.

Frank, R. M. (1978) Les stries brunes de Retzius en microscopie électronique à balayage, *Journal de Biologie Buccale*, **6**, 139–151.

Frank, R. M. & Nalbandian, J. (1989) Structure and ultrastructure of dentine, in Berkovitz, B. K. B., Boyde, A., Frank, R. M., Höhling, H. J., Moxham, B. J., Nalbandian, J. & Tonge, C. H. (eds.) *Teeth*, Handbook of Microscopic Anatomy, New York, Berlin & Heidelberg: Springer Verlag, Vol. V/6, pp. 173–247.

Frayer, D. W. (1978) *Evolution of the dentition in Upper Palaeolithic and Mesolithic Europe*, University of Kansas Publications in Anthropology 10, Lawrence: University of Kansas.

Frayer, D. W. (1980) Sexual dimorphism and cultural evolution in the Late Pleistocene and Holocene of Europe, *Journal of Human Evolution*, **9**, 399–415.

Frayer, D. W. (1984) Biological and cultural change in the European Late Pleistocene and Early Holocene, in Smith, F. H. & Spencer, F. (eds.) *The origin of modern humans*, New York: Alan R. Liss, pp. 211–250.

Frayer, D. W. (1991) On the etiology of interproximal grooves, *American Journal of Physical Anthropology*, **85**, 299–304.

Frayer, D. W. & Russell, M. D. (1987) Artificial grooves on the Krapina Neanderthal teeth, *American Journal of Physical Anthropology*, **74**, 393–405.

Fremlin, J. H. (1961) The preparation of thin sections of dental enamel, *Archives of Oral Biology*, **5**, 55–60.

Friskopp, J. (1983) Ultrastructure of nondecalcified supragingival and subgingival calculus, *Journal of Periodontology*, **54**, 542–550.

Friskopp, J. & Hammarstrom, L. (1980) A comparative, scanning electron microscopic study of supragingival and subgingival calculus, *Journal of Periodontology*, **51**, 553–562.

Fujita, T. (1939) Neue Feststellungen uber Retzius'schen Parallelstreifung des Zahnschmelzes, *Anatomischer Anzeiger*, **87**, 350–355.

Fyfe, D. M., Chandler, N. P. & Wilson, N. H. F. (1994) Alveolar bone status of some pre-seventeenth Century inhabitants in Taumako, Solomon Islands, *International Journal of Osteoarchaeology*, **3**, 29–36.

Gantt, D. G. (1982) Neogene hominid evolution: a tooth's inside view, in Kurtén, B. (ed.) *Teeth: form, function, and evolution*, New York: Columbia University Press, pp. 93–108.

Garn, S. M., Cole, P. E. & van Astine, W. L. (1979a) Sex discriminatory effectiveness using combinations of root lengths and crown diameters, *American Journal of Physical Anthropology*, **50**, 115–118.

Garn, S. M., Cole, P. E., Wainwright, R. L. & Guire, K. E. (1977) Sex discriminatory effectiveness using combinations of permanent teeth, *Journal of Dental Research*, **56**, 697.

Garn, S. M., Dahlberg, A. A., Lewis, A. B. & Kerewsky, R. S. (1966a) Groove pattern, cusp size, and tooth size, *Journal of Dental Research*, **45**, 970.

Garn, S. M., Kerewsky, R. S. & Lewis, A. B. (1966b) Extent of sex influence on Carabelli's polymorphism, *Journal of Dental Research*, **45**, 1823.

Garn, S. M., Koski, K. & Lewis, A. B. (1957) Problems in determining the tooth eruption sequence in fossil and modern man, *American Journal of Physical Anthropology*, **15**, 313–332.

Garn, S. M. & Lewis, A. B. (1963) Phylogenetic and intra-specific variations in tooth sequence polymorphism, in Brothwell, D. R. (ed.) *Dental anthropology*, London: Pergamon Press, pp. 53–74.

Garn, S. M. & Lewis, A. B. & Bonné, B. (1961) Third molar polymorphism and the timing of tooth formation, *Nature*, **192**, 989.

Garn, S. M., Lewis, A. B. & Bonné, B. (1962a) Third molar formation and its development course, *Angle Orthodontist*, **32**, 270–279.

Garn, S. M., Lewis, A. B. & Kerewsky, R. S. (1964) Sex differences in tooth size, *Journal of Dental Research*, **43**, 306.

Garn, S. M., Lewis, A. B. & Kerewsky, R. S. (1965a) Size interrelationships of the mesial and distal teeth, *Journal of Dental Research*, **44**, 350–354.

Garn, S. M., Lewis, A. B. & Kerewsky, R. S. (1966c) Bilateral asymmetry and concordance in cusp number and crown morphology of the mandibular first molar, *Journal of Dental Research*, **45**, 1820.

Garn, S. M., Lewis, A. B. & Kerewsky, R. S. (1966d) The meaning of bilateral asymmetry in the permanent dentition, *Angle Orthodontist*, **36**, 55–62.

Garn, S. M., Lewis, A. B. & Kerewsky, R. S. (1966e) Sexual dimorphism in the buccolingual tooth diameter, *Journal of Dental Research*, **45**, 1819.

Garn, S. M., Lewis, A. B. & Kerewsky, R. S. (1967a) Buccolingual size asymmetry and its developmental meaning, *Angle Orthodontist*, **37**, 186–193.

Garn, S. M., Lewis, A. B. & Kerewsky, R. S. (1967b) Sex difference in tooth shape, *Journal of Dental Research*, **46**, 1470.

Garn, S. M., Lewis, A. B. & Kerewsky, R. S. (1967c) Shape similarities throughout the dentition, *Journal of Dental Research*, **46**, 1481.

Garn, S. M., Lewis, A. B. & Kerewsky, R. S. (1968a) The magnitude and implications of the relationship between tooth size and body size, *Archives of Oral Biology*, **13**, 129–131.

Garn, S. M., Lewis, A. B. & Kerewsky, R. S. (1968b) Relationship between buccolingual and mesiodistal crown diameters, *Journal of Dental Research*, **47**, 495.

Garn, S. M., Lewis, A. B. & Kerewsky, S. (1965b) Genetic, nutritional, and maturational correlates of dental development, *Journal of Dental Research*, **44**, 228–242.

Garn, S. M., Lewis, A. B., Koski, K. & Polachek, D. L. (1958) The sex difference in tooth calcification, *Journal of Dental Research*, **37**, 561–567.

Garn, S. M., Lewis, A. B., Swindler, D. R. & Kerewsky, R. S. (1967d) Genetic control of dimorphism in tooth size, *Journal of Dental Research*, **46**, 963–972.

Garn, S. M., Lewis, A. B. & Vicinus, J. H. (1962b) Third molar agenesis and reduction in the number of other teeth, *Journal of Dental Research*, **41**, 717.

Garn, S. M., Lewis, A. B. & Walenga, A. J. (1968c) Crown size profile pattern comparisons of 14 human populations, *Archives of Oral Biology*, **13**, 1235–1242.

Garn, S. M., Lewis, A. B. & Walenga, A. J. (1968d) Genetic basis of the crown-size profile pattern, *Journal of Dental Research*, **47**, 503.

Garn, S. M., Lewis, A. B. & Walenga, A. J. (1969) Crown-size profile patterns and presumed evolutionary trends, *American Anthropologist*, **71**, 79–84.

Garn, S. M., Nagy, J. M., Sandusky, S. T. & Trowbridge, F. (1973a) Economic impact on tooth emergence, *American Journal of Physical Anthropology*, **39**, 233–238.

Garn, S. M., Osborne, R. H. & McCabe, K. D. (1979b) The effect of prenatal factors on crown dimensions, *American Journal of Physical Anthropology*, **51**, 665–678.

Carn, S. M., Sandusky, S. T., Nagy, J. M. & Trowbridge, F. L. (1973b) Negro–Caucasoid differences in permanent tooth emergence at a constant income level, *Archives of Oral Biology*, **18**, 609–615.

Garn, S. M. & Smith, B. H. (1980) Developmental communalities in tooth emergence timing, *Journal of Dental Research*, **59**, 1178.

Gibson, H. L. (1971) Multi-spectrum investigation of prehistoric teeth, *Dental Radiography and Photography*, **44**, 57–64.

Gill, P., Ivanov, P. L., Kimpton, C., Piercy, R., Benson, N., Tully, G., Evett, I., Hagelberg, E. & Sullivan, K. (1994) Identification of the remains of the Romanov family by DNA analysis, *Nature Genetics*, **6**, 130–135.

Gillard, R. D., Pollard, A. M., Sutton, P. A. & Whittaker, D. K. (1990) An improved method for age at death determination from the measurement of D-Aspartic acid in dental collagen, *Archaeometry*, **32**, 61–70.

Gingerich, P. D. (1977) Correlation of tooth size and body size in living hominoid Primates, with a note on the relative brain size in *Aegyptopithecus* and *Proconsul*, *American Journal of Physical Anthropology*, **47**, 395–398.

Gleiser, I. & Hunt, E. E. (1955) The permanent mandibular first molar: its calcification, eruption and decay, *American Journal of Physical Anthropology*, **13**, 253–284.

Glimcher, M. J., Cohensolal, L., Kossiva, D. & Dericqles, A. (1990) Biochemical analyses of fossil enamel and dentin, *Paleobiology*, **16**, 219–232.

Goaz, P. W. & White, S. C. (1994) *Oral radiology, Principles and interpretation*, 3rd edn, St Louis: C V Mosby.

Goldberg, H. J. V., Weintraub, J. A., Roghmann, K. J. & Cornwell, W. S. (1976)

Measuring periodontal disease in ancient populations: root and wear indices in study of American Indian skulls, *Journal of Periodontology*, **47**, 348–351.

Goldstein, J. I., Newbury, D. E., Echlin, P., Joy, D. C., Romig, A. D., Lyman, C. E., Fiori, C. & Lifshin, E. (1992) *Scanning electron microscopy and X-ray microanalysis*, 2nd edn, New York: Plenum Press.

Goodman, A. H., Armelagos, G. J. & Rose, J. C. (1980) Enamel hypoplasias as indicators of stress in three prehistoric populations from Illinois, *Human Biology*, **52**, 515–528.

Goodman, A. H., Armelagos, G. J. & Rose, J. C. (1984a) The chronological distribution of enamel hypoplasias from prehistoric Dickson Mounds populations, *American Journal of Physical Anthropology*, **65**, 259–266.

Goodman, A. H., Lallo, J., Armelagos, G. J. & Rose, J. C. (1984b) Health changes at Dickson Mounds, Illinois (A.D. 950–1300), in Cohen, M. N. & Armelagos, G. J. (eds.) *Palaeopathology at the origins of agriculture*, New York: Academic Press, pp. 271–306.

Goodman, A. H., Martinez, C. & Clavez, A. (1991) Nutritional supplementation and the development of linear enamel hypoplasias in children from Tezonteopan, Mexico, *American Journal of Clinical Nutrition*, **53**, 773–781.

Goodman, A. H. & Rose, J. C. (1990) Assessment of sytemic physiological perturbations from dental enamel hypoplasias and associated histological structures, *Yearbook of Physical Anthropology*, **33**, 59–110.

Goodman, A. H. & Rose, J. C. (1991) Dental enamel hypoplasias as indicators of nutritional status, in Kelly, M. A. & Larsen, C. S. (eds.) *Advances in dental anthropology*, New York: Wiley–Liss, pp. 279–293.

Goodman, A. H., Thomas, R. B., Swedlund, A. C. & Armelagos, G. J. (1988) Biocultural perspectives of stress in prehistoric, historical and contemporary population research, *Yearbook of Physical Anthropology*, **31**, 169–202.

Goose, D. H. (1963) Dental measurement: an assessment of its value in anthropological studies, in Brothwell, D. R. (ed.) *Dental anthropology*, London: Pergamon Press, pp. 125–148.

Goose, D. H. (1971) The inheritance of tooth size in British families, in Dahlberg, A. A. (ed.) *Dental morphology and evolution*, Chicago: University of Chicago Press, pp. 263–270.

Gordon, K. D. (1984a) Hominoid dental microwear: complications in the use of microwear analysis to detect diet, *Journal of Dental Research*, **63**, 1043–1046.

Gordon, K. D. (1984b) Orientation of occlusal contacts in the chimpanzee, *Pan troglodytes verus*, deduced from scanning electron microscopic analysis of dental microwear patterns, *Archives of Oral Biology*, **29**, 783–787.

Gordon, K. D. (1988) A review of methodology and quantification of dental microwear analysis, *Scanning Microscopy*, **2**, 1139–1147.

Grandjean, P., Nielsen, O. V. & Shapiro, I. M. (1979) Lead retention in ancient Nubian and contemporary populations, *Journal of Environmental Pathology & Toxicology*, **2**, 781–787.

Gray, S. W. & Lamons, F. P. (1959) Skeletal development and tooth eruption in Atlanta children, *American Journal of Orthodontics*, **45**, 272–277.

Greene, D. L. (1967) *Dentition of Meroitic, X-Group, and Christian populations from Wadi Halfa, Sudan*, Department of Anthropology, University of Utah, Anthropological Papers 85, Salt Lake City: Department of Anthropology, University of Utah Press.

Greene, D. L. (1972) Dental anthropology of early Egypt and Nubia, *Journal of Human Evolution*, **13**, 315–324.

Greene, D. L. (1984) Fluctuating dental asymmetry and measurement error, *American Journal of Physical Anthropology*, **65**, 283–289.

Greene, D. L., Ewing, G. H. & Armelagos, G. J. (91967) Dentition of a Mesolithic population from Wadi Haifa, Sudan, *American Journal of Physical Anthropology*, **27**, 41–55.

Gregory, W. K. (1922) *The origin and evolution of the human dentition*, Baltimore: Williams & Wilkins.

Gregory, W. K. & Hellman, M. (1926) The crown patterns of fossil and recent human molar teeth and their meaning, *Natural History*, **26**, 300–309.

Grine, F. E. (1981) Relative sizes of the maxillary deciduous canine and central incisor teeth in the Kalahari San and South African Negro, *Annals of the South African Museum*, **7**, 229–245.

Grine, F. E. (1985) Australopithecine evolution: the deciduous dental evidence, in Delson, E. A. (ed.) *Ancestors: the hard evidence*, New York: A Liss, pp. 153–167.

Grine, F. E. (1986) Dental evidence for dietary differences in *Australopithecus* and *Paranthropus*: a quantitative analysis of permanent molar microwear, *Journal of Human Evolution*, **15**, 783–822.

Grine, F. E. (1987) Quantitative analysis of occlusal microwear in *Australopithecus* and *Paranthropus, Scanning Microscopy*, **1**, 647–656.

Grine, F. E. (1988) *Evolutionary history of the 'robust' australopithecines*, New York: Aldine de Gruyter.

Grine, F. E., Gwinnett, A. J. & Oaks, J. H. (1990) Early hominid dental pathology: interproximal caries in 1.5 million-year-old *Paranthropus robustus* from Swartkrans, *Archives of Oral Biology*, **35**, 381–386.

Grine, F. E. & Kay, R. F. (1988) Early hominid diets from quantitative image analysis of dental microwear, *Nature*, **333**, 765–768.

Grün, R. & Stringer, C. B. (1991) Electron spin resonance dating and the evolution of modern humans, *Archaeometry*, **33**, 153–199.

Grüneberg, H. (1963) *The pathology of development*, Oxford: Blackwell Scientific Publications.

Gustafson, A.-G. (1955) The similarity between contralateral pairs of teeth, *Odontologisk Tidskrift*, **63**, 245–248.

Gustafson, A.-G. (1959) A morphologic investigation of certain variations in the structure and mineralisation of human dental enamel, *Odontologisk Tidskrift*, **67**, 361–472.

Gustafson, G. (1947) Microscopic examination of teeth as a means of identification in forensic medicine, *Journal of the American Dental Associations*, **35**, 720–724.

Gustafson, G. (1950) Age determination on teeth, *Journal of the American Dental Association*, **41**, 45–54.

Gustafson, G. & Gustafson, A.-G. (1967) Microanatomy and histochemistry of enamel, in Miles, A. E. W. (ed.) *Structural and chemical organization of teeth*, London: Academic Press, Vol. 2, pp. 135–162.

Gustafson, G. & Koch, G. (1974) Age estimation up to 16 years of age based on dental development, *Odontologisk Revy*, **25**, 297–306.

Gustafson, B. G., Quensel, C. E., Swedlander, L. L., Lundquist, C., Granen, H., Bonow, B. E. & Krasse, B. (1954) The Vipeholm dental caries study. The effect of different levels of carbohydrate intake on caries activity in 436 individuals observed for five years, *Acta Odontologica Scandinavica*, **11**, 232–364.

Gwinnett, A. J. (1966) Histology of normal enamel. III. Phase contrast study, *Journal of Dental Research*, **45**, 865–869.

Gysi, A. (1931) Metabolism in adult enamel, *Dental Digest*, **37**, 661–668.

Haavikko, K. (1970) The formation and the alveolar and clinical eruption of the permanent teeth, *Proceedings of the Finnish Dental Society*, **66**, 101–170.

Haavikko, K. (1973) The physiological resorption of the roots of deciduous teeth in Helsinki children, *Proceedings of the Finnish Dental Society*, **69**, 93–98.

Haavikko, K. (1974) Tooth formation age estimate on a few selected teeth: a simple method for clinical use, *Proceedings of the Finnish Dental Society*, **70**, 15–19.

Haeussler, A. M. & Turner, C. G. (1992) The dentition of Soviet Central Asians and the quest for New World ancestors, in Lukacs, J. R. (ed.) *Culture, Ecology & Dental Anthropology*, Journal of Human Ecology Special Issue No 2, Delhi: Kamla-Raj Enterprises, pp. 273–297.

Haeussler, A. M., Turner II, C. G. & Irish, J. D. (1988) Concordance of American and Soviet methods in dental anthropology, *American Journal of Physical Anthropology*, **75**, 218.

Hagelberg, E. (1994) Analysis of genetic information from archaeological and forensic bone, in Eglinton, C. & Kay, R. L. F. (eds.) *Biomolecular palaeontology*, NERC Earth Sciences Directorate Special Publication No 94/1, Swindon: Natural Environment Research Council, pp. 39–41.

Hagelberg, E., Bell, L. S., Allen, T., Boyde, A., Jones, S. J. & Clegg, J. B. (1991a) Analysis of ancient bone DNA: techniques and applications, in Eglinton, G. & Curry, G. B. (eds.) *Molecules through time: fossil molecules and biochemical systematics*, London: The Royal Society, pp. 399–408.

Hagelberg, E. & Clegg, J. B. (1991) Isolation and characterisation of DNA from archaeological bone, *Philosophical Transactions of the Royal Society of London, Series B*, **244**, 45–50.

Hagelberg, E., Gray, I. C. & Jeffreys, A. J. (1991b) Identification of the skeletal remains of a murder victim by DNA analysis, *Nature*, **352**, 427–429.

Hagelberg, E., Sykes, B. & Hedges, R. (1989) Ancient bone DNA amplified, *Nature*, **342**, 485.

Hägg, U. & Matsson, L. (1985) Dental maturity as an indicator of chronological age: the accuracy and precision of three methods, *European Journal of Orthodontics*, **7**, 24–34.

Haikel, Y., Frank, R. M. & Voegel, J. C. (1983) Scanning electron microscopy of the human enamel surface layer of incipient carious lesions, *Caries Research*, **17**, 1–13.

Hanihara, K. (1963) Crown characteristics of the deciduous dentition of the Japanese–American hybrids, in Brothwell, D. R. (ed.) *Dental anthropology*, London: Pergamon Press, pp. 105–124.

Hanihara, K. (1967) Racial characteristics in the dentition, *Journal of Dental Research*, **46**, 923–926.

Hanihara, K. (1969) Mongoloid dental complex in the permanent dentition, *VIIIth Congress of Anthropological and Ethnological Sciences 1968*, Tokyo and Kyoto, 298–300.

Hanihara, T. (1990) Affinities of the Phillipine Negritos with Japanese and the Pacific populations based on dental measurements: the basic populations in East Asia, *Journal of the Anthropological Society of Nippon*, **98**, 13–28.

Hanihara, T. (1991) Dentition of Nansei Islanders and the peopling of the Japanese Archipelago: the basic populations in East Asia, IX, *Journal of the Anthropological Society of Nippon*, **99**, 399–409.

Hanihara, T. (1992a) Biological relationships among Southeast Asians, Jomonese, and the Pacific populations as viewed from dental characters – the basic populations in East Asia, X, *Journal of the Anthropological Society of Nippon*, **100**, 53–67.

Hanihara, T. (1992b) Dental and cranial affinities among populations of East Asia and the Pacific: the basic populations in East Asia, IV, *American Journal of Physical Anthropology*, **88**, 163–182.

Hanihara, T. (1992c) Dental variation of the Polynesian populations, *Journal of the Anthropological Society of Nippon*, **100**, 291–302.

Hanihara, T. (1992d) Negritos, Australian aborigines, and the 'Proto-Sundadont' dental pattern: the basic populations in East Asia, V, *American Journal of Physical Anthropology*, **88**, 183–196.

Hanihara, T., Ishida, H., Oshima, N., Kondo, O. & Masuda, T. (1994) Dental calculus and other dental disease in a human skeleton of the Okhotsk culture unearthed at Hamanaka-2 site, Rebun-Island, Hokkaido, Japan, *International Journal of Osteoarchaeology*, **4**, 343–351.

Hardwick, J. L. (1960) The incidence and distribution of caries throughout the ages in relation to the Englishman's diet, *British Dental Journal*, **108**, 9–17.

Hare, P. E. (1980) Organic chemistry of bone and its relation to the survival of bone in the natural environment, in Behrensmeyer, A. K. & Hill. A. P. (eds.) *Fossils in the making. Vertebrate taphonomy and paleoecology*, Prehistory Archaeology and Ecology Series, Chicago: University of Chicago Press, pp. 208–219.

Harris, E. F. (1992) Laterality in human odontometrics: analysis of a contemporary American White series, in Lukacs, J. R. (ed.) *Culture, ecology & dental anthropology*, Journal of Human Ecology Special Issue No 2, Delhi: Kamla-Raj Enterprises, pp. 157–170.

Harris, E. F. & Bailit, H. L. (1980) The metaconule: a morphologic and familial analysis of a molar cusp in humans, *American Journal of Physical Anthropology*, **53**, 349–358.

Harris, E. F. & Bailit, H. L. (1988) A principal components analysis of human odontometrics, *American Journal of Physical Anthropology*, **75**, 87–99.

Harris, E. F. & McKee, J. H. (1990) Tooth mineralization standards for blacks and whites from the middle southern United States, *Journal of Forensic Sciences*, **35**, 859–872.

Harris, E. F. & Nweeia, M. T. (1980) Tooth size of Ticuna Indians, Colombia, with phenetic comparisons to other Amerindians, *American Journal of Physical Anthropology*, **53**, 81–91.

Harris, E. F. & Rathbun, T. A. (1991) Ethnic differences in the apportionment of tooth sizes, in Kelley, M. A. & Larsen, C. S. (eds.) *Advances in Dental Anthropology*, New York: Wiley–Liss, pp. 121–142.

Harris, E. F., Woods, M. A. & Robinson, Q. C. (1993) Dental health patterns in an urban Midsouth population: race, sex and age changes, *Quintessence International*, **24**, 45–52.

Harris, R. (1963) Biology of the children of Hopewood House, Bowral, Australia. Observations on dental caries experience extending over five years (1957–1961), *Journal of Dental Research*, **42**, 1387–1399.

Hartley, W. G. (1979) *Hartley's microscopy*, Charlbury: Senecio Publishing Co Ltd.

Hartman, S. E. (1988) A cladistic analysis of hominoid molars, *Journal of Human Evolution*, **17**, 489–502.

Hartman, S. E. (1989) Stereophotogrammetric analysis of occlusal morphology of extant hominoid molars: phenetics and function, *American Journal of Physical Anthropology*, **80**, 145–166.

Hayashi, Y. (1993) High resolution electron microscopy of the junction between enamel and dental calculus, *Scanning Microscopy*, **7**, 973–978.

Hedges, R. E. M. & Sykes, B. C. (1992) Biomolecular archaeology: past, present and future, in Pollard, A. M. (ed.) *New developments in archaeological science*, Proceedings of the British Academy 77, Oxford: Oxford University Press, pp. 267–283.

Hellman, M. (1928) Racial characters in the human dentition, *Proceedings of the American Philosophical Society*, **67**, 157–174.

Helm, S. & Prydso, U. (1979) Prevalence of malocclusion in medieval and modern Danes contrasted, *Scandinavian Journal of Dental Research*, **87**, 91–97.

Henderson, A. M. & Corruccini, R. S. (1976) Relationship between tooth size and body size in American Blacks, *Journal of Dental Research*, **54**, 94–96.

Henderson, P., Marlow, C. A., Molleson, T. I. & Williams, C. T. (1983) Patterns of chemical change during fossilization, *Nature*, **306**, 358–360.

Hiiemae, K. M. (1978) Mammalian mastication: a review of the activity of the jaw muscles and the movements they produce in chewing, in Butler, P. M. & Joysey, K. A. (ed.) *Development, function and evolution of teeth*, London: Academic Press, pp. 359–398.

Hildebolt, C. F. & Molnar, S. (1991) Measurement and description of periodontal disease in anthropological studies, in Kelley, M. A. & Larsen, C. S. (eds.) *Advances in dental anthropology*, New York: Wiley–Liss, pp. 225–240.

Hillson, S. W. (1979) Diet and dental disease, *World Archaeology*, **11**, 147–162.

Hillson, S. W. (1986a) *Teeth*, Cambridge Manuals in Archaeology, Cambridge: Cambridge University Press.

Hillson, S. W. (1986b) Teeth, age, growth and archaeology, in David, R. A. (ed.) *Science in Egyptology. Proceedings of the 'Science in Egyptology' Symposia*, Manchester: Manchester University Press, pp. 475–484.

Hillson, S. W. (1992a) Dental enamel growth, perikymata and hypoplasia in ancient tooth crowns, *Journal of the Royal Society of Medicine*, **85**, 460–466.

Hillson, S. W. (1992b) Impression and replica methods for studying hypoplasia and perikymata on human tooth crown surfaces from archaeological sites, *International Journal of Osteoarchaeology*, **2**, 65–78.

Hillson, S. W. (1992c) *Mammal bones and teeth. An introductory guide to methods of identification*, London: Institute of Archaeology, University College London.

Hillson, S. W. (1992d) Studies of growth in dental tissues, in Lukacs, J. R. (ed.) *Culture, ecology & dental anthropology*, Journal of Human Ecology Special Issue No 2, Delhi: Kamla-Raj Enterprises, pp. 7–23.

Hinrichsen, C. F. L. & Engel, M. B. (1966) Fine structure of partially demineralised enamel, *Archives of Oral Biology*, **11**, 65–93.

Hinton, R. J. (1981) Form and patterning of anterior tooth wear among aboriginal human groups, *American Journal of Physical Anthropology*, **54**, 555–564.

Hinton, R. J. (1982) Differences in interproximal and occlusal tooth wear among prehistoric Tennessee Indians: implications for masticatory function, *American Journal of Physical Anthropology*, **57**, 103–115.

Hofman-Axthelm, W. (1981) *History of dentistry*, Chicago: Quintessence Publishing Co. Inc.

Hojo, M. (1954) On the pattern of the dental abrasion, *Okajimas Folia Anatomica Japonica*, **26**, 11–30.

Hollinger, J. O., Lorton, L., Krantz, W. A. & Connelly, M. (1984) A clinical and laboratory comparison of irreversible hydrocolloid impression techniques, *Journal of Prosthetic Dentistry*, **51**, 304–309.

Horowitz, H. S. (1989) Fluoride and enamel defects, *Advances in Dental Research*, **3**, 143–146.

Horowitz, H. S., Driscoll, W. S., Meyers, R. J., Heifetz, S. B. & Kingman, A. (1984) A new method for assessing the prevalence of dental fluorosis – the tooth surface index of fluorosis, *Journal of the American Dental Association*, **109**, 37–41.

Howells, W. W. (1973) *Cranial variation in man. A study by multivariate analysis of patterns of difference among recent human populations*, Papers of the Peabody Museum of Archaeology and Ethnology 67, Cambridge, Massachusetts: Harvard University.

Howells, W. W. (1989) *Skull shapes and the map. Craniometric analyses in the dispersion of modern Homo*, Papers of the Peabody Museum of Archaeology and Ethnology 79, Cambridge, Massachusetts: Harvard University.

Howells W. W. (1995) *Ethnic identification of crania from measurements*, Papers of the Peabody Museum of Archaeology and Ethnology 82, Cambridge, Massachusetts: Harvard University.

Howie, F. M. P. (1984) Materials used for conserving fossil specimens since 1930: a review, in Brommelle, N. S., Pye, E. M., Smith, P. & Thomson, G. (eds.) *Adhesives and consolidants. Preprints of the Contributions to the Paris Congress, 2–8 September 1984*, London: International Institute for Conservation of Historic and Artistic Works, pp. 92–97.

Hrdlicka, A. (1920) Shovel-shaped teeth, *American Journal of Physical Anthropology*, **3** (Old Series), 429–465.

Hrdlicka, A. (1952) *Practical anthropometry*, 4th edn, Philadelphia: Wistar Institute.

Huda, T. F. J. & Bowman, J. E. (1994) Variation in cross-striation number between striae in an archaeological population, *International Journal of Osteoarchaeology*, **4**, 49–52.

Huda, T. F. J. & Bowman, J. E. (1995) Age determination from dental microstructure in juveniles, *American Journal of Physical Anthropology*, **97**, 135–150.

Hunter, J. (1771) *The natural history of the teeth*, London.

Hunter, J. (1778) *A practical treatise on the diseases of the teeth*, London.

Hunter, W. S. & Priest, W. R. (1960) Errors and discrepancies in measurement of tooth size, *Journal of Dental Research*, **39**, 405–408.

Hurme, V. O. (1949) Ranges of normalcy in the eruption of the permanent teeth, *Journal of Dentistry for Children*, **16**, 11–15.

Hurme, V. O. (1951) Time and sequence of tooth eruption, *Journal of Forensic Sciences*, **2**, 317–388.

Hurme, V. O. & van Wagenen, G. (1961) Basic data on the emergence of permanent teeth in the rhesus monkey (*Macaca mulatta*), *Proceedings of the American Philosophical Society*, **105**, 105–140.

Hutchinson, D. L. & Larsen, C. S. (1990) Stress and lifeway changes: the evidence from enamel hypoplasias', in Larsen, C. S. (ed.) *The archaeology of Mission Santa Catalina de Guale: 2. Biocultural interpretations of a population in transition*, Anthropological Papers of the American Museum of Natural History 68, New York: American Museum of Natural History, pp. 50–65.

Hutchinson, J. (1857) On the influence of hereditary syphilis on the teeth, *Transactions of the Odontological Society of Great Britain*, **2**, 95–106.

Ikeya, M. (1993) *New applications of electron spin response: ESR dating, dosimetry and microscopy*, World Science Publications.

Infante, P. F. & Gillespie, G. M. (1977) Enamel hypoplasia in relation to caries in Guatemalan children, *Journal of Dental Research*, **56**, 493–498.

Irish, J. D. & Turner, C. G. (1987) More lingual surface attrition of the maxillary anterior teeth in American Indians: prehistoric Panamanians, *American Journal of Physical Anthropology*, **73**, 209–213.

Irish, J. D. & Turner II, C. G. (1990) West African dental affinity of Late Pleistocene Nubians. Peopling of the Eurafrican–South Asian triangle II, *Homo*, **41**, 42–53.

Israel, H. & Lewis, A. B. (1971) Radiographically determined linear permanent tooth growth from age 6 years, *Journal of Dental Research*, **50**, 334–342.

Jackson, D. & Weidmann, S. M. (1959) The relationship between age and the fluorine content of human dentine and enamel: a regional survey, *British Dental Journal*, **107**, 303–306.

Jacobi, K. P., Collins Cook, D., Corruccini, R. S. & Handler, J. S. (1992) Congenital syphilis in the past: slaves at Newton Plantation, Barbados, West Indies, *American Journal of Physical Anthropology*, **89**, 145–158.

James, W. W. (1960) *The jaws and teeth of primates*, London: Pitman Medical Publishing Co. Ltd.

Jaspers, M. T. & Witkop, C. J. (1980) Taurodontism, an isolated trait associated with syndromes and X-chromosomal aneuploidy, *American Journal of Human Genetics*, **32**, 396–413.

Jaswal, S. (1983) Age and sequence of permanent-tooth emergence among Khasis, *American Journal of Physical Anthropology*, **62**, 177–186.

Johanson, D. C. (1979) A consideration of the 'Dryopithecus pattern', *Ossa*, **6**, 125–138.

Johanson, G. (1971) Age determination from human teeth, *Odontologisk Revy*, **22**, 1–126.

Johansson, A. (1992) A cross-cultural study of occlusal tooth wear, *Swedish Dental Journal*, **86**, 1–59.

Johansson, A., Fareed, K. & Omar, R. (1991) Analysis of possible factors influencing the occurrence of occlusal tooth wear in a young Saudi population, *Acta Odontologica Scandinavica*, **49**, 139–145.

Johansson, A., Haraldson, T., Omar, R., Kiliaridis, S. & Carlsson, G. E. (1993) A system for assessing the severity and progression of occlusal tooth wear, *Journal of Oral Rehabilitation*, **20**, 125–131.

Jones, S. J. (1981) Cement, in Osborn, J. W. (ed.) *Dental anatomy and embryology*, A Companion to Dental Studies, Oxford: Blackwell Scientific Publications, Vol. 1 (2), pp. 193–205.

Jones, S. J. (1987) The root surface: an illustrated review of some scanning electron microscope studies, *Scanning Microscopy*, **1**, 2003–2018.

Jones, S. J. & Boyde, A. (1972) A study of human root cementum surfaces as prepared for and examined in the scanning electron microscope, *Zeitschrift für Zellforschung und Microskopische Anatomie*, **130**, 318–337.

Jones, S. J. & Boyde, A. (1984) Ultrastructure of dentin and dentinogenesis, in Linde A. (ed.) *Dentin and dentinogenesis*, Boca Raton: CRC Press Inc., Vol. 1, pp. 81–134.

Jones, S. J. & Boyde, A. (1987) Scanning microscopic observations on dental caries, *Scanning Microscopy*, **1**, 1991–2002.

Jordan, R. E., Abrams, L. & Kraus, B. S. (1992) *Kraus' dental anatomy and occlusion*, 2nd edn, St Louis: Mosby Year Book.

Jørgensen, K. D. (1955) The Dryopithecus pattern in recent Danes and Dutchmen, *Journal of Dental Research*, **34**, 195–208.

Kaidonis, J. A., Townsend, G. C. & Richards, L. C. (1992) Brief communication: interproximal tooth wear: a new observation, *American Journal of Physical Anthropology*, **88**, 105–107.

Kambe, T., Yonemitsu, K., Kibayashi, K. & Tsunenari, S. (1991) Application of a computer assisted image analyzer to the assessment of area and number of sites of dental attrition and its use for age estimation, *Forensic Science International*, **50**, 97–109.

Kanazawa, E., Morris, D. H., Sekikawa, M. & Ozaki, T. (1988) Comparative study of the upper molar occlusal table morphology among seven human populations, *American Journal of Physical Anthropology*, **77**, 271–278.

Kanazawa, E., Sekikawa, M. & Ozaki, T. (1984) Three-dimensional measurements of the occlusal surfaces of upper molars in a Dutch population, *Journal of Dental Research*, **63**, 1298–1301.

Kanazawa, E., Sekikawa, M. & Ozaki, T. (1990) A quantitative investigation of irregular cuspules in human maxillary permanent molars, *American Journal of Physical Anthropology*, **83**, 173–180.

Karn, K. W., Shockett, H. P., Moffitt, W. C. & Gray, J. L. (1984) Topographic classification of deformities of the alveolar process, *Journal of Periodontology*, **55**, 336–340.

Karnosh, L. J. (1926) Histopathology of syphilitic hypoplasia of teeth, *Archives of Dermatology & Syphilology*, **13**, 25–42.

Katz, D. & Suchey, J. M. (1986) Age determination of the male *Os pubis, American Journal of Physical Anthropology*, **69**, 427–436.

Katzenberg, M. A. (1992) Advances in stable isotope analysis of prehistoric bones, in Saunders, S. R. & Katzenberg, M. A. (eds.) *Skeletal biology of past peoples: research methods*, New York: Wiley–Liss, pp. 105–120.

Kaul, S. S. & Corruccini, R. S. (1992) Dental arch length reduction through interproximal attrition in modern Australian aborigines, in Lukacs, J. R. (ed.) *Culture, Ecology & Dental Anthropology*, Journal of Human Ecology Special Issue No 2, Delhi: Kamla-Raj Enterprises, pp. 195–200.

Kawasaki, K., Tanaka, S. & Isikawa, T. (1980) On the daily incremental lines in human dentine, *Archives of Oral Biology*, **24**, 939–943.

Kay, R. F. (1977) The evolution of molar occlusion in the Cercopithecidae and early catarrhines, *American Journal of Physical Anthropology*, **46**, 327–352.

Kay, R. F. (1978) Molar structure and diet in extant Ceropithecidae, in Butler, P. M. & Joysey, K. A. (eds.) *Development, function and evolution of teeth*, London: Academic Press, pp. 309–339.

Kay, R. F. (1987) Analysis of primate dental microwear using image processing techniques, *Scanning Microscopy*, **1**, 657–662.

Kay, R. F. & Hiiemae, K. M. (1974) Jaw movement and tooth use in recent and fossil primates, *American Journal of Physical Anthropology*, **40**, 27–256.

Kay, R. F., Rasmussen, D. T. & Beard, K. C. (1984) Cementum annulus counts provide a means for age determination in *Macaca mulatta* (Primates, Anthropidea), *Folia Primatologia,*, **42**, 85–95.

Keegan, W. F. (1989) Stable isotope analysis of prehistoric diet, in Iscan, M. Y. & Kennedy, K. A. R. (eds.) *Reconstruction of life from the skeleton*, New York: Alan R Liss, pp. 223–236.

Keene, H. J. (1967) Australopithecine dental dimensions in a contemporary population, *American Journal of Physical Anthropology*, **27**, 379–384.

Keith, A. (1913) Problems relating to the earlier forms of prehistoric man, *Proceedings of the Royal Society of Medicine (Odontology)*, **6**, 103–119.

Kelley, M. A. & Larsen, C. S. (eds.) (1991) *Advances in dental anthropology*, New York: Wiley–Liss.

Kerr, N. W. (1990) The prevalence and pattern of distribution of root caries in a Scottish medieval population, *Journal of Dental Research*, **69**, 857–860.

Kerr, N. W. (1991) Prevalence and natural history of periodontal disease in Scotland – the medieval period (900–1600 AD), *Journal of Periodontal Research*, **26**, 346–354.

Kerr, N. W., Bruce, M. F. & Cross, J. F. (1990) Caries experience in Mediaeval Scots, *American Journal of Physical Anthropology*, **83**, 69–70.

Kieser, J. A. (1990) *Human adult odontometrics*, Cambridge Studies in Biological Anthropology 4, Cambridge: Cambridge University Press.

Kieser, J. A. & Groenveld, H. T. (1988) Patterns of variability in the South African Negro dentition, *Journal of the Dental Association of South Africa*, **43**, 105–110.

Kieser, J. A., Preston, C. B. & Evans, W. G. (1983) Skeletal age at death: an evaluation of the Miles method of ageing, *Journal of Archaeological Science*, **10**, 9–12.

Kilgore, L. (1995) Patterns of dental decay in African great apes, *Palaeopathology Association 22nd Annual Meeting*, Oakland, 6.

Kilian, J. & Vlcek, E. (1989) Age determination from teeth in the adult, in Iscan, M. Y. (ed.) *Age markers in the human skeleton*, Springfield: Charles C Thomas, pp. 255–275.

Klein, H. (1945) Etiology of enamel hypoplasia in rickets as determined by studies on rats and swine, *Journal of the American Dental Association*, **18**, 866–884.

Klein, H., Palmer, C. E. & Knutson, J. W. (1938) Studies on dental caries. I. Dental status and dental needs of elementary schoolchildren, *Public Health Reports*, **53**, 751–765.

Klineberg, I. (1991) *Occlusion: principles and assessment*, Oxford: Wright.

Kolakowski, D., Harris, E. F. & Bailit, H. L. (1980) Complex segregation analysis of Carabelli's Trait in a Melanesian population, *American Journal of Physical Anthropology*, **53**, 301–308.

Koob, S. P. (1984) The consolidation of archaeological bone, in Brommelle, N. S., Pye, E. M., Smith, P. & Thomson, G. (eds.) *Adhesives and consolidants. Preprints of the Contributions to the Paris Congress, 2–8 September 1984*, London: International Institute for Conservation of Historic and Artistic Works, pp. 98–102.

Koski, K. & Garn, S. M. (1957) Tooth eruption sequence in fossil and modern man, *American Journal of Physical Anthropology*, **15**, 469–488.

Kraus, B. S. (1959) Calcification of the human deciduous teeth, *Journal of the American Dental Association*, **59**, 1128–1136.

Kraus, B. S. & Jordan, R. E. (1965) *The human dentition before birth*, Philadelphia: Lea & Febiger.

Krejci, I., Reich, T., Bucher, W. & Lutz, F. (1994) Eine neue Methode zur dreidimensionalen Verschleissmessung, *Schweizerische Monatsschrift für Zahnmedizin*, **104**, 160–169.

Kreshover, S. J. (1944) The pathogenesis of enamel hypoplasia: an experimental study, *Journal of Dental Research*, **23**, 231–238.

Kreshover, S. J. (1960) Metabolic disturbances in tooth formation, *Annals of the New York Academy of Sciences*, **85**, 161–167.

Kreshover, S. J. & Clough, O. W. (1953) Prenatal influences on tooth development II. Artificially induced fever in rats, *Journal of Dental Research*, **32**, 565–572.

Kreshover, S. J., Clough, O. W. & Bear, D. M. (1953) Prenatal influences on tooth development I. Alloxan diabetes in rats, *Journal of Dental Research*, **32**, 246–261.

Kreshover, S. J., Clough, O. W. & Hancock, J. A. (1954) Vaccinia infection in pregnant rabbits and its effect on maternal and fetal dental tissues, *Journal of the American Dental Association*, **49**, 549–562.

Kreshover, S. J. & Hancock, J. A. (1956) The effect of lymphocytic choriomeningitis on pregnancy and dental tissue in mice, *Journal of Dental Research*, **35**, 467–483.

Kronfeld, R. (1935a) Development and calcification of the human deciduous dentition, *The Bur*, **15**, 18–25.

Kronfeld, R. (1935b) First permanent molar: its condition at birth and its postnatal development, *Journal of the American Dental Association*, **22**, 1131–1155.

Kronfeld, R. (1935c) Postnatal development and calcification of the anterior permanent teeth, *Journal of the American Dental Association*, **22**, 1521–1536.

Kronfeld, R. & Schour, I. (1939) Neonatal dental hypoplasia, *Journal of the American Dental Association*, **26**, 18–32.

Krueger, H. W. (1991) Exchange of carbon with biological apatite, *Journal of Archaeological Science*, **18**, 355–361.

Kuhnlein, H. V. & Calloway, D. H. (1977) Minerals in human teeth: differences between preindustrial and contemporary Hopi Indians, *American Journal of Clinical Nutrition*, **30**, 883–886.

Kurtén, B. (ed.) (1982) *Teeth: form, function, and evolution*, New York: Columbia University Press.

Kuttler, V. (1959) Classification of dentin into primary, secondary and tertiary, *Oral Surgery, Oral Medicine, Oral Pathology*, **12**, 906–1001.

Kvaal, S. I., Sellevold, B. J. & Solheim, T. (1994) A comparison of different non-destructive methods of age estimation in skeletal material, *International Journal of Osteoarchaeology*, **4**, 363–370.

Langlais, R. P., Langland, O. E. & Nortjé, C. J. (1995) *Diagnostic imaging of the jaws*, Malvern: Williams & Wilkins.

Larato, D. C. (1970) Intrabony defects in the dry human skull, *Journal of Periodontology*, **41**, 496–498.

Larsen, C. S. (1983) Behavioural implications of temporal change in cariogenesis, *Journal of Archaeological Science*, **10**, 1–8.

Larsen, C. S. (1985) Dental modifications and tool use in the western Great Basin, *American Journal of Physical Anthropology*, **67**, 393–402.

Larsen, C. S. (1995) Biological changes in human populations with agriculture, *Annual Review of Anthropology*, **24**, 185–213.

Larsen, C. S., Shavit, R. & Griffin, M. C. (1991) Dental caries evidence for dietary change: an archaeological context, in Kelley, M. A. & Larsen, C. S. (eds.) *Advances in dental anthropology*, New York: Wiley–Liss, pp. 179–202.

Larsen, E. B. (1979) *Moulding and casting of museum objects using siliconerubber and epoxyresin*, Privately published.

Lasker, G. W. (1951) Genetic analysis of racial traits of the teeth, *Cold Spring Harbor Symposia on Quantitative Biology*, XV, 191–203.

Lavelle, C. L. B. (1970) Analysis of attrition in adult human molars, *Journal of Dental Research*, **49**, 822–828.

Lavelle, C. L. B. (1977) Relationship between tooth and long bone size, *American Journal of Physical Anthropology*, **46**, 423–426.

Lavelle, C. L. B. (1984) A metrical comparison of maxillary first premolar form, *American Journal of Physical Anthropology*, **63**, 397–403.

Lavelle, C. L. B. & Moore, W. J. (1969) Alveolar bone resorption in Anglo-Saxon and seventeenth century mandibles, *Journal of Periodontal Research*, **4**, 70–73.

Lavelle, C. L. B. & Moore, W. J. (1973) The incidence of agenesis and polygenesis in the Primate dentition, *American Journal of Physical Anthropology*, **38**, 671–680.

Le Bot, P. & Salmon, D. (1980) Congenital defects of the upper lateral incisors (ULI) and the morphology of other teeth in man, *American Journal of Physical Anthropology*, **53**, 479–486.

Leakey, M. C., Feibel, C. S., McDougall, I. & Walker, A. (1995) New four-million-year-old hominid species from Kanapoi and Allia Bay, Kenya, *Nature*, **376**, 565–571.

Lee, G. T. R. & Goose, D. H. (1972) The inheritance of dental traits in a Chinese population in the United Kingdom, *Journal of Medical Genetics*, **9**, 336–339.

Lee-Thorpe, J. A. & van der Merwe, N. J. (1991) Aspects of the chemistry of modern and fossil biological apatites, *Journal of Archaeological Science*, **18**, 343–354.

Lee-Thorpe, J. A. & van der Merwe, N. J. & Brain, C. K. (1989) Isotopic evidence for dietary differences between two extinct baboon species from Swartkrans, *Journal of Archaeological Science*, **18**, 183–190.

Leek, F. F. (1972) Teeth and bread in ancient Egypt, *Journal of Egyptian Archaeology*, **58**, 126–132.

Lehner, T. (1992) *Immunology of oral disease*, 3rd edn, Oxford: Blackwell Scientific Publications.

Leigh, R. W. (1925) Dental pathology of Indian tribes of varied environmental and food conditions, *American Journal of Physical Anthropology*, **8**, 179–199.

Leinonen, A., Wasz-Höckert, B. & Vuorinen, P. (1972) Usefulness of the dental age obtained by orthopantomography as an indicator of the physical age, *Proceedings of the Finnish Dental Society*, **68**, 235–242.

Leutenegger, W. & Kelley, J. T. (1977) Relationship of sexual dimorphism in canine size and body size to social, behavioural and ecological correlates in anthropoid primates, *Primates*, **18**, 117–136.

Levers, B. G. H. & Darling, A. I. (1983) Continuous eruption of some adult human teeth of ancient populations, *Archives of Oral Biology*, **28**, 401–408.

Levesque, G.-Y., Demirjian, A. & Tanguay, R. (1981) Sexual dimorphism in the development, emergence and agenesis of the mandibular third molar, *Journal of Dental Research*, **60**, 1735–1741.

Lewis, D. W. & Grainger, R. M. (1967) Sex-linked inheritance of tooth size: a family study, *Archives of Oral Biology*, **12**, 539–544.

Lieberman, D. E. (1993) Life history variables preserved in dental cementum micro-structure, *Science*, **261**, 1162–1164.

Lieberman, D. E., Deacon, T. W. & Meadow, R. H. (1990) Computer image enhance-ment and analysis of cementum increments as applied to teeth of *Gazella gazella*, *Journal of Archaeological Science*, **17**, 519–533.

Lieberman, D. E. & Meadow, R. H. (1992) The biology of cementum increments (with an archaeological application), *Mammal Review*, **22**, 57–78.

Limbrey, S. (1975) *Soil science and archaeology*, Studies in Archaeological Science, London: Academic Press.

Lingström, P., Birkhed, D., Granfeldt, Y. & Björck, K. (1993) pH measurements of human dental plaque after consumption of starchy foods using the microtouch and the sampling method, *Caries Research*, **27**, 394–401.

Lipsinic, F. E., Paunovitch, E., Houston, G. D. & Robinson, S. F. (1986) Correlation of age and incremental lines in the cementum of human teeth, *Journal of Forensic Sciences*, **31**, 982–989.

Littleton, J. & Frohlich, B. (1993) Fish-eaters and farmers: dental pathology in the Arabian Gulf, *American Journal of Physical Anthropology*, **92**, 427–447.

Liversidge, H. M. (1994) Accuracy of age estimation from developing teeth of a population of known age (0 to 5.4 years), *International Journal of Osteoarchaeol-ogy*, **4**, 37–46.

Liversidge, H. M., Dean, M. C. & Molleson, T. I. (1993) Increasing human tooth length between birth and 5.4 years, *American Journal of Physical Anthropology*, **90**, 307–313.

Loe, H., Anerud, A. & Boysen, H. (1992) The natural history of periodontal disease in man: prevalence, severity, and extent of gingival recession, *Journal of Periodontology*, **63**, 489–495.

Loevy, H. (1983) Maturation of permanent teeth in black and latino children, *Acta de Odontologia Pediatrica*, **4**, 59–62.

Logan, W. H. G. (1935) A histologic study of the anatomic structures forming the oral cavity, *Journal of the American Dental Association*, **22**, 3–30.

Logan, W. H. G. & Kronfeld, R. (1933) Development of the human jaws and surrounding structures from birth to the age of fifteen years, *Journal of the American Dental Association*, **20**, 379–427.

Lombardi, A. V. (1978) A factor analysis of morphogenetic fields in the human dentition, in Butler, P. M. & Joysey, K. A. (eds.) *Development, function and evolution of teeth*, London: Academic Press, pp. 203–214.

Lovejoy, C. O. (1985) Dental wear in the Libben population: its functional pattern and role in the determination of adult skeletal age at death, *American Journal of Physical Anthropology*, **68**, 47–56.

Lovejoy, C. O., Meindl, R. S., Mensforth, R. P. & Barton, T. J. (1985) Multifactorial determination of skeletal age at death: a method and blind tests of its accuracy, *American Journal of Physical Anthropology*, **68**, 1–14.

Lowenstein, J. M. & Scheuenstuhl, G. (1991) Immunological methods in molecular palaeontology, in Eglinton, G. & Curry, G. B. (eds.) *Molecules through time: fossil molecules and biochemical systematics*, London: The Royal Society, pp. 375–380.

Lubell, D., Jackes, M., Schwarcz, H., Knyf, M. & Meiklejohn, C. (1994) The Mesolithic–Neolithic transition in Portugal: isotopic and dental evidence of diet, *Journal of Archaeological Science*, **21**, 201–216.

Lucas, P. W. (1982) An analysis of the canine tooth size of old world higher primates in relation to mandibular length and body weight, *Archives of Oral Biology*, **27**, 493–496.

Lucy, D., Pollard, A. M. & Roberts, C. A. (1994) A comparison of three dental techniques for estimating age at death in humans, *Journal of Archaeological Science*, **22**, 151–156.

Lucy D., Aykroyd R. G., Pollard A. M. & Solheim T. (1996) A Bayesian approach to adult human age estimation from dental observations by Johanson's age changes, *Journal of Forensic Sciences*, **41**, 5–10.

Lucy D. & Pollard A. M. (1995) Further comments on the estimation of error associated with the Gustafson dental age estimation method, *Journal of Forensic Sciences*, **40**, 222–227.

Lukacs J. R. (1992) Dental paleopathology and agricultural intensification in south Asia: new evidence from Bronze Age Harappa, *American Journal of Physical Anthropology*, **87**, 133–150.

Lukacs, J. R. & Hemphill, B. E. (1991) The dental anthropology of prehistoric Baluchistan: a morphometric approach to the peopling of South Asia, in Kelley, M. A. & Larsen, C. S. (eds.) *Advances in Dental Anthropology*, New York: Wiley–Liss, pp. 77–119.

Lukacs, J. R. & Pastor, R. F. (1988) Activity-induced patterns of dental abrasion in prehistoric Pakistan: evidence from Mehrgarh and Harappa, *American Journal of Physical Anthropology*, **76**, 377–398.

Lukacs, J. R., Retief, D. H. & Jarrige, J.-F. (1985) Dental disease in prehistoric Balu-chistan, *National Georgraphic Research*, Spring 1985, 184–197.

Luke, D. A., Stack, M. V. & Hey, E. N. (1978) A comparison of morphological and gravimetric methods of estimation human foetal age from the dentition, in Butler, P. M. & Joysey, K. A. (eds.) *Development, function and evolution of teeth*, London: Academic Press, pp. 511–518.

Lundström, A. (1948) *Tooth size and occlusion in twins*, Basel: S Karger.

Lundström, A. (1963) Tooth morphology as a basis for distinguishing monozygotic and dizygotic twins, *American Journal of Human Genetics*, **15**, 34–43.

Lunt, D. A. (1954) A case of taurodontism in a modern European molar, *Dental Record*, **74**, 307–312.

Lunt, D. A. (1974) The prevalence of dental caries in the permanent dentition of Scottish prehistoric and medieval Danes, *Archives of Oral Biology*, **19**, 431–437.

Lunt, D. A. (1978) Molar attrition in Medieval Danes, in Butler, P. M. & Joysey, K. A. (eds.) *Development, function and evolution of teeth*, London: Academic Press, pp. 465–482.

Lunt, R. C. & Law, D. B. (1974) A review of the chronology of calcification of deciduous teeth, *Journal of the American Dental Association*, **89**, 599–606.

Lustmann, J., Lewin-Epstein, J. & Shteyer, A. (1976) Scanning electron microscopy of dental calculus, *Calcified Tissue Research*, **21**, 47–55.

Lynch, E. & Beighton, D. (1994) A comparison of primary root caries lesions classi-fied according to colour, *Caries Research*, **28**, 233–239.

Maas, M. C. (1991) Enamel structure and microwear: an experimental study of the response of enamel to shearing force, *American Journal of Physical Anthropology*, **85**, 31–49.

Maas, M. C. (1994) A hierarchical view of the relationship of enamel microstructure and wear, *American Journal of Physical Anthropology*, Supplement 18, 132–133.

Macho, G. A. & Berner, M. E. (1994) Enamel thickness and the helicoidal occlusal plane, *American Journal of Physical Anthropology*, **94**, 327–338.

MacPhee, T. & Cowley, G. (1975) *Essentials of periodontology and periodontics*, 2nd edn, Oxford: Blackwell Scientific Publications.

Mangion, J. J. (1962) Two cases of taurodontism in modern human jaws, *British Dental Journal*, **113**, 309–312.

Manji, F., Fejerskov, O., Baelum, V. & Nagelkerke, N. (1989) Dental calculus and caries experience in 14–65 year olds with no access to dental care, in Ten Cate, J. M. (ed.) *Recent advances in the study of dental calculus*, Oxford: IRL Press at Oxford University Press, pp. 223–234.

Mann, A. (1988) The nature of Taung dental maturation, *Nature*, **333**, 123.

Mann, A., Lampl, M. & Monge, J. (1990a) Patterns of ontogeny in human evolution: evidence from dental development, *Yearbook of Physical Anthropology*, **33**, 111–150.

Mann, A., Monge, J. & Lampl, M. (1990b) Dental caution, *Nature*, **348**, 202.

Mann, A. E. (1975) *Some paleodemographic aspects of the South African australopi-thecines*, University of Pennsylvania Publications in Anthropology No 1, Philadel-phia: University of Pennsylvania.

Mann, A. E., Lampl, M. & Monge, J. (1987) Maturational patterns in early hominids, *Nature*, **328**, 673–674.

Mann, A. E., Monge, J. M. & Lampl, M. (1991) Investigation into the relationship between perikymata counts and crown formation times, *American Journal of Physical Anthropology*, **86**, 175–188.

Maples, W. R. (1978) An improved technique using dental histology for the estimation of adult age, *Journal of Forensic Sciences*, **23**, 764–770.

Maples, W. R. & Rice, P. M. (1979) Some difficulties in the Gustafson dental age estimations, *Journal of Forensic Sciences*, **24**, 118–172.

Marsh, P. & Martin, M. (1992) *Oral microbiology*, 3rd edn, London: Chapman & Hall.

Martin, L. B. (1985) Significance of enamel thickness in hominoid evolution, *Nature*, **314**, 260–263.

Martin, L. B. (1986) Relationships among extant and extinct great apes and humans, in Wood, B. A., Martin, L. & Andrews, P. (eds.) *Major topics in primate and human evolution*, Cambridge: Cambridge University Press, pp. 161–187.

Martin, L. B. & Boyde, A. (1984) Rates of enamel formation in relation to enamel thickness in hominoid primates, in Fearnhead, R. W. & Suga, S. (eds.) *Tooth enamel IV*, Amsterdam, Elsevier Science Publications BV, pp. 447–451.

Martin, L. B., Boyde, A. & Grine, F. E. (1988) Enamel structure in primates: a review of scanning electron microscope studies, *Scanning Microscopy*, **2**, 1503–1526.

Massler, M. & Schour, I. (1946) The appositional life span of the enamel and dentin-forming cells. I. Human deciduous teeth and first permanent molars, *Journal of Dental Research*, **25**, 145–150.

Massler, M., Schour, I. & Poncher, H. (1941) Developmental pattern of the child as reflected in the calcification pattern of the teeth, *American Journal of Diseases of Children*, **62**, 33–67.

Masters, P. M. (1984) Stereochemical age determinations from the Barrow Eskimo remains, *Arctic Anthropology*, **21**, 77–82.

Masters, P. M. (1986a) Age determination of living mammals using aspartic acid racemization in structural proteins, in Zimmerman, M. R. & Angel, J. L. (eds.) *Dating and the age determination of biological materials*, London: Croom Helm, pp. 270–283.

Masters, P. M. (1986b) Amino acid racemization dating – a review, in Zimmerman, M. R. & Angel, J. L. (eds.) *Dating and the age determination of biological materials*, London: Croom Helm, pp. 39–58.

Masters P. M. (1987) Preferential preservation of noncollagenous protein during bone diagenesis: implications for chronometric and stable isotope measurements, *Geochimica et Cosmochimica Acta*, **51**, 3209–3214.

Masters, P. M. & Bada, J. L. (1978) Amino acid racemization dating of bone and shell, in Carter, G. F. (ed.) *Archaeological chemistry II*, Advances in Chemistry Series 171, Washington DC: American Chemical Society, pp. 117–138.

Masters, P. M. & Zimmerman, M. R. (1978) Age determination of an Alaskan mummy: morphological and biochemical correlation, *Science*, **201**, 811–812.

Matsikidis, G. & Schulz, P. (1982) Altersbestimmung nach dem Gebiß mit Hilfe des Zahnfilms, *Zahnärzliche Mitteilungen*, **72**, 2524–2528.

Mayhall, J. T. (1992) Techniques for the study of dental morphology, in Saunders, S. R. & Katzenberg, M. A. (eds.) *Skeletal biology of past peoples: research methods*, New York: Wiley–Liss, pp. 59–78.

Mayhall, J. T. & Alvesalo, L. (1992) Sexual dimorphism in the three-dimensional determinations of the maxillary first molar: cusp height, area, volume and position, in Smith, P. & Tchernov, E. (eds.) *Structure, function and evolution of teeth*, London & Tel Aviv: Freund Publishing House Ltd, pp. 425–436.

Mayhall, J. T. & Kanazawa, E. (1989) Three-dimensional analysis of the maxillary first molar crowns of Canadian Inuit, *American Journal of Physical Anthropology*, **78**, 73–78.

Mayhall, J. T. & Saunders, S. R. (1986) Dimensional and discrete dental trait asymmetry relationships, *American Journal of Physical Anthropology*, **69**, 403–411.

Mayhall, J. T., Saunders, S. R. & Belier, P. L. (1982) The dental morphology of North American whites: a reappraisal, in Kurtén, B. (ed.) *Teeth: form, function, and evolution*, New York: Columbia University Press, pp. 245–258.

McHenry, H. M. & Corruccini, R. S. (1980) Late Tertiary hominoids and human origins, *Nature*, **285**, 397–398.

McKee, J. K. & Molnar, S. (1988) Measurements of tooth wear among Australian aboriginees: II. Intrapopulational variation in patterns of dental attrition, *American Journal of Physical Anthropology*, **76**, 125–136.

McKinley, J. I. (1994) *The Anglo-Saxon cemetery at Spong Hill, North Elmham. Part VIII: the cremations*, East Anglian Archaeology Report No 69, Dereham: Field Archaeology Division, Norfolk Museums Service.

Mellanby, M. (1927) The structure of human teeth, *British Dental Journal*, **48**, 737–751.

Mellanby, M. (1929) *Diet and teeth: an experimental study. Part I. Dental structure in dogs*, Medical Research Council, Special Report Series, No 140, London: His Majesty's Stationery Office.

Mellanby, M. (1930) *Diet and teeth: an experimental study. Part II. A. Diet and dental disease. B. Diet and dental structure in mammals other than the dog*, Medical Research Council, Special Report Series, No 153, London: His Majesty's Stationery Office.

Mellanby, M. (1934) *Diet and teeth: an experimental study. Part III. The effect of diet on the dental structure and disease in man*, Medical Research Council, Special Report Series, No 191, London: His Majesty's Stationery Office.

Menaker, L. (ed.) (1980) *The biologic basis of dental caries. An oral biology textbook*, Hagerstown: Harper & Row.

Mendis, B. R. R. N. & Darling, A. I. (1979) A scanning electron microscope and microradiographic study of closure of human coronal dentinal tubules related to occlusal attribution and caries, *Archives of Oral Biology*, **24**, 725–733.

Meskin, L. H. & Corlin, R. J. (1963) Agenesis and peg-shaped permanent maxillary lateral incisors, *Journal of Dental Research*, **42**, 1476–1479.

Metzger, Z., Buchner, A. & Gorsky, M. (1980) Gustafson's method for age determination from teeth – a modification for the use of dentists in identification teams, *Journal of Forensic Sciences*, **25**, 742–749.

Miles, A. E. W. (1958) The assessment of age from the dentition, *Proceedings of the Royal Society of Medicine*, **51**, 1057–1050.

Miles, A. E. W. (1962) Assessment of the ages of a population of Anglo-Saxons from their dentitions, *Proceedings of the Royal Society of Medicine*, **55**, 881–886.

Miles, A. E. W. (1963) Dentition and the estimation of age, *Journal of Dental Research*, **42**, 255–263.

Miles, A. E. W. (1978) Teeth as an indicator of age in man, in Butler, P. M. & Joysey, K. A. (eds.) *Development, function and evolution of teeth*, London: Academic Press, pp. 455–462.

Miles, A. E. W. & Grigson, C. (eds.), 1990 *Colyer's variations and diseases of the teeth of animals*, Revised Edition edn, Cambridge: Cambridge University Press.

Millard, A. R. & Hedges, R. E. M. (1995) The role of the environment in uranium uptake by buried bone, *Journal of Archaeological Science*, **22**, 239–250.

Miller, C. S., Dove, S. B. & Cottone, J. A. (1988) Failure of use of cemental annulations in teeth to determine the age of humans, *Journal of Forensic Sciences*, **33**, 137–143.

Milner, G. R. & Larsen, C. S. (1991) Teeth as artifacts of human behavior: intentional mutilation and accidental modification, in Kelley, M. A. & Larsen, C. S. (eds.) *Advances in dental anthropology*, New York: Wiley–Liss, pp. 357–378.

Mimura, T. (1939) Horoshitsu ni mirareru Seicho-sen no shuki (The periodicity of growth lines seen in the enamel), *Kobyo-shi*, **13**, 545–455.

Mincer, H. H., Harris, E. F. & Berryman, H. E. (1993) The A.B.F.O. study of third molar development and its use as an estimator of chronological age, *Journal of Forensic Sciences*, **38**, 379–390.

Mizoguchi, Y. (1985) *Shovelling: a statistical analysis of its morphology*, Tokyo: Tokyo University Press.

Mizoguchi, Y. (1988) Degree of bilateral asymmetry of nonmetric tooth crown characters quantified by the tetrachoric correlation method, *Bulletin of the National Science Museum (Tokyo) Series D (Anthropology)*, **14**, 29–49.

Moggi-Cecchi, J. (ed.) (1995) *Aspects of dental biology: paleontology, anthropology and evolution*, Florence: International Institute for the Study of Man.

Moggi-Cecchi, J., Pacciani, E. & Pinto-Cisternas, J. (1994) Enamel hypoplasia and age at weaning in 19th-Century Florence, Italy, *American Journal of Physical Anthropology*, **93**, 299–306.

Møller, I. J. (1982) Fluorides and dental fluorosis, *International Dental Journal*, **32**, 135–147.

Møller, I. J., Poulsen, S. & Nielson, V. O. (1972) The prevalence of dental caries in Godhavn and Scoresbysund districts, Greenland, *Scandinavian Journal of Dental Research*, **80**, 169–180.

Molleson, T., Jones, K. & Jones, S. (1993) Dietary change and the effects of food preparation on microwear patterns in the Late Neolithic of abu Hureyra, northern Syria, *Journal of Human Evolution*, **24**, 455–468.

Molnar, S. (1971) Human tooth wear, tooth function and cultural variability, *American Journal of Physical Anthropology*, **34**, 175–190.

Molnar, S., McKee, J. K. & Molnar, I. (1983a) Measurements of tooth wear among Australian aborigines: I. Serial loss of the enamel crown, *American Journal of Physical Anthropology*, **61**, 51–65.

Molnar, S., McKee, J. K., Molnar, I. M. & Przybeck, T. R. (1983b) Tooth wear rates among contemporary Australian Aborigines, *Journal of Dental Research*, **62**, 562–565.

Molnar, S. & Molnar, I. M. (1985) The incidence of enamel hypoplasia among the Krapina Neanderthals, *American Anthropologist*, **87**, 536–549.

Molnar, S., Przybeck, T. R., Gantt, D. G., Elizondo, R. S. & Wilkerson, J. E. (1981) Dentin apposition rates as markers of primate growth, *American Journal of Physical Anthropology*, **55**, 443–450.

Moon, H. (1877) On irregular and defective tooth development, *Transactions of the Odontological Society of Great Britain*, **9**, 223–243.

Moore, W. J. & Corbett, M. E. (1971) Distribution of dental caries in ancient British populations: I Anglo-Saxon period, *Caries Research*, **5**, 151–168.

Moore, W. J. & Corbett, M. E. (1973) Distribution of dental caries in ancient British populations: II Iron Age, Romano-British and Medieval periods, *Caries Research*, **7**, 139–153.

Moore, W. J. & Corbett, M. E. (1975) Distribution of dental caries in ancient British populations: III The 17th Century, *Caries Research*, **9**, 163–175.

Moorrees, C. F. A. (1957a) The Aleut dentition, *Tijdschrift voor Tandheelkunde*, LXIV, 3–15.

Moorrees, C. F. A. (1957b) *The Aleut dentition*, Cambridge, Massachusetts: Harvard University Press.

Moorrees, C. F. A. & Chandha, J. M. (1962) Crown diameters of corresponding tooth groups in the deciduous and permanent dentition, *Journal of Dental Research*, **41**, 466–470.

Moorrees, C. F. A., Fanning, E. A. & Hunt, E. E. (1963a) Age variation of formation stages for ten permanent teeth, *Journal of Dental Research*, **42**, 1490–1502.

Moorrees, C. F. A., Fanning, E. A. & Hunt, E. E. (1963b) Formation and resorption of three deciduous teeth in children, *American Journal of Physical Anthropology*, **21**, 205–213.

Moorrees, C. F. A. & Kent, R. L. (1978) A step function model using tooth counts to assess the developmental timing of the dentition, *Annals of Human Biology*, **5**, 55–68.

Moorrees, C. F. A. & Reed, R. B. (1964) Correlations among crown diameters of human teeth, *Archives of Oral Biology*, **9**, 685–697.

Moorrees, C. F. A., Thomsen, S. O., Jensen, E. & Yen, P. K. J. (1957) Mesiodistal crown diameters of deciduous and permanent teeth, *Journal of Dental Research*, **36**, 39–47.

Mörmann, J. E. & Mühlemann, H. R. (1981) Oral starch degradation and its influence on acid production in human dental plaque, *Caries Research*, **15**, 166–175.

Morris, D. H. (1970) On deflecting wrinkles and the *Dryopithecus* pattern in human mandibular molars, *American Journal of Physical Anthropology*, **32**, 97–104.

Morris, D. H. (1973) Bushman maxillary canine polymorphism, *South African Journal of Science*, **71**, 333–335.

Morris, D. H., Glasstone Hughes, S. & Dahlberg, A. A. (1978) Uto-Aztecan premolar: the anthropology of a dental trait, in Butler, P. M. & Joysey, K. A. (eds.) *Development, function and evolution of teeth*, London: Academic Press, pp. 59–67.

Morris, P. (1978) The use of teeth for estimating the age of wild mammals, in Butler, P. M. & Joysey, K. A. (eds.) *Development, function and evolution of teeth*, London: Academic Press, pp. 483–494.

Moss, M. L. & Moss-Salentijn, L. (1977) Analysis of developmental processes related to human dental sexual dimorphism in permanent and deciduous canines, *American Journal of Physical Anthropology*, **46**, 407–414.

Muller, D. & Perizonius, W. R. K. (1980) The scoring of defects of the alveolar process in human crania, *Journal of Human Evolution*, **9**, 113–116.

Mundorff, S. A., Featherstone, J. D. B., Bibby, B. G., Curzon, M. E. J., Eisenberg, A. D. & Espeland, M. A. (1990) Cariogenic potential of foods. I. Caries in the rat model, *Caries Research*, **24**, 344–355.

Mundorff-Shrestha, S. A., Featherstone, J. D. B., Eisenberg, A. D., Cowles, E., Curzon, M. E. J., Espeland, M. A. & Shields, C. P. (1994) Cariogenic potential of foods. II. Relationship of food composition, plaque microbial counts, and salivary perameters to caries in the rat model, *Caries Research*, **28**, 106–115.

Murphy, T. (1959a) The changing pattern of dentine exposure in human tooth attrition, *American Journal of Physical Anthropology*, **17**, 167–178.

Murphy, T. (1959b) Gradients of dentine exposure in human molar tooth attrition, *American Journal of Physical Anthropology*, **17**, 179–186.

Murphy, T. R. (1964) Reduction of the dental arch by approximal attrition. A quantitative assessment, *British Dental Journal*, **116**, 483–488.

Murray, S. R. & Murray, K. A. (1989) Computer software for hypoplasia analysis, *American Journal of Physical Anthropology*, **78**, 277–278.

Nalbandian, J. & Soggnaes, R. F. (1960) Structural age changes in human teeth, in Shock, N. W. (ed.) *Ageing – some social and biological aspects. Symposia presented at the Chicago meeting, December 29–30, 1959*, American Association for the Advancement of Science Publication No 65, Washington DC: American Association for the Advancement of Science, pp. 367–382.

Navia, J. M. (1994) Carbohydrates and dental health, *American Journal of Clinical Nutrition*, **59**, 719s–727s.

Newman, H. N. & Poole, D. F. G. (1974) Observations with scanning and transmission electron microscopy on the structure of human surface enamel, *Archives of Oral Biology*, **19**, 1135–1143.

Nichol, C. R. (1989) Complex segregation analysis of dental morphological variants, *American Journal of Physical Anthropology*, **78**, 37–59.

Nichol, C. R. & Turner II, C. G. (1986) Intra- and interobserver concordance in classifying dental morphology, *American Journal of Physical Anthropology*, **69**, 299–315.

Nortjé, C. J. (1983) The permanent mandibular third molar, its value in age determination, *Journal of Forensic Odonto-Stomatology*, **1**, 27–31.

Nowell, G. W. (1978) An evaluation of the Miles method of ageing using the Tepe Hissar dental sample, *American Journal of Physical Anthropology*, **49**, 271–276.

Nyström, M., Haataja, J., Kaataja, M., Evälahti, E., Peck, L. & Kleemola-Kujala, E. (1986) Dental maturity in Finnish children, estimated from the development of seven permanent mandibular teeth, *Acta Odontologica Scandinavica*, **44**, 193–198.

Nyvad, B. & Fejerskov, O. (1982) Root surface caries: clinical, histopathological and microbiological features and clinical implications, *International Dental Journal*, **32**, 311–326.

O'Sullivan, E. A., Williams, S. A., Wakefield, R. C., Cape, J. E. & Curzon, M. E. (1993) Prevalence and site characteristics of dental caries in primary molar teeth from prehistoric times to the 18th century in England, *Caries Research*, **27**, 147–153.

Oakley, K. P. (1969) Analytical methods of dating bones, in Brothwell, D. R. & Higgs, E. S. (eds.) *Science in Archaeology*, London: Thames & Hudson, pp. 35–45.

Ogilvie, M. D., Curran, B. K. & Trinkaus, E. (1989) Incidence and patterning of dental enamel hypoplasia among the Neandertals, *American Journal of Physical Anthropology*, **79**, 25–41.

Osborn, J. W. (1970) The mechanism of ameloblast movement: a hypothesis, *Calcified Tissue Research*, **5**, 344–359.

Osborn, J. W. (1973) Variations in structure and development of enamel, *Oral Science Reviews*, **3**, 3–83.

Osborn, J. W. (1978) Morphogenetic gradients: fields versus clones, in Butler, P. M. & Joysey, K. A. (eds.) *Development, function and evolution of teeth*, London: Academic Press, pp. 171–199.

Osborn, J. W. (ed.) (1981) *Dental anatomy and embryology*, A Comparison to Dental Studies, Oxford: Blackwell Scientific Publications.

Osborn, J. W. (1982) Helicoidal plane of dental occlusion, *American Journal of Physical Anthropology*, **57**, 273–281.

Osborne, R. H., Horowitz, S. L. & de George, F. V. (1958) Genetic variation of tooth dimensions: a twin study of the permanent anterior teeth, *American Journal of Human Genetics*, **10**, 350–356.

Owen, C. P., Wilding, R. J. & Morris, A. G. (1991) Changes in mandibular condyle morphology related to tooth wear in a prehistoric human population, *Archives of Oral Biology*, **36**, 799–804.

Owen, R. (1845) *Odontography or a treatise on the comparative anatomy of the teeth: their physiological relations, mode of development and microscopic structure in the vertebrate animals*, London: Hyppolyte Baillière.

Ozaki, T., Satake, T. & Kanazawa, E. (1987) Morphological significance of root length variability in comparison with other crown dimensions, *J Nihon Univ Sch Dent*, **29**, 233–240.

Pal, A. (1971) Gradients of dentine exposure in human molars, *Journal of the Indian Anthropological Society*, **6**, 67–73.

Palomino, H., Chakraborty, R. & Rothhammer, F. (1977) Dental morphology and population diversity, *Human Biology*, **49**, 61–70.

Pantke, H. (1957) Untersuchungen über Retziusstreifen und Perikymantien, *Stoma*, **10**, 32–40.

Parker, R. B. & Toots, H. (1980) Trace elements in bones as paleobiological indicators, in Behrensmeyer, A. K. & Hill, A. P. (eds.) *Fossils in the making. Vertebrate taphonomy and paleoecology*, Prehistory Archaeology and Ecology Series, Chicago: University of Chicago Press, pp. 197–297.

Pastor, R. F. (1992) Dietary adaptations and dental microwear in Mesolithic and Chalcolithic South Asia, in Lukacs, J. R. (ed.) *Culture, ecology & dental anthropology*, Journal of Human Ecology Special Issue No 2, Delhi: Kamla-Raj Enterprises, pp. 215–228.

Pastor, R. F. (1994) A multivariate dental microwear analysis of prehistoric groups from the Indian subcontinent (abstract), *American Journal of Physical Anthropology*, Supplement 18, 158–159.

Pedersen, P. O. (1949) The East Greenland Eskimo dentition, *Meddelelser om Grønland*, **142**, 1–244.

Pedersen, P. O. (1966) Nutritional aspects of dental caries, *Odontologisk Revy*, **17**, 91–100.

Pederson, P. O., Dahlberg, A. A. & Alexandersen, V. (1967) International Symposium on Dental Morphology, Fredensborg, Denmark, September 27–29, 1965, *Journal of Dental Research*, 46 Supplement, 769–992.

Pedersen, P. O. & Scott, D. B. (1951) Replica studies of the surfaces of teeth from Alaskan Eskimo, West Greenland natives, and American whites, *Acta Odontologica Scandinavica*, **9**, 261–292.

Penning, C., van Amerongen, J. P., Seef, R. E. & ten Cate, J. M. (1992) Validity of probing for fissure caries diagnosis, *Caries Research*, **26**, 445–449.

Perrin, W. F. & Myrick, A. C. (eds.) (1980) *Growth of odontocetes and sirenians: problems in age determination. Proceedings of the International Conference on determining age of odontocete Ceteans (and Sirenians). La Jolla, California, September 5–19, 1978*, Report of the International Whaling Commission, Special Issue No 3, Cambridge: International Whaling Commission.

Perzigian, A. J. (1976) The dentition of the Indian Knoll skeletal population: odontometrics and cusp number, *American Journal of Physical Anthropology*, **44**, 113–122.

Perzigian, A. J. (1977) Fluctuating dental asymmetry: variation among skeletal populations, *American Journal of Physical Anthropology*, **47**, 81–88.

Perzigian, A. J. (1981) Allometric analysis of dental variation in a human population, *American Journal of Physical Anthropology*, **54**, 341–345.

Pfeiffer, S. & Fairgrieve, S. I. (1994) Evidence from ossuaries: the effect of contact on the health of Iroquoians, in Larsen, C. S. & Milner, G. R. (eds.) *In the wake of contact. Biological responses to conquest*, New York: Wiley–Liss, pp. 47–62.

Pickerill, H. P. (1912) The structure of the enamel, *Dental Cosmos*, **55**, 959–988.

Pindborg, J. J. (1970) *Pathology of the dental hard tissues*, Philadelphia: W B Saunders.

Pindborg, J. J. (1982) Aetiology of developmental enamel defects not related to fluorosis, *International Dental Journal*, **32**, 123–134.

Poole, D. F. G. & Tratman, E. K. (1978) Post-mortem changes on human teeth from late upper Palaeolithic/Mesolithic occupants of an English limestone cave, *Archives of Oral Biology*, **23**, 1115–1120.

Portin, P. & Alvesalo, L. (1974) The inheritance of shovel shape in maxillary central incisors, *American Journal of Physical Anthropology*, **41**, 59–62.

Potter, R. H. & Nance, W. E. (1976) A twin study of dental dimension. I. Discordance, asymmetry, and mirror imagery, *American Journal of Physical Anthropology*, **44**, 391–396.

Potter, R. H., Nance, W. E., Yu, P.-L. & Davis, W. B. (1976) A twin study of dental dimension. II. Independent genetic determinants, *American Journal of Physical Anthropology*, **44**, 397–412.

Potter, R. H. Y., Corruccini, R. S. & Green, L. J. (1981) Variance of occlusion traits in twins, *Journal of Craniofacial Genetics & Developmental Biology*, **1**, 217–227.

Potter, R. H. Y., Rice, J. P., Dahlberg, A. A. & Dahlberg, T. (1983) Dental size traits within families: path analysis for first molar and lateral incisors, *American Journal of Physical Anthropology*, **61**, 283–289.

Potter, R. H. Y., Yu, P. L., Dahlberg, A. A., Merritt, A. D. & Connelly, P. M. (1968) Genetic studies of tooth size factors in Pima Indian families, *American Journal of Human Genetics*, **20**, 89–100.

Powell, B. & Garnick, J. J. (1978) The use of extracted teeth to evaluate clinical measurements of periodontal disease, *Journal of Periodontology*, **49**, 621–624.

Powell, M. L. (1985) The analysis of dental wear and caries for dietary reconstruction, in Gilbert, R. I. & Mielke, J. H. (eds.) *Analysis of prehistoric diets*, New York: Academic Press, pp. 307–338.

Pöyry, M., Nyström, M. & Ranta, R. (1986) Comparison of two tooth formation rating methods, *Proceedings of the Finnish Dental Society*, **82**, 127–133.

Preiswerk, G. (1895) *Beiträge zur Kentniss der Schmelzstructur bei Säugetieren mit besonderer Berüksichtigung der Ungulaten*, University of Basel, Phil. Diss.

Pullinger, A. G. & Seligman, D. A. (1993) The degree to which attrition characterizes differentiated patient groups of temporomandibular disorders, *Journal of Orofacial Pain*, **7**, 196–208.

Radlanski,R. J. & Renz, H. (ed.) (1995) *Proceedings of the 10th International Symposium on Dental Morphology*, Berlin: "M" Marketing Services.

Rathbun, T. A. (1987) Health and disease at a South Carolina Plantation: 1840–1870, *American Journal of Physical Anthropology*, **74**, 239–253.

Reid, C., van Reenen, J. R. & Groeneveld, H. T. (1991) Tooth size and the Carabelli trait, *American Journal of Physical Anthropology*, **84**, 427–432.

Reid, C., van Reenen, J. F. & Groeneveld, H. T. (1992) The Carabelli trait and maxillary molar cusp and crown base areas, in Smith, P. & Tchernov, E. (eds.) *Structure, function and evolution of teeth*, London & Tel Aviv: Freund Publishing House Ltd, pp. 451–466.

Remane, A. (1930) Zur Mesztechnik der Primatenzähne, in Abderhanden, E. (ed.) *Handbuch der biologischen Arbeitsmethoden*, Berlin: Urban & Schwarzenberg, Vol. 7, pp. 609–635.

Retzius, A. (1836) Mikroskopiska undersökningar öfver Tändernes, särddeles Tandbenets, struktur, *Konelige Vetensskaps Academiens Handlingar För År 1836*, 52–140.

Retzius, A. (1837) Bemerkungen über den inneren Bau der Zähne, mit besonderer Rücksicht auf den im Zahnknochen vorkommenden Röhrenbau, *(Müllers) Archiv Anat. Phys.*, Year 1837, 486–566.

Rice, P. M. & Maples, W. R. (1979) Some difficulties in the Gustafson dental age estimations, *Journal of Forensic Sciences*, **24**, 118–172.

Richards, A., Fejerskov, O. & Baelum, V. (1989) Enamel fluoride in relation to severity of human dental fluorosis, *Advances in Dental Research*, **3**, 147–153.

Richards, L. & Brown, T. (1981a) Dental attrition and age relationships in Australian aboriginals, *Archaeology of Oceania*, **16**, 94–98.

Richards, L. C. (1984) Principal axis analysis of dental attrition data from two Australian Aboriginal populations, *American Journal of Physical Anthropology*, **65**, 5–13.

Richards, L. C. (1990) Tooth wear and temporomandibular joint change in Australian aboriginal populations, *American Journal of Physical Anthropology*, **82**, 377–384.

Richards, L. C. & Brown, T. (1981b) Dental attrition and degenerative arthritis of the temporomandibular joint, *Journal of Oral Rehabilitation*, **8**, 293–307.

Richards, L. C. & Miller, S. L. (1991) Relationships between age and dental attrition in Australian aboriginals, *American Journal of Physical Anthropology*, **84**, 159–164.

Richards, M. B., Sykes, B. C. & Hedges, R. E. M. (1995) Authenticating DNA extracted from ancient skeletal remains, *Journal of Archaeological Science*, **22**, 291–299.

Risnes, S. (1984) Rationale for consistency in the use of enamel surface terms: perikymata and imbrications, *Scandinavian Journal of Dental Research*, **92**, 1–5.

Risnes, S. (1985a) Circumferential continuity of perikymata in human dental enamel investigated by scanning electron microscopy, *Scandinavian Journal of Dental Research*, **93**, 185–191.

Risnes, S. (1985b) A scanning electron microscope study of the three-dimensional extent of Retzius lines in human dental enamel, *Scandinavian Journal of Dental Research*, **93**, 145–152.

Risnes, S. (1989) Ectopic tooth enamel. An SEM study of the structure of enamel in enamel pearls, *Advances in Dental Research*, **3**, 258–264.

Rixon, A. E. (1976) *Fossil animal remains: their preparation and conservation*, London: The Athlone Press.

Robinow, M., Richards, T. W. & Anderson, M. (1942) The eruption of deciduous teeth, *Growth*, **6**, 127–133.

Robinson, C. & Kirkham, J. (1982) Dynamics of amelogenesis as revealed by protein compositional studies, in Butler, W. T. (ed.) *The chemistry and biology of mineralized tissues*, Birmingham, Alabama: Ebsco Media Inc, pp. 248–263.

Robinson, C., Kirkham, J., Weatherell, J. A. & Strong, M. (1986) Dental enamel – a living fossil, in Cruwys, E. & Foley, R. A. (eds.) *Teeth and anthropology*, B. A. R. International Series No 291, Oxford: British Achaeological Reports, pp. 31–54.

Robinson, C., Lowe, N. R. & Weatherell, J. A. (1975) Amino acid composition, distribution and origin of 'tuft' protein in human and bovine dental enamel, *Archives of Oral Biology*, **20**, 29–42.

Robinson, J. T. (1952) Some hominid features of the ape-man dentition, *Journal of the Dental Association of South Africa*, **7**, 1–12.

Robinson, J. T. (1956) *The dentition of the Australopithecinae*, Transvaal Museum Memoir No 9, Pretoria: Transvaal Museum.

Robinson, J. T. & Allin, E. F. (1966) On the Y of the Dryopithecus pattern of mandibular molar teeth, *American Journal of Physical Anthropology*, **25**, 323–324.

Roche, A. F. (1992) *Growth, maturation, and body composition. The Fels Longitudinal Study 1929–1991*, Cambridge Studies in Biological Anthropology, Cambridge: Cambridge University Press.

Romero, J. (1958) *Mutilaciones dentarias prehispanicas de Mexico y America en general*, Investigaciones 3, Mexico: Instituto Nacional de Antropologica e Historia.

Romero, J. (1970) Dental mutilation, trephination, and cranial deformation, in Stewart, T. D. (ed.) *Physical anthropology*, Handbook Med Am Indians 9, pp. 50–67.

Rose, J. C. (1977) Defective enamel histology of prehistoric teeth from Illinois, *American Journal of Physical Anthropology*, **46**, 439–446.

Rose, J. C. (1979) Morphological variations of human enamel prisms within abnormal striae of Retzius, *Human Biology*, **51**, 139–151.

Rose, J. C., Armelagos, G. J. & Lallo, J. W. (1978) Histological enamel indicator of childhood stress in prehistoric skeletal samples, *American Journal of Physical Anthropology*, **49**, 511–516.

Rose, J. C. & Harmon, A. (1986) Enamel microwear and prehistoric North American diets, *American Journal of Physical Anthropology*, **69**, 257.

Rose J. C., Marks M. K. & Tieszen L. L. (1991) Bioarchaeology and subsistence in the central and lower portions of the Mississippi valley, in Powell M. L., Bridges P. S. & Mires A. M. (eds.) *What mean these bones? Studies in southeastern bioarchaeology*, Tuscasloosa & London: University of Alabama Press, pp. 7–21.

Rose, J. C. & Tucker, T. L. (1994) Identification of dietary components using microwear patterns: a bioarcheological method, *American Journal of Physical Anthropology*, Supplement 18, 172–173.

Rönnholm, D., Markén, K.-E. & Arwill, T. (1951) Record systems for dental caries and other conditions of the teeth and surrounding tissues, *Odontologisk Tidskrift*, **59**, 34–56.

Rösing, F. W. (1983) Sexing immature human skeletons, *Journal of Human Evolution*, **12**, 149–155.

Rowles, S. L. (1967) Chemistry of the mineral phase of dentine, in Miles, A. E. W. (ed.) *Structural and chemical organization of teeth*, London: Academic Press, Vol. 2. pp. 201–246.

Rudney, J. D. (1983) The age-related distribution of dental indicators of growth disturbance in ancient Lower Nubia: an etiological model from the ethnographic record, *Journal of Human Evolution*, **12**, 535–543.

Rushton, M. A. (1933) Fine contour-lines of enamel of milk teeth, *Dental Record*, **53**, 170.

Russell, D. E., Santoro, J.-P. & Sigogneau-Russell, D. (eds.) (1988) *Teeth revisited: Proceedings of the VIIth International Symposium on Dental Morphology, Paris 1986*, Memoires National Hist nat Paris (Serie C) Volume 53.

Saari, J. T., Hurt, W. C. & Biggs, N. L. (1968) Periodontal bony defects in the dry skull, *Journal of Periodontology*, **39**, 278–283.

Saheki, M. (1958) On the heredity of the tooth crown configuration studied in twins, *Japanese Journal of Anatomy*, **33**, 456–470.

Saleemi, M. A., Hägg, U., Jalil, F. & Zamani, S. (1994) Timing of emergence of individual primary teeth, *Swedish Dental Journal*, **18**, 107–112.

Salido, E. C., Yen, P. H., Koprivnikar, K., Yu, L. C. & Shapiro, L. J. (1992) The human enamel protein gene amelogenin is expressed from both the X-chromosomes and the Y-chromosomes, *American Journal of Human Genetics*, **50**, 303–316.

Sandford, M. K. (1992) A reconsideration of trace element analysis in prehistoric bone, in Saunders, S. R. & Katzenberg, M. A. (eds.) *Skeletal biology of past peoples: research methods*, New York: Wiley–Liss, pp. 79–104.

Sandford, M. K. (1993) Understanding the biogenic–diagenetic continuum: interpreting elemental concentrations of archaeological bone, in Sandford, M. K. (ed.) *Investigations of ancient human tissue. Chemical analyses in anthropology*, Food and Nutrition in History and Anthropology Volume 10, Langhorne, Pennsylvania: Gordon & Breach, pp. 3–57.

Santini, A., Land, M. & Raab, G. M. (1990) The accuracy of simple ordinal scoring of tooth attrition in age assessment, *Forensic Science International*, **48**, 175–184.

Sarnat, B. G. & Schour, I. (1941) Enamel hypoplasia (chronologic enamel aplasia) in relation to systemic disease: a chronologic, morphologic and etiologic classification, *Journal of the American Dental Association*, **28**, 1989–2000.

Sarnat, B. G. & Schour, I. (1942) Enamel hypoplasia (chronologic enamel aplasia) in relation to systemic disease: a chronologic, morphologic and etiologic classification, *Journal of the American Dental Association*, **29**, 397–418.

Saunders, E. (1837) *The teeth as a test of age, considered with reference to the factory children: addressed to the members of both Houses of Parliament*, London: H Renshaw.

Saunders, S., DeVito, C., Herring, A., Southern, R. & Hoppa, R. (1993) Accuracy tests of tooth formation age estimations for human skeletal remains, *American Journal of Physical Anthropology*, **92**, 173–188.

Saunders, S. R. (1992) Subadult skeletons and growth related studies, in Saunders, S. R. & Katzenberg, M. A. (eds.) *Skeletal biology of past peoples: research methods*, New York: Wiley–Liss, pp. 1–20.

Scheie, A. A. (1989) The role of plaque in dental calculus formation, in Ten Cate, J. M. (ed.) *Recent advances in the study of dental calculus*, Oxford: IRL Press at Oxford University Press, pp. 47–56.

Schenk, R. K., Olah, A. J. & Herrmann, N. (1984) Preparation of calcified tissues for light microscopy, in Dickson, G. R. (ed.) *Methods of calcified tissue preparation*, Amsterdam: Elsevier, pp. 1–56.

Schluger, S., Yuodelis, R., Page, R. C. & Johnson, R. H. (1990) *Periodontal diseases. Basic phenomena, clinical management, and occlusal and restorative interrelationships*, 2nd edn, Philadelphia: Lea & Febiger.

Schmidt, W. J. & Keil, A. (1971) *Polarizing microscopy of dental tissues. Theory, methods and results from the structural analysis of normal and diseased hard dental tissues and tissues associated with them in man and other vertebrates*, English translation by D. F. G. Poole & A. I. Darling, Oxford: Pergamon Press.

Schour, I. (1936) Neonatal line in enamel and dentin of human deciduous teeth and first permanent molar, *Journal of the American Dental Association*, **23**, 1946–1955.

Schour, I. & Hoffman, M. M. (1939a) Studies in Tooth Development. I. The 16 μ calcification rhythm in the enamel and dentin from fish to man, *Journal of Dental Research*, **18**, 91–102.

Schour, I. & Hoffman, M. M. (1939b) Studies in tooth development. II. The rate of apposition of enamel and dentin in man and other animals, *Journal of Dental Research*, **18**, 161–175.

Schour, I. & Kronfeld, R. (1938) Tooth ring analysis: IV. Neonatal dental hypoplasia analysis of the teeth of an infant with injury of the brain at birth, *Archives of Pathology*, **26**, 471–490.

Schour, I. & Massler, M. (1940) Studies in tooth development: the growth pattern of human teeth, *Journal of the American Dental Association*, 27, 1778–1792, 1918–1931.

Schour, I. & Massler, M. (1941) The development of the human dentition, *Journal of the American Dental Association*, **28**, 1153–1160.

Schour, I. & Massler, M. (1944) *Development of the human dentition*, 2nd edn, Chicago: American Dental Association.

Schour, I. & Poncher, H. G. (1937) Rate of apposition of human enamel and dentin as measured by the effects of acute fluorosis, *American Journal of Diseases of Children*, **54**, 757–776.

Schreger, D. (1800) Beitrag zur Geschichte der Zähne, *Beitr Vergleich Zergliderungskunst*, **1**, 1–7.

Schultz, P. D. & McHenry, H. M. (1975) Age distribution of enamel hypoplasia in prehistoric California Indians, *Journal of Dental Research*, **54**, 913.

Schulz, P. D. (1977) Task activity and anterior tooth grooving in prehistoric California Indians, *American Journal of Physical Anthropology*, **46**, 87–91.

Schuman, E. L. & Sognnaes, R. F. (1956) Developmental microscopic defects in the teeth of sub-human primates, *American Journal of Physical Anthropology*, **14**, 193–214.

Schüpbach, P., Guggenheim, B. & Lutz, F. (1989) Human root caries: histopathology of initial lesions in cementum and dentin, *Journal of Oral Pathology & Medicine*, **3**, 146–156.

Schüpbach, P., Guggenheim, B. & Lutz, F. (1990) Human root caries: histopatholoty of advanced lesions, *Caries Research*, **24**, 145–158.

Schüpbach, P., Lutz, F. & Guggenheim, B. (1992) Human root caries: histopathology of arrested lesions, *Caries Research*, **26**, 153–164.

Schwarcz, H. P. & Schoeninger, M. J. (1991) Stable isotope analyses in human nutritional ecology, *Yearbook of Physical Anthropology*, **34**, 283–321.

Sciulli, P. W. (1977) A descriptive and comparative study of the deciduous dentition of prehistoric Ohio Valley Amerindians, *American Journal of Physical Anthropology*, **47**, 71–80.

Sciulli, P. W. (1978) Developmental abnormalities of the permanent dentition in prehistoric Ohio Valley Amerindians, *American Journal of Physical Anthropology*, **48**, 193–198.

Sciulli, P. W., Doyle, W. J., Kelley, C., Siegel, P. & Siegel, M. I. (1979) The interaction of stressors in the induction of increased levels of fluctuating asymmetry in the laboratory rat, *American Journal of Physical Anthropology*, **50**, 279–284.

Scott, D. B., Kaplan, H. & Wyckoff, R. W. G. (1949) Replica studies of changes in tooth surfaces with age, *Journal of Dental Research*, **28**, 31–47.

Scott, D. B. & Wyckoff, R. W. G. (1949) Studies of tooth surface structure by optical and electron microscopy, *Journal of the American Dental Association*, **39**, 275–282.

Scott, E. C. (1979a) Dental wear scoring technique, *American Journal of Physical Anthropology*, **51**, 213–218.

Scott, E. C. (1979b) Principal axis analysis of dental attrition data, *American Journal of Physical Anthropology*, **51**, 203–212.

Scott, G. R. (1977) Classification, sex dimorphism, association, and population variation of the canine distal accessory ridge, *Human Biology*, **49**, 453–469.

Scott, G. R. (1978) The relationship between Carabelli's trait and the protostylid, *Journal of Dental Research*, **57**, 570.

Scott, G. R. (1979c) Association between the hypocone and Carabelli's trait of the maxillary molars, *Journal of Dental Research*, **58**, 1403–1404.

Scott, G. R. (1980) Population variation of Carabelli's trait, *Human Biology*, **52**, 63–78.

Scott, G. R. & Dahlberg, A. A. (1982) Microdifferentiation in tooth crown morphology among Indians of the American Southwest, in Kurtén, B. (ed.) *Teeth: form, function, and evolution*, New York: Columbia University Press, pp. 259–291.

Scott, J. H. & Symons, N. B. B. (1974) *Introduction to dental anatomy*, 7th edn, Edinburgh: Churchill Livingstone.

Scott, W. A. & Hillson, S. (1988) An application of the EM algorithm to archaeological data analysis, in Rahtz S. (ed.) *Computer and quantitative methods in archaeology 1988*, B. A. R. International Series 446(ii), Oxford: British Archaeological Reports, pp. 43–52.

Scott, W. A., Whittaker J., Green, M. & Hillson, S. (1991) Graphical modelling of archaeological data, in Lockyear, K. & Rahtz, S. (eds) *Computer applications and quantitative methods in archaeology 1990*, International Series 565, Oxford: British Archaeological Reports, pp. 111–116.

Seeto, E. & Seow, W. K. (1991) Scanning electron microscopic analysis of dentin in vitamin D – resistant rickets–assessment of mineralization and correlation with clinical findings, *Pediatric Dentistry*, **13**, 43–48.

Seligman, D. A., Pullinger, A. G. & Solberg, W. K. (1988) The prevalence of dental attrition and its association with factors of age, gender, occlusion, and TMJ symptomatology, *Journal of Dental Research*, **67**, 1323–1333.

Semal, P. (1988) *Evolution et variabilite des dimensions dentaires chez Homo sapiens neanderthalensis*, Viroinval, Belgique: Centre d'Etudes et de Documentation Archeologiques.

Shapiro, I. M., Mitchell, G., Davidson, I. & Katz, S. H. (1975) The lead content of teeth, *Archives of Environmental Health*, **30**, 483–486.

Sharma, J. C. (1992) Dental morphology and odontometry of twins and the heritability of dental variation, in Lukacs, J. R. (ed.) *Culture, ecology & dental anthropology*, Journal of Human Ecology Special Issue No 2, Delhi: Kamla-Raj Enterprises, pp. 49–60.

Sharma, K. & Corruccini, R. S. (1986) Genetic basis of dental occlusal variation in northwest Indian twins, *European Journal of Orthodontics*, **8**, 91–97.

Shear, M. (1992) *Cysts of the oral regions*, 3rd edn, Oxford: Wright.

Sheiham, A. (1983) Sugars in dental decay, *Lancet*, **1**, 282–284.

Sheridan, S. G., Mittler, D. M., Van Gerven, D. P. & Covert, H. H. (1991) Biomechanical association of dental and temporomandibular pathology in a medieval Nubian population, *American Journal of Physical Anthropology*, **85**, 201–205.

Shipman, P., Foster, G. & Schoeninger, M. (1984) Burnt bones and teeth: an experimental study of color, morphology, crystal structure and shrinkage, *Journal of Archaeological Science*, **11**, 307–325.

Shrestha, B. M. (1980) Use of ultraviolet light in early detection of smooth surface carious lesions in rats, *Caries Research*, **14**, 448–451.

Sillen, A. & Kavanagh, M. (1992) Strontium and paleodietary research: a review, *Yearbook of Physical Anthropology*, **25**, 67–90.

Silverstone, L. M., Johnson, N. W., Hardie, J. M. & Williams, R. A. D. (1981) *Dental caries. Aetiology, pathology and prevention*, London: Macmillan.

Skinner, M. & Goodman, A. H. (1992) Anthropological uses of developmental defects of enamel, in Saunders, S. R. & Katzenberg, M. A. (eds.) *Skeletal biology of past peoples: research methods*, New York: Wiley–Liss, pp. 153–175.

Smith, B. C., Fisher, D. L., Weedn, V. W., Warnock, G. R. & Holland, M. M. (1993) A systematic approach to the sampling of dental DNA, *Journal of Forensic Sciences*, **38**, 1194–1209.

Smith, B. H. (1984) Patterns of molar wear in hunter-gatherers and agriculturalists, *American Journal of Physical Anthropology*, **63**, 39–56.

Smith, B. H. (1986a) Dental development in *Australopithecus* and early *Homo*, *Nature*, **323**, 327–330.

Smith, B. H. (1986b) Development and evolution of the helicoidal plane of dental occlusion, *American Journal of Physical Anthropology*, **69**, 21–35.

Smith, B. H. (1987) Reply to 'Maturational patterns in early hominids' by A. E. Mann, M. Lampl and J. Monge, *Nature*, **328**, 674–675.

Smith, B. H. (1991a) Dental development and the evolution of life history in Hominidae, *American Journal of Physical Anthropology*, **86**, 157–174.

Smith, B. H. (1991b) Standards of human tooth formation and dental age assessment, in Kelly, M. A. & Larsen, C. S. (eds.) *Advances in dental anthropology*, New York: Wiley–Liss, pp. 143–168.

Smith, B. H., Crummett, T. L. & Brandt, K. L. (1994) Ages of eruption of primate teeth: a compendium for ageing individuals and comparing life histories, *Yearbook of Physical Anthropology*, **37**, 177–232.

Smith, B. H. & Garn, S. M. (1987) Polymorphisms in eruption sequence of permanent teeth in American children, *American Journal of Physical Anthropology*, **74**, 289–303.

Smith, F. H., Falsetti, A. B. & Donnelly, S. M. (1989) Modern human origins, *Yearbook of Physical Anthropology*, **32**, 35–68.

Smith, P., Bar-Yosef, O. & Sillen, A. (1984) Archaeological and skeletal evidence for dietary change during the late Pleistocene/early Holocene in the Levant, in Cohen, M. N. & Armelagos, G. J. (eds.) *Palaeopathology at the origins of agriculture*, New York: Academic Press, pp. 101–136.

Smith, P. & Tchernov, E. (eds.) (1992) *Structure, function and evolution of teeth*, London & Tel Aviv: Freund Publishing House Ltd.

Smith, R. J. & Bailit, H. J. (1977) Variation in dental occlusion and arches among Melanesians of Bougainville Island, Papua New Guinea: I Methods, age changes, sex differences and population comparison, *American Journal of Physical Anthropology*, **47**, 195–208.

Smith, R. J., Kolakowski, D. & Bailit, H. J. (1978) Variation in dental occlusion and arches among Melanesians of Bougainville Island, Papua New Guinea: II Clinal variation, geographic microdifferentiation and synthesis, *American Journal of Physical Anthropology*, **48**, 331–342.

Soames, J. V. & Southam, J. C. (1993) *Oral pathology*, 2nd edn, Oxford: Oxford University Press.

Sofaer, J. A. (1969) The genetics and expression of a dental morphological variant in the mouse, *Archives of Oral Biology*, **14**, 1213–1223.

Sofaer, J. A. (1970) Dental morphological variation and the Hardy–Weinburg law, *Journal of Dental Research*, **49**, 1505–1508.

Sofaer, J. A., MacLean, C. J. & Bailit, H. L. (1972a) Heredity and morphological variation in early and late developing teeth of the same morphological class, *Archives of Oral Biology*, **17**, 811–816.

Sofaer, J. A., Niswander, J. D. & MacLean, C. J. (1972b) Population studies on Southwestern Indian tribes. V. Tooth morphology as an indicator of biological distance, *American Journal of Physical Anthropology*, **37**, 357–366.

Soggnaes, R. F. (1950) Histological studies of ancient and recent teeth with special regard to differential diagnosis between intra-vitam and post-mortem characteristics, *American Journal of Physical Anthropology*, **8**, 269–270.

Soggnaes, R. F. (1956) Histological evidence of developmental lesions in teeth originating from paleolithic, prehistoric, and ancient man, *American Journal of Pathology*, **32**, 547–577.

Solheim, T. (1990) Dental cementum apposition as an indicator of age, *Scandinavian Journal of Dental Research*, **98**, 510–519.

Solheim, T. (1992) Amount of secondary dentin as an indicator of age, *Scandinavian Journal of Dental Research*, **100**, 193–199.

Spriggs, J. A. & Van Byeren, T. (1984) A practical approach to the excavation and recording of ancient Maya burials, *The Conservator* **8**, 41–47.

Staaf, V., Mornstad, H. & Welander, U. (1991) Age estimation based on tooth development: a test of reliability and validity, *Scandinavian Journal of Dental Research*, **99**, 281–286.

Stack, M. V. (1964) A gravimetric study of crown growth rate of the human deciduous dentition, *Biology of the Neonate*, **6**, 197–224.

Stack, M. V. (1967) Vertical growth rates of the deciduous teeth, *Journal of Dental Research*, **46**, 879–882.

Stack, M. V. (1971) Relative rates of weight gain in human deciduous teeth, in Dahlberg, A. A. (ed.) *Dental morphology and evolution*, Chicago: University of Chicago Press, pp. 59–62.

Stack, M. V. (1986) Trace elements in teeth of Egyptians and Nubians, in David, R. A. (ed.) *Science in Egyptology. Proceedings of the 'Science in Egyptology' Symposia*, Manchester: Manchester University Press, pp. 219–222.

Stead, I. M., Bourke, J. B. & Brothwell, D. R. (eds.) (1986) *Lindow Man. The body in the bog*, London: British Museum Publications.

Steadman, L. T., Brudevold, F., Smith, F. A., Gardner, D. E. & Little, M. F. (1959) Trace elements in ancient Indian teeth, *Journal of Dental Research*, **38**, 285–292.

Stephan, R. M. (1944) Intra-oral hydrogen ion concentrations associated with dental caries activity, *Journal of Dental Research*, **23**, 257–266.

Stermer Beyer-Olsen, E. M. & Risnes, S. (1994) Radiographic analysis of dental development used in age determination of infant and juvenile skulls from a Medieval archaeological site in Norway, *International Journal of Osteoarchaeology*, **4**, 299–304.

Stone A. C., Milner, G. R., Pääbo, S. & Stoneking, M. (1996) Sex determination of ancient human skeletons using DNA, *American Journal of Physical Anthropology*, **99**, 231–228.

Stoner, K. E. (1995) Dental pathology in *Pongo satyrus borneensis, American Journal of Physical Anthropology*, **98**, 307–322.

Stott, G. G., Sis, R. F. & Levy, B. M. (1980) Cemental annulation as an age criterion in the common marmoset (*Callithrix jaculus*), *Journal of Medical Primatology*, **9**, 274–285.

Stott, G. G., Sis, R. F. & Levy, B. M. (1982) Cemental annulation as an age criterion in forensic dentistry, *Journal of Dental Research*, **61**, 814–817.

Strachan, T. & Read, A. P. (1996) *Human molecular genetics*, Oxford: BIOS Scientific Publishers.

Stringer, C. B. (1990) The emergence of modern humans, *Scientific American*, December 1990, 68–74.

Stringer, C. B., Dean, M. C. & Martin, R. D. (1990) A comparative study of cranial and dental development within a recent British sample and among Neanderthals, in de Rousseau, J. (ed.) *Primate life history and evolution*, New York: Wiley–Liss, pp. 115–152.

Suarez, B. K. (1974) Neanderthal dental asymmetry and the probable mutation effect, *American Journal of Physical Anthropology*, **41**, 411–416.

Suarez, B. K. & Spence, M. A. (1974) The genetics of hypodontia, *Journal of Dental Research*, **53**, 781.

Suckling, G. W., Nelson, D. G. A. & Patel, M. J. (1989) Macroscopic and scanning electron microscopic appearance and hardness values of developmental defects in human permanent tooth enamel, *Advances in Dental Research*, **3**, 219–233.

Suga, S. (1989) Enamel hypopmineralization viewed from the pattern of progressive mineralization of human and monkey developing enamel, *Advances in Dental Research*, **3**, 188–189.

Sunderland, E. P., Smith, C. J. & Sunderland, R. (1987) A histological study of the chronology of initial mineralization in the human deciduous dentition, *Archives of Oral Biology*, **32**, 167–174.

Suwa, G. (1986) Dental metric assessment of the Omo fossils: implications for the phylogenetic position of *Australopithecus africanus*, in Grine, F. E. (ed.) *Evolutionary history of the 'robust' australopithecines*, New York: Aldine de Gruyter, pp. 199–222.

Suwa, G., Wood, B. A. & White, T. D. (1994) Further analysis of mandibular molar crown and cusp areas in Pliocene and Early Pleistocene hominids, *American Journal of Physical Anthropology*, **93**, 407–426.

Swärdstedt, T. (1966) *Odontological aspects of a medieval population in the province of Jämtland/Mid Sweden*, University of Lund, Sweden, Akademisk Avhandling som med vederbörligt tillstand av Odontologiska Fakulteten vid Lunds Universitet för vinnande av Odontologie Doktorgrad offentilgen försvaras i Tandläkarhöskolans Aula, Malmö, Fredagen den 9 December 1966 K19 CT.

Sweeney, E. A., Saffir, A. J. & Leon, R. D. (1971) Linear hypoplasia of deciduous incisor teeth in malnourished children, *American Journal of Clinical Nutrition*, **24**, 29–31.

Swindler, D. R. (1976) *Dentition of living primates*, London: Academic Press.

Tal, H. (1985) Periodontal bone loss in dry mandibles of South African blacks: a biometric study, *Journal of Dental Research*, **64**, 925–929.

Tattersall, I. (1968) Dental paleopathology of Mediaeval Britain, *Journal of the History of Medicine*, **23**, 380–385.

Teaford, M. F. (1988a) A review of dental microwear and diet in modern mammals, *Scanning Microscopy*, **2**, 1149–1166.

Teaford, M. F. (1988b) Scanning electron microscope diagnosis of wear patterns versus artefacts on fossil teeth, *Scanning Microscopy*, **2**, 1167–1175.

Teaford, M. F. (1991) Dental microwear: what can it tell us about diet and dental function, in Kelley, M. A. & Larsen, C. S. (eds.) *Advances in dental anthropology*, New York: Wiley–Liss, pp. 341–356.

Teaford, M. F. & Walker, A. (1984) Quantitative differences in dental microwear between primate species with different diets and a comment on the presumed diet of *Sivapithecus*, *American Journal of Physical Anthropology*, **64**, 191–200.

Ten Cate, A. R. (1985) *Oral histology: development, structure and function*, St Louis: C V Mosby.

Thomson, H. (1990) *Occlusion* 2nd edn, London: Wright.

Thylstrup, A., Chironga, L., Carvalho, J. D. & Ekstrand, K. R. (1989) The occurrence of dental calculus in occlusal fissures as an indication of caries activity, in Ten Cate, J. M. (ed.) *Recent advances in the study of dental calculus*, Oxford: IRL Press at Oxford University Press, pp. 211–222.

Thylstrup, A. & Fejerskov, O. (1978) Clinical appearance and surface distribution of dental fluorosis in permanent teeth in relation to histological changes, *Community Dentistry & Oral Epidemiology*, **6**, 315–328.

Tobias, P. V. (1980) The natural history of the helicoidal occlusal plane and its evolution in early *Homo*, *American Journal of Physical Anthropology*,, **53**, 173–187.

Tomenchuk, J. & Mayhall, J. T. (1979) A correlation of tooth wear and age among modern Igloolik Eskimos, *American Journal of Physical Anthropology*, **51**, 67–78.

Tomes, C. S. (1984) *A manual of dental anatomy. Human and comparative*, 4th edn, London: J & A Churchill.

Tóth, K. (1970) *The epidemiology of dental caries in Hungary*, English translation by E. Kerner, Budapest: Akadémiai Kiadó.

Townsend, G. C. (1980) Heritability of deciduous tooth size in Australian aboriginals, *American Journal of Physical Anthropology*, **53**, 297–300.

Townsend, G. C. & Alvesalo, L. (1985) Tooth size in 47 XYY males: evidence for a direct effect of the Y chromosome on growth, *Australian Dental Journal*, **30**, 268–272.

Townsend, G. C. & Brown, T. (1978) Heritability of permanent tooth size, *American Journal of Physical Anthropology*, **49**, 497–505.

Townsend, G. C. & Brown, T. (1980) Dental asymmetry in Australian aboriginals, *Human Biology*, **52**, 661–673.

Trautz, O. R. (1967) Crystalline organization of dental mineral, in Miles, A. E. W. (ed.) *Structural and chemical organization of teeth*, London: Academic Press, Vol. 2, pp. 165–200.

Turner II, C. G. (1967) Dental genetics and microevolution in prehistoric and living Koniag Eskimo, *Journal of Dental Research*, **46**, 911–917.

Turner II, C. G. (1979) Dental anthropological indications of agriculture among the Jomon people of Central Japan, *American Journal of Physical Anthropology*, **51**, 619–636.

Turner II, C. G. (1987) Late Pleistocene and Holocene population history of east Asia based on dental variation, *American Journal of Physical Anthropology*, **73**, 305–321.

Turner II, C. G. (1989) Teeth and prehistory in Asia, *Scientific American*, February 1989, 88–96.

Turner II, C. G. (1990) Major features of Sundadonty and Sinodonty, including suggestions about East Asian microevolution, population hsitory, and late Pleistocene relationships with Australian aboriginals, *American Journal of Physical Anthropology*, **82**, 295–317.

Turner II, C. G. & Markowitz, M. A. (1990) Dental discontinuity between Late Pleistocene and recent Nubians. Peopling of the Eurafrican-South Asian traingle I, *Homo*, **41**, 32–41.

Turner II, C. G., Nichol, C. R. & Scott, G. R. (1991) Scoring procedures for key morphological traits of the permanent dentition: the Arizona State University Dental Anthropology System, in Kelley, M. A. & Larsen, C. S. (eds.) *Advances in dental anthropology*, New York: Wiley–Liss, pp. 13–31.

Ubelaker, D. H. (1978) *Human skeletal remains: excavation, analysis, interpetation*, Chicago: Aldine.

Ubelaker, D. H. (1984) Prehistoric human biology of Equador, in Cohen, M. N. & Armelagos, G. J. (eds.) *Palaeopathology at the origins of agriculture*, New York: Academic Press, pp. 491–513.

Ubelaker, D. H. (1987) Estimating age at death from immature human skeletons: an overview, *Journal of Forensic Sciences*, **32**, 1254–1263.

Ubelaker, D. H. (1994) The biological impact of European contact in Ecuador, in Larsen, C. S. & Milner, G. R. (eds.) *In the wake of contact. Biological responses to conquest*, New York: Wiley–Liss, pp. 147–160.

Ubelaker, D. H., Phenice, T. W. & Bass, W. M. (1969) Artificial interproximal grooving of the teeth in American Indians, *American Journal of Physical Anthropology*, **30**, 145–149.

Ungar, P. S. (1995) A semiautomated image analysis procedure for the quantification of dental microwear II, *Scanning*, **17**, 57–59.

Ungar, P. S., Simon, J. C. & Cooper, J. W. (1990) A semiautomated image analysis procedure for the quantification of dental microwear, *Scanning*, **13**, 31–36.

van Amerongen, J. P., Penning, C., Kidd, E. A. M. & ten Cate, J. M. (1992) An in vitro assessment of the extent of caries under small occlusal cavities, *Caries Research*, **26**, 89–93.

van Beek, G. C. (1983) *Dental morphology. An illustrated guide*, 2nd edn, Bristol: Wright.

van der Merwe, N. J. (1992) Light stable isotopes and the reconstruction of prehistoric diets, in Pollard, A. M. (ed.) *New developments in archaeological science*, Proceedings of the British Academy 77, Oxford: Oxford University Press, pp. 247–264.

van der Velden, U., Abbas, F., Armand, S., de Graaf, J., Timmerman, M. F., van der Weijden, G. A., van Winkelhoff, A. J. & Winkel, E. G. (1993) The effect of sibling relationship on the periodontal condition, *Journal of Clinical Periodontology*, **20**, 683–690.

Van Reenen, J. F. (1982) The effects of attrition on tooth dimensions of San

(Bushmen), in Kurtén, B. (ed.) *Teeth: form, function, and evolution*, New York: Columbia University Press, pp. 182–203.

van Reenen, J. F. (1992) Dental wear in San (Bushmen), in Lukacs, J. R. (ed.) *Culture, ecology & dental anthropology*, Journal of Human Ecology Special Issue No 2, Delhi: Kamla-Raj Enterprises, pp. 201–213.

Varrela, T. M. (1991) Prevalence and distribution of dental caries in a late medieval population in Finland, *Archives of Oral Biology*, **36**, 553–559.

Vasiliadis, L., Darling, A. I. & Levers, B. G. H. (1983a) The amount and distribution of sclerotic human root dentine, *Archives of Oral Biology*, **28**, 645–649.

Vasiliadis, L., Darling, A. I. & Levers, B. G. H. (1983b) The histology of sclerotic human root dentine, *Archives of Oral Biology*, **28**, 693–700.

Vehkalahti, M. & Paunio, L. (1994) Association between root caries occurrence and periodontal state, *Caries Research*, **28**, 301–306.

Vitzthum, V. J. & Wikander, R. (1988) Incidence and correlates of enamel hypoplasia in non-human primates, *American Journal of Physical Anthropology*, **75**, 284.

von Ebner, V. (1906) Uber die Entwicklung der leimgebenden Fibrillen im Zahnbein, *Sitzungsberichte der Akademie der Wissenschaften in Wien*, **115**, 281–349.

Vrba, E. S. & Grine, F. E. (1978) Australopithecine prism patterns, *Science*, **202**, 890–892.

Wada, K., Ohtaishi, N. & Hachiya, N. (1978) Determination of age in the Japanese monkey from growth layers in the dental cementum, *Primates*, **19**, 775–784.

Waldron, H. A. (1983) On the post-mortem accumulation of lead by skeletal tissues, *Journal of Archaeological Science*, **10**, 35–40.

Walker, A., Hoeck, H. N. & Perez, L. (1978) Microwear of mammalian teeth as an indicator of diet, *Science*, **201**, 908–910.

Walker, P. L. (1978) A quantitative analysis of dental attrition rates in the Santa Barbara Channel area, *American Journal of Physical Anthropology*, **48**, 101–106.

Walker, P. L., Dean, G. & Shapiro, P. (1991) Estimating age from tooth wear in archaeological populations, in Kelley, M. A. & Larsen, C. S. (eds.) *Advances in dental anthropology*, New York: Wiley–Liss, pp. 169–178.

Walker, P. L. & Erlandson, J. (1986) Dental evidence for prehistoric dietary change on the northern Channel Islands, *American Antiquity*, **51**, 375–383.

Walker, P. L. & Hagen, E. H. (1994) A topographical approach to dental microwear analysis, *American Journal of Physical Anthropology*, Supplement 18, 203.

Wallington, E. A. (1972) *Histological methods for bone*, Butterworths Laboratory Aids, London: Butterworths.

Waters, P. H. (1983) A review of the moulding and casting materials and techniques in use at the Palaeontology Laboratory, British Museum (Natural History), *The Conservator*, **7**, 37–43.

Watkinson, D. (ed.) (1978) *First aid for finds. Archaeology section of the United Kingdom Institute for Conservation*, Hereford: Rescue, The British Archaeological Trust.

Watson, P. J. C. (1986) A study of the pattern of alveolar recession, in Cruwys, E. & Foley, R. A. (eds.) *Teeth and anthropology*, B. A. R. International Series No 291, Oxford: British Archaeological Reports, pp. 123–132.

Watt, I. M. (1985) *The principles and practice of electron microscopy*, Cambridge: Cambridge University Press.

Webb, S. (1995) *Palaeopathology of aboriginal Australians*, Cambridge: Cambridge University Press.

Weber, D. F. & Ashrafi, S. H. (1979) Structure of Retzius lines in partially demineralized human enamel, *Anatomical Record*, **194**, 563–570.

Weber, D. F. & Eisenmann, D. (1971) Microscopy of the neonatal line in developing human enamel, *American Journal of Anatomy*, **132**, 375–392.

Weber, D. F. & Glick, P. L. (1975) Correlative microscopy of enamel prism orientation, *American Journal of Anatomy*, **144**, 407–420.

Weidenreich, F. (1937) *The dentition of Sinanthropus pekinensis: a comparative odontography of the hominids*, Palaeontologica Sinica, New Series D, No 1 (Whole Series No 101), Peking

Werelds, R. J. (1961) Observations macroscopiques et microscopiques sur certains altérations *post-mortem* des dents, *Bulletin du Groupement International pour les Recherches Scientifique en Stomatologie*, **4**, 7–60.

Werelds, R. J. (1962) Nouvelles observations sur les dégredations post-mortem de la dentine et due cément des dents inhumées, *Bulletin du Groupement International pour les Recherches Scientifique en Stomatologie*, **5**, 559–591.

Whaites, E. (1992) *Essentials of dental radiography and radiology*, Dental Series, Edinburgh: Churchill Livingstone.

White, T. D. (1978) Early hominid enamel hypoplasia, *American Journal of Physical Anthropology*, **49**, 79–84.

White, T. D., Johanson, D. C. & Kimbel, W. H. (1981) *Australopithecus africanus*: its phyletic position reconsidered, *South African Journal of Science*, **77**, 445–470.

White, T. D., Suwa, G. & Asfaw, B. (1994) *Australopithecus ramidus*, a new species of early hominid from Aramis, Ethiopia, *Nature*, **371**, 306–312.

White, T. D., Suwa, G. & Asfaw, B. (1995) Corrigendum, *Nature*, **275**, 88.

Whittaker, D. (1992) Quantitative studies on age changes in the teeth and surrounding structures in archaeological material: a review, *Journal of the Royal Society of Medicine*, **85**, 97–101.

Whittaker, D. K., Davies, G. & Brown, M. (1985a) Tooth loss, attrition and temporomandibular joint changes in a Romano-British population, *Journal of Oral Rehabilitation*, **12**, 407–419.

Whittaker, D. K., Griffiths, S., Robson, A., Roger Davies, P., Thomas, G. & Molleson, T. (1990) Continuing tooth eruption and alveolar crest height in an eighteenth-century population from Spitalfields, east London, *Archives of Oral Biology*, **35**, 81–85.

Whittaker, D. K. & MacDonald, D. G. (1989) *Colour atlas of forensic dentistry*, Wolfe Medical.

Whittaker, D. K., Molleson, T., Bennett, R. B., Edwards, I., Jenkins, P. R. & Llewelyn, J. H. (1981) The prevalence and distribution of dental caries in a Romano-British population, *Archives of Oral Biology*, **26**, 237–245.

Whittaker, D. K., Molleson, T., Daniel, A. T., Williams, J. T., Rose, P. & Resteghini, R. (1985b) Quantitative assessment of tooth wear, alveolar-crest height and continuing eruption in a Romano-British population, *Archives of Oral Biology*, **30**, 493–501.

Whittaker, D. K., Parker, J. H. & Jenkins, C. (1982) Tooth attrition and continuing eruption in a Romano-British population, *Archives of Oral Biology*, **27**, 405–409.

Whittaker, D. K. & Richards, D. (1978) Scanning electron microscopy of the neonatal line in human enamel, *Archives of Oral Biology*, **23**, 45–50.

Whittaker, D. K., Ryan, S., Weeks, K. & Murphy, W. M. (1987) Patterns of approximal wear in cheek teeth of a Romano-British population, *American Journal of Physical Anthropology*, **73**, 389–396.

Williams, C. T. & Marlow, C. A. (1987) Uranium and thorium distributions in fossil bones from Olduvai Gorge, Tanzania and Kanam, Kenya, *Journal of Archaeological Science*, **14**, 297–309.

Williams, R. A. D. & Elliot, J. C. (1989) *Basic and applied dental biochemistry* 2nd edn, Dental Series, Edinburgh: Churchill Livingstone.

Wilson, D. F. & Shroff, F. R. (1970) The nature of the striae of Retzius as seen with the optical microscope, *Australian Dental Journal*, **15**, 3–24.

Winter, G. B. & Brook, A. H. (1975) Enamel hypoplasia and anomalies of the enamel, *Dental Clinics of North America*, **19**, 3–24.

Woelfl, J. B. (1990) *Dental anatomy: its relevance to dentistry*, 4th edn, Philadelphia: Lea & Febiger.

Wolanski, N. (1966) A new method for the evaluation of tooth formation, *Acta Genetica (Basel)*, **16**, 186–197.

Wolpoff, M. H. (1970) Interstitial wear, *American Journal of Physical Anthropology*, **34**, 205–228.

Wolpoff, M. H. (1971) *Metric trends in hominid dental evolution*, Cleveland: Case Western Reserve University Press.

Wolpoff, M. H., Monge, J. M. & Lampl, M. (1988) Was Taung human or an ape?, *Nature*, **335**, 501.

Wolpoff, M. L. (1976) Some aspects of the evolution of early hominid sexual dimorphism, *Current Anthropology*, **17**, 579–606.

Wood, B. A. (1991) *Hominid cranial remains*, Koobi Fora Research Project 4, Oxford: Clarendon Press.

Wood, B. A. & Abbott, S. A. (1983) Analysis of the dental morphology of Plio-Pleistocene hominids. I. Mandibular molars: crown area measurements and morphological traits, *Journal of Anatomy*, **136**, 197–219.

Wood, B. A., Abbott, S. A. & Graham, S. H. (1983) Analysis of the dental morphology of Plio-Pleistocene hominids. II. Mandibular molars – study of cusp areas, fissure pattern and cross sectional shape of the crown, *Journal of Anatomy*, **137**, 287–314.

Wood, B. A., Abbott, S. A. & Uytterschaut, H. (1988) Analysis of the dental morphology of Plio-Pleistocene hominids. IV. Mandibular postcanine root morphology, *Journal of Anatomy*, **156**, 107–139.

Wood, B. A. & Engelman, C. A. (1988) Analysis of the dental morphology of Plio-Pleistocene hominids. V. Maxillary postcanine tooth morphology, *Journal of Anatomy*, **161**, 1–35.

Wood, B. A. & Uytterschaut, H. (1987) Analysis of the dental morphology of Plio-Pleistocene hominids. III. Mandibular premolar crowns, *Journal of Anatomy*, **154**, 121–156.

Wood, B. F. & Green, L. J. (1969) Second premolar morphologic trait similarities in twins, *Journal of Dental Research*, **48**, 74–78.

Wyckoff, R. G. (1972) *Biochemistry of animal fossils*. Bristol: Scientechnica.

y'Edynak, G. (1978) Culture, diet, and dental reduction in Mesolithic forager-fishers of Yugoslavia, *Current Anthropology*, **19**, 616–618.

y'Edynak, G. (1989) Yugoslav Mesolithic dental reduction, *American Journal of Physical Anthropology*, **78**, 17–36.

y'Edynak, G. (1992) Dental pathology: a factor in post-Pleistocene Yugoslav dental reduction, in Lukacs, J. R. (ed.) *Culture, ecology & dental anthropology*, Journal of Human Ecology Special Issue No 2, Delhi: Kamla-Raj Enterprises, pp. 133–144.

y'Edynak, G. & Fleisch, S. (1983) Microevolution and biological adaptability in the transition from food-collecting to food-producing in the Iron Gates of Yugoslavia, *Journal of Human Evolution*, **12**, 279–296.

Yamada, H. & Brown, T. (1988) Contours of maxillary molars studied in Australian aboriginals, *American Journal of Physical Anthropology*, **76**, 399–407.

Yilmaz, S., Newman, H. N. & Poole, D. F. G. (1977) Diurnal periodicity of von Ebner growth lines in pig dentine, *Archives of Oral Biology*, **22**, 511–513.

Yoneda, M. (1982) Growth layers in dental cementum of *Saguinis* monkeys in South America, *Primates*, **23**, 460–464.

Yoon, S. H., Brudevold, F., Gardner, D. E. & Smith, F. A. (1958) Fluoride in enamel, coronal dentin, and roots of human teeth, *Journal of Dental Research*, **37**, 25.

Yoon, S. H., Brudevold, F., Gardner, D. E. & Smith, F. A. (1960) Distribution of fluorine in the teeth and fluorine levels in water, *Journal of Dental Research*, **39**, 845–856.

Zsigmondy, O. (1893) On congenital defects of the enamel, *Dental Cosmos*, **35**, 709–717.

Zubov, A. A. (1977) Odontoglyphics: the laws of variation of the human molar crown microrelief, in Dahlberg, A. A. & Graber, T. M. (eds.) *Orofacial growth and development*, The Hague: Mouton, pp. 269–282.

Zubov, A. A. & Nikityuk, B. A. (1978) Prospects for the application of dental morphology in twin type analysis, *Journal of Human Evolution*, **7**, 519–524.

Index

Ordinary references are in Roman type, main references in **bold**, illustrations in *italics*.